Undergraduate Texts in Mathematics

Editors

S. Axler
F.W. Gehring
P.R. Halmos

Springer
New York
Berlin
Heidelberg
Barcelona
Budapest
Hong Kong
London
Milan
Paris
Santa Clara
Singapore
Tokyo

Undergraduate Texts in Mathematics

Anglin: Mathematics: A Concise History and Philosophy.
Readings in Mathematics.

Anglin/Lambek: The Heritage of Thales.
Readings in Mathematics.

Apostol: Introduction to Analytic Number Theory. Second edition.

Armstrong: Basic Topology.

Armstrong: Groups and Symmetry.

Axler: Linear Algebra Done Right.

Bak/Newman: Complex Analysis.

Banchoff/Wermer: Linear Algebra Through Geometry. Second edition.

Berberian: A First Course in Real Analysis.

Brémaud: An Introduction to Probabilistic Modeling.

Bressoud: Factorization and Primality Testing.

Bressoud: Second Year Calculus.
Readings in Mathematics.

Brickman: Mathematical Introduction to Linear Programming and Game Theory.

Browder: Mathematical Analysis: An Introduction.

Cederberg: A Course in Modern Geometries.

Childs: A Concrete Introduction to Higher Algebra. Second edition.

Chung: Elementary Probability Theory with Stochastic Processes. Third edition.

Cox/Little/O'Shea: Ideals, Varieties, and Algorithms.

Croom: Basic Concepts of Algebraic Topology.

Curtis: Linear Algebra: An Introductory Approach. Fourth edition.

Devlin: The Joy of Sets: Fundamentals of Contemporary Set Theory. Second edition.

Dixmier: General Topology.

Driver: Why Math?

Ebbinghaus/Flum/Thomas: Mathematical Logic. Second edition.

Edgar: Measure, Topology, and Fractal Geometry.

Elaydi: An Introduction to Difference Equations.

Exner: An Accompaniment to Higher Mathematics.

Fischer: Intermediate Real Analysis.

Flanigan/Kazdan: Calculus Two: Linear and Nonlinear Functions. Second edition.

Fleming: Functions of Several Variables. Second edition.

Foulds: Combinatorial Optimization for Undergraduates.

Foulds: Optimization Techniques: An Introduction.

Franklin: Methods of Mathematical Economics.

Hairer/Wanner: Analysis by Its History.
Readings in Mathematics.

Halmos: Finite-Dimensional Vector Spaces. Second edition.

Halmos: Naive Set Theory.

Hämmerlin/Hoffmann: Numerical Mathematics.
Readings in Mathematics.

Iooss/Joseph: Elementary Stability and Bifurcation Theory. Second edition.

Isaac: The Pleasures of Probability.
Readings in Mathematics.

James: Topological and Uniform Spaces.

Jänich: Linear Algebra.

Jänich: Topology.

Kemeny/Snell: Finite Markov Chains.

Kinsey: Topology of Surfaces.

Klambauer: Aspects of Calculus.

Lang: A First Course in Calculus. Fifth edition.

Lang: Calculus of Several Variables. Third edition.

Lang: Introduction to Linear Algebra. Second edition.

Lang: Linear Algebra. Third edition.

Lang: Undergraduate Algebra. Second edition.

Lang: Undergraduate Analysis.

Lax/Burstein/Lax: Calculus with Applications and Computing. Volume 1.

LeCuyer: College Mathematics with APL.

Lidl/Pilz: Applied Abstract Algebra.

(continued following index)

Saber N. Elaydi

An Introduction to Difference Equations

With 64 illustrations

Springer

Saber N. Elaydi
Department of Mathematics
Trinity University
San Antonio, TX 78212-7200
USA

Mathematics Subject Classifications (1991): 39-01, 39Axx, 58Fxx, 34Dxx, 34Exx

Library of Congress Cataloging-in-Publication Data
Elaydi, Saber, 1943-
    An introduction to difference equations / Saber N. Elaydi.
      p.   cm. – (Undergraduate texts in mathematics)
    Includes bibliographical references and index.
    ISBN 0-387-94582-2 (hbk. : alk. paper)
    1. Difference equations.   I. title.   II. Series.
QA431.E43   1995
515'.625–dc20   95-37485

Printed on acid-free paper.

Production managed by Robert Wexler; manufacturing supervised by Joe Quatela.
Photocomposed copy prepared by Bytheway Typesetting Services using Springer's svsing style file.
Printed and bound by R.R Donnelley and Sons, Harrisonburg, VA.
Printed in the United States of America.

9 8 7 6 5 4 3 2 1

ISBN 0-387-94582-2 Springer-Verlag New York Berlin Heidelberg SPIN 10508791

Dedicated to Salwa, Tarek, Raed, and Ghada

Preface

This book grew out of lecture notes I used in a course on difference equations that I taught at Trinity University for the past five years. The classes were largely populated by juniors and seniors majoring in Mathematics, Engineering, Chemistry, Computer Science, and Physics.

This book is intended to be used as a textbook for a course on difference equations at the level of both advanced undergraduate and beginning graduate. It may also be used as a supplement for engineering courses on discrete systems and control theory.

The main prerequisites for most of the material in this book are calculus and linear algebra. However, some topics in later chapters may require some rudiments of advanced calculus. Since many of the chapters in the book are independent, the instructor has great flexibility in choosing topics for the first one-semester course. A diagram showing the interdependence of the chapters in the book appears following the preface.

This book presents the current state of affairs in many areas such as stability, Z-transform, asymptoticity, oscillations and control theory. However, this book is by no means encyclopedic and does not contain many important topics, such as Numerical Analysis, Combinatorics, Special functions and orthogonal polynomials, boundary value problems, partial difference equations, chaos theory, and fractals. The nonselection of these topics is dictated not only by the limitations imposed by the elementary nature of this book, but also by the research interest (or lack thereof) of the author.

Great efforts were made to present even the most difficult material in an elementary format and to write in a style that makes the book accessible to students with varying backgrounds and interests. One of the main features of the book is the inclusion of a great number of applications in economics, social sciences, biology, physics, engineering, neural network, etc. Moreover, this book contains a very extensive and highly selected set of exercises at the end of each section. The exercises form an integral part of the text. They range from routine problems designed to build basic skills to more challenging problems that produce deeper understanding and build technique. The starred problems are the most challenging and the instructor may assign them as long-term projects. Another important feature of the book is that it encourages students to make mathematical discoveries through calculator/computer experimentation. I have included many programs for the calculator TI-85. Students are encouraged to improve these programs and to develop their own.

Chapter 1 deals with first order difference equations or one-dimensional maps on the real line. It includes a thorough and complete analysis of stability for many famous maps (equations) such as the Logistic map, the Tent map, and the Baker map. A rudiment of bifurcation and chaos theory is also included in Section 1.6. This section raises more questions and gives few answers. It is intended to arouse the readers' interest in this exciting field.

In Chapter 2 we give solution methods for linear difference equations of any order. Then we apply the obtained results to investigate the stability and the oscillatory behavior of second order difference equations. At the end of the chapter we give four applications: the propagation of annual plants, the gambler's ruin, the national income, and the transmission of information.

Chapter 3 extends the study in Chapter 2 to systems of difference equations. We introduce two methods to evaluate A^n. In Section 3.1, we introduce the Putzer algorithm and in Section 3.3, the method of the Jordan form is given. Many applications are then given in Section 3.5, which include Markov chains, trade models, and the heat equation.

Chapter 4 investigates the question of stability for both scalar equations and systems. Stability of nonlinear equations are studied via linearization (Section 4.5) and by the famous method of Liapunov (Section 4.6). Our exposition here is restricted to autonomous (time-invariant) systems. I believe that the extension of the theory to nonautonomous (time-variant) systems, though technically involved, will not add much more understanding to the subject matter.

Chapter 5 delves deeply into the Z-transform theory and techniques (Sections 5.1, 5.2). Then the results are applied to study the stability of Volterra difference scalar equations (Sections 5.3, 5.4) and systems (Sections 5.5, 5.6). For readers familiar with differential equations, Section 5.7 provides a comparison between the Z-transform and the Laplace transform. Most of the results on Volterra difference equations appear here for the first time in a book.

Chapter 6 takes us to the realm of control theory. Here we cover most of the basic concepts including controllability, observability, observers, and stabilizability by feedback. Again we restrict the presentation to autonomous (time-invariant) systems since this is just an introduction to this vast and growing discipline. Moreover, most practitioners deal mainly with time-invariant systems.

In Chapter 7 we give a comprehensive and accessible study of asymptotic methods for difference equations. Starting from the Poincaré Theorem, the chapter covers most of the recent development in the subject. Section 7.4 (asymptotically diagonal systems) presents an extension of Levinson's theorem for differential equations. While in Section 7.5 we carry our study to nonlinear difference equations. Several open problems are given that would serve as topics for research projects.

Finally, Chapter 8 presents a brief introduction to oscillation theory. In Section 8.1 the basic results on oscillation for three-term linear difference equations are introduced. Extension of these results to nonlinear difference equations is presented in Section 8.2. Another approach to oscillation theory, for self-adjoint equations, is presented in Section 8.3. Here we also introduce a discrete version of Sturm's separation theorem.

Suggestions for Use of the Text

The diagram shows the interdependence of the chapters

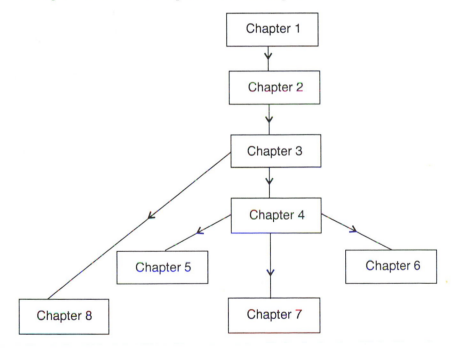

Suggestions for a First Course

i. If you want a course that emphasizes stability and control, then you may select Chapters 1, 2, and 6 and parts of 3, 4, and 5.

ii. For a course on the classical theory of difference equations, one may include Chapters 1, 2, and 8 and parts of 3, 4, and 7.

I am indebted to Gerry Ladas, who read many parts of the book and suggested many useful improvements, especially within the section on stability of scalar difference equations (Section 4.3). His influence through papers and lectures on Chapter 8 (oscillation theory) is immeasurable. My thanks go to Vlajko Kocic who thoroughly read and made many helpful comments about Chapter 4 on Stability. Jim McDonald revised the chapters on the Z-transform and Control Theory (Chapters 5 and 6) and made significant improvements. I am very grateful to him for his contributions to this book. My sincere thanks go to Paul Ehlo, who read the entire manuscript and offered valuable suggestions that led to many improvements in the final draft of the book. I am also grateful to Istvan Gyori for his comments on Chapter 8 and to Ronald Mickens for his review of the whole manuscript and for his advice and support. I would like to thank the following mathematicians who encouraged and helped me in numerous ways during the preparation of the book: Allan Peterson, Donald Bailey, Roberto Hasfura, Haydar Akça, and Shunian Zhang. I am grateful to my students Jeff Bator, Michelle MacArthur, and Nhung Tran who caught misprints and mistakes in the earlier drafts of this book. My special thanks are due to my student Julie Lundquist, who proofread most of the book and made improvements in the presentation of many topics.

My thanks go to Connie Garcia who skillfully typed the entire manuscript with its many many revised versions. And finally, it is a pleasure to thank Ina Lindemann and Robert Wexler from Springer-Verlag for their enthusiastic support of this project.

Saber Elaydi
San Antonio, TX 1995

Contents

1

Dynamics of First Order Difference Equations

1.1 Introduction

Difference equations usually describe the evolution of certain phenomena over the course of time. For example, if a certain population has discrete generations, the size of the $n + 1$st generation $x(n + 1)$ is a function of the nth generation $x(n)$. This relation expresses itself in the *difference equation*

$$x(n + 1) = f(x(n)). \qquad (1.1.1)$$

We may look at this problem from another point of view. Starting from a point x_0, one may generate the sequence

$$x_0, \, f(x_0), \, f(f(x_0)), \, f(f(f(x_0))), \, \dots .$$

For convenience we adopt the notation

$$f^2(x_0) = f(f(x_0)), \, f^3(x_0) = f(f(f(x_0))), \text{ etc.}$$

$f(x_0)$ is called the *first iterate* of x_0 under f, $f^2(x_0)$ is called the second iterate of x_0 under f, and more generally $f^n(x_0)$ is the nth iterate of x_0 under f. The set of all (positive) iterates $\{f^n(x_0) : n \geq 0\}$ is called the *(positive) orbit* of x_0, and will be denoted by $O^+(x_0)$. This iterative procedure is an example of a *discrete dynamical system*. Letting $x(n) = f^n(x_0)$, we have

$$x(n + 1) = f^{n+1}(x_0) = f[f^n(x_0)] = f(x(n))$$

and hence we recapture Eq. (1.1.1). Observe that $x(0) = f^0(x_0) = x_0$. For example, let $f(x) = x^2$ and $x_0 = 0.6$. To find the sequence of iterates $\{f^n(x_0)\}$, we key 0.6 into a calculator, and then repeatedly depress the x^2 button. We obtain the numbers

$$0.6, 0.36, 0.1296, 0.01679616, \ldots.$$

A few more key strokes on the calculator will be enough to convince the reader that the iterates $f^n(0.6)$ tend to 0. The reader is invited to verify that for all $x_0 \in (0, 1)$, $f^n(x_0)$ tends to 0 as n tends to ∞ and that $f^n(x_0)$ tends to ∞ if $x_0 \notin [-1, 1]$. Obviously, $f^n(0) = 0$, $f^n(1) = 1$ for all positive integers n, and $f^n(-1) = 1$ for $n = 1, 2, 3, \ldots$.

After this discussion one may conclude correctly that difference equations and discrete dynamical systems represent two sides of the same coin. For instance, when mathematicians talk about difference equations, they usually refer to the analytic theory of the subject and when they talk about discrete dynamical systems, they generally refer to its geometrical and topological aspects.

If the function f in Eq. (1.1.1) is replaced by a function g of two variables, that is $g: Z^+ \times R \to R$, where Z^+ is the set of positive integers and R is the set of real numbers, then we have

$$x(n + 1) = g(n, x(n)). \tag{1.1.2}$$

Equation (1.1.2) is called *nonautonomous* or time-variant whereas Eq. (1.1.1) is called *autonomous* or time-invariant. The study of Eq. (1.1.2) is much more complicated and does not lend itself to discrete dynamical system theory of first order equations. If an initial condition $x(n_0) = x_0$ is given, then for $n \geq n_0$ there is a *unique* solution $x(n) \equiv x(n, n_0, x_0)$ of Eq. (1.1.2) such that $x(n_0, n_0, x_0) = x_0$. This may be shown easily by iteration. Now

$$
\begin{aligned}
x(n_0 + 1, n_0, x_0) &= g(n_0, x(n_0)) \\
&= g(n_0, x_0), \\
x(n_0 + 2, n_0, x_0) &= g(n_0 + 1, x(n_0 + 1)) \\
&= g(n_0 + 1, f(n_0, x_0)), \\
x(n_0 + 3, n_0, x_0) &= g(n_0 + 2, x(n_0 + 2)) \\
&= g[n_0 + 2, f(n_0 + 1, f(n_0, x_0))].
\end{aligned}
$$

And inductively we get

$$x(n, n_0, x_0) = g[n - 1, x(n - 1, n_0, x_0)].$$

1.2   Linear First Order Difference Equations

In this section we study the simplest special cases of Eqs. (1.1.1) and (1.1.2), namely, linear equations. A typical linear *homogeneous* first order equation is given by

$$x(n + 1) = a(n)x(n), \qquad x(n_0) = x_0, \qquad n \geq n_0 \geq 0 \tag{1.2.1}$$

and the associated *nonhomogeneous* equation given by

$$y(n + 1) = a(n)y(n) + g(n), \qquad y(n_0) = y_0, \qquad n \geq n_0 \geq 0 \qquad (1.2.2)$$

where in both equations, it is assumed that $a(n) \neq 0$, $a(n)$ and $g(n)$ are real-valued functions defined for $n \geq n_0 \geq 0$.

One may obtain the solution of Eq. (1.2.1) by a simple iteration.

$$\begin{aligned}
x(n_0 + 1) &= a(n_0)x(n_0) = a(n_0)x_0, \\
x(n_0 + 2) &= a(n_0 + 1)x(n_0 + 1) = a(n_0 + 1)a(n_0)x_0, \\
x(n_0 + 3) &= a(n_0 + 2)x(n_0 + 2) = a(n_0 + 2)a(n_0 + 1)a(n_0)x_0.
\end{aligned}$$

And inductively, it is easy to see that

$$\begin{aligned}
x(n) &= x(n_0 + n - n_0)) \\
&= a(n - 1)a(n - 2)\ldots a(n_0)x_0 \\
&= \left[\prod_{i=n_0}^{n-1} a(i)\right] x_0. \qquad (1.2.3)
\end{aligned}$$

The unique solution of the *nonhomogeneous* Eq. (1.2.2) may be found as follows:

$$\begin{aligned}
y(n_0 + 1) &= a(n_0)y_0 + g(n_0), \\
y(n_0 + 2) &= a(n_0 + 1)y(n_0 + 1) + g(n_0 + 1) \\
&= a(n_0 + 1)a(n_0)y_0 + a(n_0 + 1)g(n_0) + g(n_0 + 1).
\end{aligned}$$

Now we use mathematical induction to show that for all $n \in Z^+$,

$$y(n) = \left[\prod_{i=n_0}^{n-1} a(i)\right] y_0 + \sum_{r=n_0}^{n-1}\left[\prod_{i=r+1}^{n-1} a(i)\right] g(r). \qquad (1.2.4)$$

To establish this, assume that Formula (1.2.4) holds for $n = k$. Then from Eq. (1.2.2), $y(k + 1) = a(k)y(k) + g(k)$, which by Formula (1.2.4) yields

$$\begin{aligned}
y(k + 1) &= a(k)\left[\prod_{i=n_0}^{k-1} a(i)\right] y_0 + \sum_{r=n_0}^{k-1}\left[a(k)\prod_{i=r+1}^{k-1} a(i)\right] g(r) + g(k) \\
&= \left[\prod_{i=n_0}^{k} a(i)\right] y_0 + \sum_{r=n_0}^{k-1}\left(\prod_{i=r+1}^{k} a(i)\right) g(r) \\
&\quad + \left(\prod_{i=k+1}^{k} a(i)\right) g(k) \text{ (see footnote 1)} \\
&= \left[\prod_{i=n_0}^{k} a(i)\right] y_0 + \sum_{r=n_0}^{k}\left(\prod_{i=r+1}^{k} a(i)\right) g(r).
\end{aligned}$$

Hence Formula (1.2.4) holds for all $n \in Z^+$.

---

[1] Notice that we have adopted the notation $\prod_{i=k+1}^{k} a(i) = 1$ and $\sum_{i=k+1}^{k} a(i) = 0$.

1.2.1 *Important Special Cases*

There are two special cases of Eq. (1.2.2) which are important in many applications. The first equation is given by

$$y(n + 1) = ay(n) + g(n), \quad y(0) = y_0. \tag{1.2.5}$$

Using Formula (1.2.4) one may establish that

$$y(n) = a^n y_0 + \sum_{k=0}^{n-1} a^{n-k-1} g(k). \tag{1.2.6}$$

The second equation is given by

$$y(n + 1) = ay(n) + b, \qquad y(0) = y_0. \tag{1.2.7}$$

Using Formula (1.2.6) we obtain

$$y(n) = \begin{cases} a^n y_0 + b \left[\frac{a^n - 1}{a - 1} \right] & \text{if } a \neq 1 \\ y_0 + bn & \text{if } a = 1 \end{cases}. \tag{1.2.8}$$

We now give some examples to practice the above formulas.

Example 1.1. Solve the equation

$$y(n + 1) = (n + 1)y(n) + 2^n(n + 1)!, \qquad y(0) = 1, \qquad n > 0.$$

Solution

$$y(n) = \prod_{i=0}^{n-1}(i + 1) + \sum_{k=0}^{n-1} \left[\prod_{i=k+1}^{n-1}(i + 1) \right] 2^k(k + 1)!$$

$$= n! + \sum_{k=0}^{n-1} n! \, 2^k$$

$$= 2^n n! \text{ (from Table 1.2).}$$

Example 1.2. Find a solution for the equation

$$x(n + 1) = 2x(n) + 3^n, \qquad x(1) = 0.5.$$

Solution From Eq. (1.2.6), we have

$$x(n) = \left(\frac{1}{2} \right) 2^{n-1} + \sum_{k=1}^{n-1} 2^{n-k-1} 3^k$$

$$= 2^{n-2} + 2^{n-1} \sum_{k=1}^{n-1} \left(\frac{3}{2} \right)^k$$

$$= 2^{n-2} + 2^{n-1} \frac{3}{2} \left(\frac{\left(\frac{3}{2} \right)^{n-1} - 1}{3/2 - 1} \right)$$

$$= 3^n - 2^{n-3}.$$

Example 1.3. A drug is administered once every four hours. Let $D(n)$ be the amount of the drug in the blood system at the nth interval. The body eliminates a certain fraction p of the drug during each time interval. If the amount administered is D_0, find $D(n)$ and $\lim_{n\to\infty} D(n)$.

Solution We first must create an equation to solve. Since the amount of drug in the patient's system at time $(n+1)$ is equal to the amount at time n minus the fraction p that has been eliminated from the body, plus the new dosage D_0, we arrive at the following equation:

$$D(n+1) = (1-p)D(n) + D_0.$$

Using Eq. (1.2.8), we solve the above equation, arriving at

$$D(n) = \left[D_0 - \frac{D_0}{p}\right](1-p)^n + \frac{D_0}{p}.$$

Hence,

$$\lim_{n\to\infty} D(n) = \frac{D_0}{p}. \tag{1.2.9}$$

Let $D_0 = 2$ cubic centimeters (cc), $p = 0.25$.

Then our original equation becomes

$$D(n+1) = 0.75 D(n) + 2, \qquad D(0) = 2.$$

Table (1.1) gives $D(n)$ for $0 \leq n \leq 10$.

It follows from Eq. (1.2.9) that $\lim_{n\to\infty} D(n) = 8$, where $D^* = 8$ cc is the equilibrium amount of drug in the body. We now enter the realm of finance for our next example.

Example 1.4 (Amortization). Amortization is the process by which a loan is repaid by a sequence of periodic payments, each of which is part payment of interest and part payment to reduce the outstanding principal.

Let $p(n)$ represent the outstanding principal after the nth payment $g(n)$. Suppose that interest charges compound at the rate r per payment period.

The formulation of our model here is based on the fact that the outstanding principal $p(n+1)$ after the $(n+1)$st payment is equal to the outstanding principal $p(n)$ after the nth payment plus the interest $rp(n)$ incurred during the $(n+1)$st period minus the nth payment $g(n)$. Hence

$$\begin{aligned} p(n+1) &= p(n) + rp(n) - g(n) \text{ or} \\ p(n+1) &= (1+r)p(n) - g(n), \qquad p(0) = p_0, \end{aligned} \tag{1.2.10}$$

Table 1.1. Values of $D(n)$

n	0	1	2	3	4	5	6	7	8	9	10
$D(n)$	2	3.5	4.62	5.47	6.1	6.58	6.93	7.2	7.4	7.55	7.66

where p_0 is the initial debt. By Eq. (1.2.8) we have

$$p(n) = (1+r)^n p_0 - \sum_{k=0}^{n-1}(1+r)^{n-k-1} g(k). \qquad (1.2.11)$$

In practice, the payment $g(n)$ is constant and, say, equal to T. In this case,

$$p(n) = (1+r)^n p_0 - \left((1+r)^n - 1\right)\left(\frac{T}{r}\right). \qquad (1.2.12)$$

If we want to pay off the loan in exactly n payments, what would be the monthly payment T? Observe first that $p(n) = 0$. Hence from Eq. (1.2.12) we have

$$T = p_0 \left[\frac{r}{1 - (1+r)^{-n}}\right].$$

Exercises 1.1 and 1.2

1. Find the solution of each difference equation:

   (a) $x(n+1) - (n+1)x(n) = 0,$ $x(0) = c.$

   (b) $x(n+1) - 3^n x(n) = 0,$ $x(0) = c.$

   (c) $x(n+1) - e^{2n} x(n) = 0,$ $x(0) = c.$

   (d) $x(n+1) - \frac{n}{n+1} x(n) = 0,$ $n \geq 1,$ $x(1) = c.$

2. Find the general solution of each difference equation:

   (a) $y(n+1) - \frac{1}{2} y(n) = 2,$ $y(0) = c.$

   (b) $y(n+1) - \frac{n}{n+1} y(n) = 4,$ $y(0) = c.$

3. Find the general solution of each difference equation:

   (a) $y(n+1) - (n+1)y(n) = 2^n (n+1)!,$ $y(0) = c.$

   (b) $y(n+1) = y(n) + e^n,$ $y(0) = c.$

4. (a) Write a difference equation that describes the number of regions created by n lines in the plane if it is required that every pair of lines meet and no more than two lines meet at one point.

   (b) Find the number of these regions by solving the difference equation in case (a).

Table 1.2. Definite sum

Number	Summation	Definite sum
1	$\displaystyle\sum_{k=0}^{n} k$	$\dfrac{n(n+1)}{2}$
2	$\displaystyle\sum_{k=0}^{n} k^2$	$\dfrac{n(n+1)(2n+1)}{6}$
3	$\displaystyle\sum_{k=0}^{n} k^3$	$\left[\dfrac{n(n+1)}{2}\right]^2$
4	$\displaystyle\sum_{k=0}^{n} k^4$	$\dfrac{n(6n^4+15n^3+10n^2-1)}{30}$
5	$\displaystyle\sum_{k=0}^{n} a^k$	$\begin{cases}(a^{n+1}-1)/a-1 & \text{if } a \neq 1 \\ n+1 & \text{if } a = 1\end{cases}$
6	$\displaystyle\sum_{k=0}^{n} ka^k,\ \ a \neq 1$	$\dfrac{(a-1)(n+1)a^{n+1}-a^{n+2}+a}{(a-1)^2}$

5. The gamma function is defined as $\Gamma(x) = \int_0^\infty t^{x-1} e^{-t} dt, \qquad x > 0.$

   (a) Show that $\Gamma(x+1) = x\Gamma(x), \qquad \Gamma(1) = 1.$
   (b) If n is a positive integer, show that $\Gamma(n+1) = n!$.
   (c) Show that $x^{(n)} = \dfrac{\Gamma(x+1)}{\Gamma(x-n+1)}.$

6. A space (three dimensional) is divided by n planes, nonparallel, and no four copunctal.

   (a) Write a difference equation that describes the number of regions created.
   (b) Find the number of these regions.

7. Verify Eq. 1.2.6.

8. Verify Eq. 1.2.8.

9. A debt of \$12,000 is to be amortized by equal payments of \$380 at the end of each month, plus a final partial payment one month after the last \$380 is paid. If the interest is at the annual rate of 12% compounded monthly, construct an amortization schedule to show the required payments.

10. Suppose that a loan of \$80,000 is to be amortized by equal monthly payments. If the interest rate is at the rate of 10% compounded monthly, find the monthly payment required to pay off the loan in 30 years.

11. Suppose the constant sum T is deposited at the end of each fixed period in a bank which pays interest rate r per period. Let $A(n)$ be the amount accumulated in the bank after n periods.

     (a) Write a difference equation that describes $A(n)$.

     (b) Solve the difference equation obtained in (a), when $A(0) = 0$, $T = \$200$, and $r = 0.008$.

12. The temperature of a body is measured as $110°F$. It is observed that the amount the temperature changes for each period of two hours is -0.3 times the difference between the previous period's temperature and the room temperature, which is $70°F$.

     (a) Write a difference equation that describes the temperature $T(n)$ of the body at the end of n periods.

     (b) Find $T(n)$.

13. Suppose that you can get a 30 year mortgage at 8% interest. How much can you afford to borrow if you can afford to make a monthly payment of $\$1,000$?

14. Radium decreases at the rate of 0.04% per year. What is its half-life? (A half-life of radioactive material is defined to be the time needed for half of the material to dissipate.)

15. Carbon Dating—It has been observed that the proportion of Carbon-14 in plants and animals is the same as in the atmosphere as long as the plant or the animal is alive. When an animal or plant dies the Carbon-14 in their tissue starts decaying at the rate r.

     (a) The half-life of a radioactive material is defined to be the time needed for half of the material to dissipate. If the half-life of Carbon-14 is 5,700 years, find r.

     (b) If the amount of Carbon-14 present in a bone of an animal is 70% of the original amount of Carbon-14, how old is the bone?

1.3   Equilibrium Points

The notion of equilibrium points (states) is central in the study of the dynamics of any physical system. In many applications in biology, economics, physics, engineering, etc., it is desirable that all states (solutions) of a given system tend to its equilibrium state (point). This is the subject of study of Stability Theory, a topic of great importance to scientists and engineers. We now give the formal definition of an equilibrium point.

Definition 1.5. A point $x*$ in the domain of f is said to be an equilibrium point of Eq. (1.1.1) if it is a fixed point of f, i.e., $f(x*) = x*$.

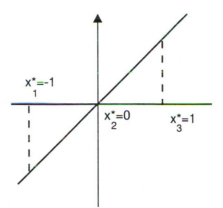

Figure 1.1. Fixed points of $f(x) = x^3$.

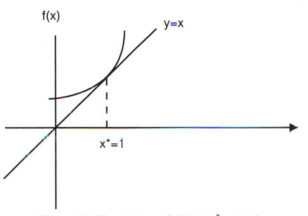

Figure 1.2. Fixed points of $f(x) = x^2 - x + 1$.

In other words x^* is a *constant solution* of Eq. (1.1.1), since if $x(0) = x^*$ is an initial point, then $x(1) = f(x^*) = x^*$ and $x(2) = f(x(1)) = f(x^*) = x^*$, and so on.

Graphically, an equilibrium point is the x coordinate of the point where the graph of f intersects the diagonal line $y = x$ (Figs. 1.1 and 1.2). For example there are three equilibrium points for the equation

$$x(n + 1) = x^3(n)$$

where $f(x) = x^3$. To find these equilibrium points, we let $f(x^*) = x^*$ or $x^3 = x$ and solve for x. Hence there are three equilibrium points $-1, 0, 1$ (Fig. 1.1).

Figure 1.2 illustrates another example where $f(x) = x^2 - x + 1$ and the difference equation is given by

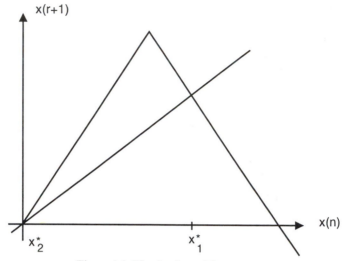

Figure 1.3. Fixed points of the tent map.

$$x(n + 1) = x^2(n) - x(n) + 1.$$

Letting $x^2 - x + 1 = x$ we find that 1 is the only equilibrium point.

There is a phenomenon that is unique to difference equations and cannot possibly occur in differential equations. It is possible in difference equations that a solution may not be an equilibrium point but may reach one after finitely many iterations. In other words, a nonequilibrium state may go to an equilibrium state in a finite time. This leads to the following definition.

Definition 1.6. Let x be a point in the domain of f. If there exists a positive integer r and an equilibrium point $x*$ of Eq. (1.1.1) such that $f^r(x) = x*$, $f^{r-1}(x) \neq x*$, then x is an eventually equilibrium point.

Example 1.7. (The tent map). Consider the equation

$$x(n + 1) = T(x(n))$$

where

$$T(x) = \begin{cases} 2x & \text{for } 0 \leq x \leq \dfrac{1}{2} \\ 2(1 - x) & \text{for } \dfrac{1}{2} < x \leq 1 \end{cases}.$$

There are two equilibrium points 0, and 2/3. The search for eventually equilibrium points is not as simple algebraically. If $x(0) = \frac{1}{4}$, then $x(1) = \frac{1}{2}$, $x(2) = 1$, and $x(3) = 0$. Thus $\frac{1}{4}$ is an eventually equilibrium point. The reader is asked to show that if $x = k/2^n$, where k and n are positive integers with $0 < k/2^n \leq 1$, then x is an eventually fixed point (Exercise 1.3, Problem 12).

One of the main objectives of dynamical systems is to study the behavior of its solutions near an equilibrium point. This study constitutes the so-called Stability Theory. Next we introduce the basic definitions of stability.

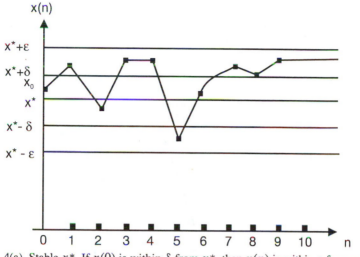

Figure 1.4(a). Stable x^*. If $x(0)$ is within δ from x^*, then $x(n)$ is within ε from $x(n)$ for all $n > 0$.

Definition 1.8. (a) The equilibrium point x^* of Eq. (1.1.1) is stable (Fig. 1.4a) if given $\varepsilon > 0$ there exists $\delta > 0$ such that $|x_0 - x^*| < \delta$ implies $|f^n(x_0) - x^*| < \varepsilon$ for all $n > 0$. If x^* is not stable, then it is called *unstable* (Fig. 1.4b).

(b) The point x^* is said to be a repelling (source) equilibrium point if there exists $\varepsilon > 0$ such that

$$0 < |x_0 - x^*| < \varepsilon \text{ implies } |f(x_0) - x^*| > |x_0 - x^*|$$

or equivalently

$$0 < |x(0) - x^*| < \varepsilon \text{ implies } |x(1) - x^*| > |x(0) - x^*| \text{ (Fig.1.4c)}$$

(c) The point x^* is an asymptotically stable (attracting) equilibrium point[2] if it is stable and there exists $\eta > 0$ such that

$$|x(0) - x^*| < \eta \text{ implies } \lim_{n \to \infty} x(n) = x^* \text{ (Fig.1.4d)}$$

If $\eta = \infty$, x^* is said to be globally asymptotically stable (Fig. 1.4e)

To determine the stability of an equilibrium point from the above definitions may prove to be a mission impossible in many cases. This is due to the fact that we may not be able to find the solution in a closed form even for the deceivingly simple-looking Eq. (1.1.1). In this section we present some of the most simple but powerful tools of the trade to help us understand the behavior of solutions of Eq. (1.1.1) in the vicinity of equilibrium points, namely, the graphical techniques. A hand-held calculator may now fulfill all your graphical needs in this section.

---

[2]In some of the literature of dynamical systems an attracting point is called a sink.

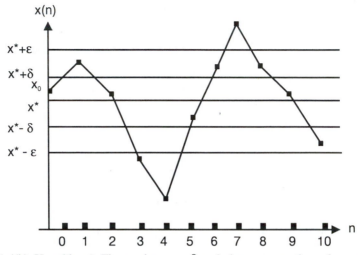

Figure 1.4(b). Unstable x^*. There exists $\varepsilon > 0$ such that no matter how close $x(0)$ is to x^*, there will be an N such that $x(N)$ is at least ε from x^*.

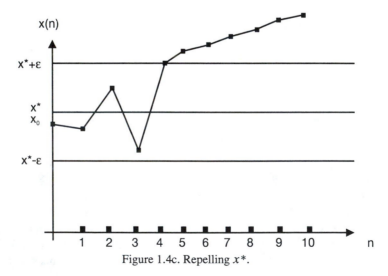

Figure 1.4c. Repelling x^*.

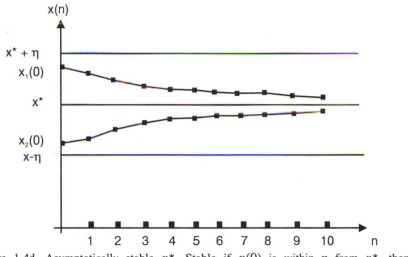
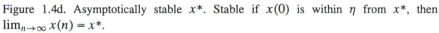

Figure 1.4d. Asymptotically stable x^*. Stable if $x(0)$ is within η from x^*, then $\lim_{n \to \infty} x(n) = x^*$.

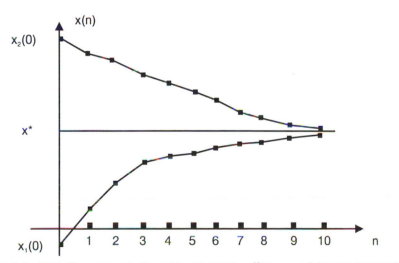

Figure 1.4e. Globally asymptotically stable x^*. Stable + $\lim_{n \to \infty} x(n) = x^*$ for all $x(0)$.

Example 1.9. (Numerical Solutions of Differential Equations). As we solve a differential equation numerically (by approximation), we actually use an associated difference equation, whether we realize it or not. Let us illustrate this development with *Euler's Method*, one of the simplest techniques for approximating the solutions of a differential equation.

Consider the first order differential equation

$$x'(t) = g(x(t)), \qquad x(t_0) = x_0, \qquad t_0 \le t \le b. \tag{1.3.1}$$

Let us divide the interval $[t_0, b]$ into N equal subintervals. The size of each subinterval is called the *step size* of the method and is denoted by $h = (b - t_0)/N$. This step size defines the *node* points $t_0, t_1, t_2, \ldots, t_N$, where $t_j = t_0 + jh$. Euler's method approximates $x'(t)$ by $(x(t + h) - x(t))/h$.

Substituting this value into Eq. (1.3.1) gives

$$x(t + h) = x(t) + hg(x(t)),$$

and for $t = t_0 + nh$, we obtain

$$x[t_0 + (n + 1)h] = x(t_0 + nh) + hg[x(t_0 + nh)] \tag{1.3.2}$$

for $n = 0, 1, 2, \ldots, N - 1$.

Adapting the difference equation notation, and replacing $x(t_0 + nh)$ by $x(n)$ gives

$$x(n + 1) = x(n) + hg[x(n)]. \tag{1.3.3}$$

Equation (1.3.3) defines the *Euler's Algorithm*, which approximates the solutions of the differential Eq. (1.3.1) at the node points.

Note that x^* is an equilibrium point of Eq. (1.3.3) if and only if $g(x^*) = 0$. Thus the differential Eq. (1.3.1) and the difference Eq. (1.3.3) have the same equilibrium points.

Let us now apply Euler's method to the differential equation:

$$x'(t) = 0.7x^2(t) + 0.7, \qquad x(0) = 1, \qquad t \in [0, 1] \qquad (DE) \text{ (see footnote 3)}$$

The exact solution of this equation is given by $x(t) = \tan\left(0.7t + \frac{\pi}{4}\right)$.

The corresponding difference equation using Euler's method is

$$x(n + 1) = x(n) + 0.7h(x^2(n) + 1), \qquad x(0) = 1 \qquad (\Delta E) \text{ (see footnote 4)}$$

Table 1.3 shows the Euler's approximations for $h = 0.2$ and 0.1, as well as the exact values. Figure 1.5 depicts the $(n, x(n))$ diagram. Notice that the smaller the step size we use, the better approximations we have.

---

[3]$DE \equiv$ Differential Equation.
[4]$\Delta E \equiv$ Difference Equation.

Table 1.3.

n	t	(ΔE) Euler $(h = 0.2)$ $x(n)$	(ΔE) Euler $(h = 0.1)$ $x(n)$	Exact (DE) $x(t)$
0	0	1	1	1
1	0.1		1.14	1.150
2	0.2	1.28	1.301	1.328
3	0.3		1.489	1.542
4	0.4	1.649	1.715	1.807
5	0.5		1.991	2.150
6	0.6	2.170	2.338	2.614
7	0.7		2.791	3.286
8	0.8	2.969	3.406	4.361
9	0.9		4.288	6.383
10	1	4.343	5.645	11.681

1.3.1   The Stair Step (Cobweb) Diagrams

We now give, in excruciating detail, another important graphical method for analyzing the stability of equilibrium (and periodic) points for Eq. (1.1.1). Since $x(n + 1) = f(x(n))$, we may draw a graph of f in the $(x(n), x(n + 1))$ plane. Then, given $x(0) = x_0$, we pinpoint the value $x(1)$ by drawing a vertical line through x_0 so that it also intersects the graph of f at $(x_0, x(1))$. Next, draw a horizontal line from $(x_0, x(1))$ to meet the diagonal line $y = x$ at the point $(x(1), x(1))$. A vertical line drawn from the point $(x(1), x(1))$ will meet the graph of f at the point $(x(1), x(2))$. Continuing this process, one may find $x(n)$ for all $n > 0$.

Example 1.10. The Logistic Equation. Let $y(n)$ be the size of a population at time n. If μ is the rate of growth of the population from one generation to another, then we may consider a mathematical model in the form

$$y(n + 1) = \mu y(n), \qquad \mu > 0. \tag{1.3.4}$$

If the initial population $y(0) = y_0$, then by simple iteration we find that

$$y(n) = \mu^n y_0 \tag{1.3.5}$$

is the solution of Eq. (1.3.4). If $\mu > 1$, then $y(n)$ increases indefinitely and $\lim_{n \to \infty} y(n) = \infty$. If $\mu = 1$, then $y(n) = y_0$ for all $n > 0$ which means that the size of the population is constant for the indefinite future. However, for $\mu < 1$, $\lim_{n \to \infty} y(n) = 0$ and the population becomes eventually extinct.

For most biological species, however, none of the above cases is valid — the population increases until it reaches a certain upper limit. Then, due to the limitations of available resources, the creatures will become testy, and engage in competition for those limited resources. This competition is proportional to the number of squabbles among them, given by $y^2(n)$. A more reasonable model would allow b,

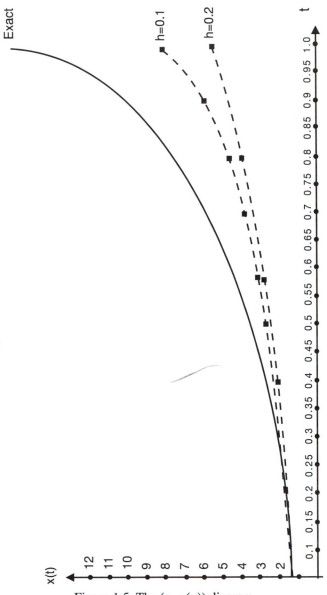

Figure 1.5. The $(n, x(n))$ diagram.

the proportionality constant, to be greater than 0,

$$y(n + 1) = \mu y(n) - by^2(n). \tag{1.3.6}$$

If, in Eq. (1.3.6), we let $x(n) = \frac{b}{\mu} y(n)$ we obtain

$$x(n + 1) = \mu x(n)(1 - x(n)) = f(x(n)). \tag{1.3.7}$$

This equation is the simplest nonlinear first order difference equation, commonly referred to as the (discrete) logistic equation. However, a closed form solution of Eq. (1.3.7) is not available (except for certain values of μ). In spite of its simplicity, this equation exhibits a rather rich and complicated dynamics. To find the equilibrium points of Eq. (1.3.7) we let $f(x^*) = \mu x^* (1 - x^*) = x^*$. Thus, we pinpoint two equilibrium points: $x^* = 0$ and $x^* = (\mu - 1)/\mu$.

Figure 1.6 gives the stair step diagram of $(x(n), x(n + 1))$ when $\mu = 2.5$ and $x(0) = 0.1$. In this case, we also have two equilibrium points. One, $x^* = 0$, is unstable, and the other, $x^* = 0.6$, is asymptotically stable.

Example 1.11. The Cobweb Phenomena (Economics applications). Here we study the pricing of a certain commodity. Let $S(n)$ be the number of units supplied in period n, $D(n)$ = number of units demanded in period n, and $p(n)$ = price per unit in period n.

For simplicity, we assume that $D(n)$ depends only linearly on $p(n)$ and is denoted by

$$D(n) = -m_d p(n) + b_d, \qquad m_d > 0, b_d > 0. \tag{1.3.8}$$

This equation is referred to as the price-demand curve. The constant m_d represents the sensitivity of consumers to price. We also assume that the price-supply curve relates the supply in any period with the price one period before, i.e.,

$$S(n + 1) = m_s p(n) + b_s, \qquad m_s > 0, \qquad b_s > 0. \tag{1.3.9}$$

The constant m_s is the sensitivity of suppliers to price. The slope of the demand curve is negative because an increase of one unit in price produces a decrease of m_d units in demand. Correspondingly, an increase of one unit in price causes an increase of m_s units in supply, creating a positive slope for that curve.

A third assumption we make here is that the market price is the price at which the quantity demanded and the quantity supplied are equal, that is when $D(n + 1) = S(n + 1)$.

Thus

$$-m_d p(n + 1) + b_d = m_s p(n) + b_s$$

or

$$p(n + 1) = Ap(n) + B = f(p(n)) \tag{1.3.10}$$

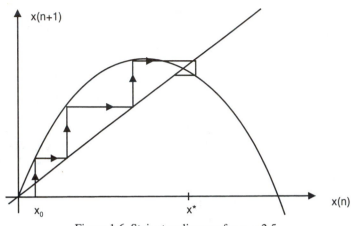

Figure 1.6. Stair step diagram for $\mu = 2.5$.

where
$$A = -\frac{m_s}{m_d}, \qquad B = \frac{b_d - b_s}{m_d}. \qquad (1.3.11)$$

This equation is a first order linear difference equation. The equilibrium price p^* is defined in economics as the price which results in an intersection of the supply $S(n+1)$ and demand $D(n)$ curves. Also, since p^* is the unique fixed point of $f(p)$ in (1.3.10), $p^* = B/(1 - A)$. (This proof arises later as Exercise 1.3, Problem 6.) Because A is the ratio of the slopes of the supply and demand curves, this ratio determines the behavior of the price sequence. There are three cases to be considered:

(a) $-1 < A < 0$

(b) $A = -1$

(c) $A < -1$.

The three cases are now depicted graphically using our old stand-by, the stair step diagram.

(i) In case a, prices alternate above and below, but converge to the equilibrium price p^*. In economics lingo, the price p^* is considered "stable"; in mathematics, we refer to it as "asymptotically stable" (Fig. 1.7a).

(ii) In case b, prices oscillate between two values only. If $p(0) = p_0$, then $p(1) = -p_0 + B$ and $p(2) = p_0$. Hence the equilibrium point p^* is stable (Fig. 1.7b).

(iii) In case c, prices oscillate infinitely about the equilibrium point p^* but progressively move further away from it. Thus, the equilibrium point is considered unstable (Fig. 1.7c).

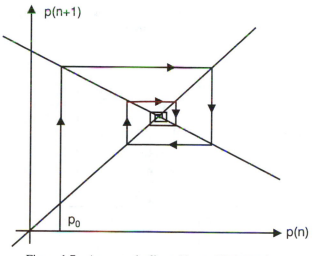
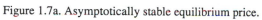

Figure 1.7a. Asymptotically stable equilibrium price.

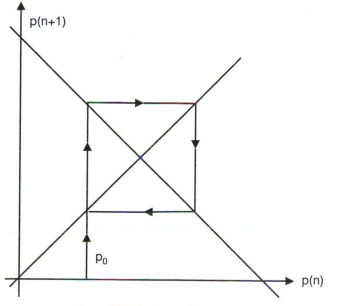

Figure 1.7b. Stable equilibrium price.

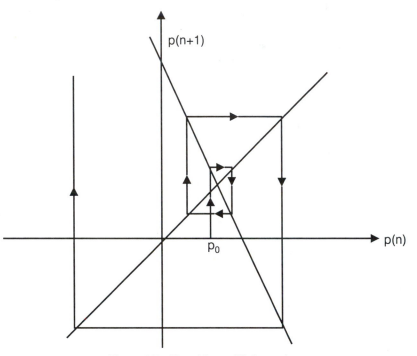

Figure 1.7c. Unstable equilibrium price.

An explicit solution of Eq. (1.3.10) with $p(0) = p_0$ is given by

$$p(n) = \left(p_0 - \frac{B}{1-A} \right) A^n + \frac{B}{1-A} \quad \text{(Exercise 1.3, Problem 9).} \quad (1.3.12)$$

This explicit solution allows us to restate Cases a and b as follows.

1.3.2   The Cobweb Theorem of Economics

If the suppliers are less sensitive to price than the consumers (i.e., $m_s < m_d$), the market will then be stable. If the suppliers are more sensitive than the consumers, the market will be unstable.

One might also find the closed form solution (1.3.12) by using a computer algebra program, such as Maple. One would enter this program:

$$\text{rsolve}(\{p(n + 1) = a * p(n) + b, \ p(0) = p_0\}, \ p(n));$$

Exercise 1.3.

1. Contemplate the equation $x(n + 1) = f(x(n))$, where $f(0) = 0$.

    (a) Prove that $x(n) \equiv 0$ is a solution of the equation.
    (b) Show that the function depicted in the $(n, x(n))$ diagram cannot possibly be a solution of the equation.

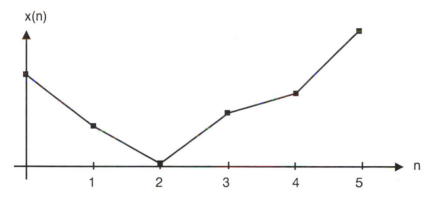

2. Newton's Method of Computing the Square Root of a Positive Number:

    The equation $x^2 = a$ can be written in the form $x = \frac{1}{2}(x + a/x)$. This form leads to Newton's method

    $$x(n + 1) = \frac{1}{2}\left[x(n) + \frac{a}{x(n)}\right].$$

    (a) Show that this difference equation has two equilibrium points $-\sqrt{a}$ and $\sqrt{a}$.
    (b) Sketch a stair step diagram for $a = 3$, $x(0) = 1$, and $x(0) = -1$
    (c) What can you conclude from (b)?

3. Pielou's Logistic Equation:

    E.C. Pielou [1] referred to the following equation as the discrete logistic equation:

    $$x(n + 1) = \frac{\alpha x(n)}{1 + \beta x(n)}, \qquad \alpha > 1, \qquad \beta < 0.$$

    (a) Find the positive equilibrium point.
    (b) Demonstrate, using the stair step diagram, that the positive equilibrium point is asymptotically stable, taking $\alpha = 2$ and $\beta = 1$.

4. Find the equilibrium points and determine their stability for the equation

    $$x(n + 1) = 5 - \frac{6}{x(n)}.$$

5. (a) Draw a stair step diagram for the Eq. (1.3.7) for $\mu = 0.5$, 3, and 3.3. What can you conclude from these diagrams?

    (b) Determine whether these values for μ given rise to periodic solutions of period 2.

6. The Cobweb Phenomenon [Eq. (1.3.10)] Economists define the equilibrium price $p*$ of a commodity as the price at which the demand function $D(n)$ is equal to the supply function $S(n+1)$, in Eqs. (1.3.8) and (1.3.9), respectively.

    (a) Show that $p* = \frac{B}{1-A}$, where A and B are defined as in Eq. (1.3.11).

    (b) Let $m_s = 2$, $b_s = 3$, $m_d = 1$, and $b_d = 15$. Find the equilibrium price $p*$. Then draw a stair step diagram for $p(0) = 2$.

7. Continuation of Problem 6:

    Economists use a different stair step diagram as we will explain in the following steps:

    (i) Let the x axis represent the price $p(n)$ and the y axis represent $s(n+1)$ or $D(n)$. Draw the supply line and the demand line and find their point of intersection $p*$.

    (ii) Starting with $p(0) = 2$ we find $s(1)$ by moving vertically to the supply line, then move horizontally to find $D(1)$ (since $D(1) = S(1)$), which determines $p(1)$ on the price axis. The supply $S(2)$ is found on the supply line directly above $p(1)$, and then $D(2)(= S(2))$ is found by moving horizontally to the demand line, etc.

    (iii) is $p*$ stable?

8. Repeat Exercises 6 and 7 for

    (a) $m_s = m_d = 2$, $b_d = 10$, and $b_s = 2$, and

    (b) $m_s = 1$, $m_d = 2$, $b_d = 14$, and $b_s = 2$.

9. Verify that Formula (1.3.12) is a solution of Eq. (1.3.10).

10. Use Formula (1.3.12) to show that

    (a) if $-1 < A < 0$, then $\lim_{n \to \infty} p(n) = B/1 - A$;

    (b) if $A < -1$, then $p(n)$ is unbounded;

    (c) if $A = -1$, then $p(n)$ takes only two values:

$$
p(n) = \begin{cases} p(0) \text{ if } n \text{ is even} \\ p(1) = \dfrac{2B}{1 - A} - p_0 \text{ if } n \text{ is odd} \end{cases}
$$

11. Suppose that the supply and demand equations are given by $D(n) = -2p(n) + 3$ and $S(n + 1) = p^2(n) + 1$.

  (a) Assuming the market price is the price at which supply equals demand, find a difference equation that relates $p(n + 1)$ to $p(n)$.

  (b) Find the positive equilibrium value of this equation.

  (c) Use the stair step diagrams to determine the stability of the positive equilibrium value.

12. Consider the Baker's map defined by

$$B(x) = \begin{cases} 2x & \text{for } 0 \leq x \leq \frac{1}{2} \\ 2x - 1 & \text{for } \frac{1}{2} < x \leq 1 \end{cases}.$$

  (i) Draw the function $B(x)$ on $[0,1]$.

  (ii) Show that $x \in [0, 1]$ is an eventually fixed point if and only if it is of the form $x = k/2^n$, where k and n are positive integers,[5] with $0 \leq k \leq 2^n - 1$.

13. Find the fixed points and the eventually fixed points of $x(n + 1) = f(x(n))$, where $f(x) = x^2$.

14. Find an eventually fixed point of the tent map of Example 1.7 which is not in the form of $k/2^n$.

15. Consider the tent map of Example 1.7. Show that if $x = k/2^n$, where k and n are positive integers with $0 < k/2^n \leq 1$, then x is an eventually fixed point.

In Nos. 16 through 18,

  (a) Find the associated difference equation.

  (b) Draw an $(n, x(n))$ diagram.

  (c) Find, if possible, the exact solution of the differential equation and draw its graph on the same plot in (b).

16. $y' = -y^2$, $y(0) = 1$, $0 \leq t \leq 1$, $h = 0.2, 0.1$.

17. $y' = -y + \frac{4}{y}$, $y(0) = 1$, $0 \leq t \leq 1$, $h = 0.25$.

18. $y' = -y + 1$, $y(0) = 1$, $0 \leq t \leq 1$, $h = 0.25$.

---

[5] A number $x \in [0,1]$ is called a dyadic rational if it has the form $k/2^n$ from some nonnegative integers k and n, with $0 \leq k \leq 2^{n-1}$.

1.4   Criteria for Asymptotic Stability of Equilibrium Points

In this section, we give a simple but powerful criterion for the asymptotic stability of equilibrium points. The following theorem is our main tool in this section.

Theorem 1.12. Let x^* be an equilibrium point of the difference equation

$$x(n + 1) = f(x(n)) \tag{1.4.1}$$

where f is continuously differentiable at x^*. The following statements then hold true:

(i) if $|f'(x^*)| < 1$, then x^* is an asymptotically stable *(attracting) point*.

(ii) if $|f'(x^*)| > 1$, then x^* is not stable. In fact, x^* is a *repelling point*.

Proof

(i) Suppose that $|f'(x^*)| < M < 1$. Then there is an interval $J = (x^*-\gamma, x^*+\gamma)$ containing x^* such that $|f'(x)| \le M < 1$ for all $x \in J$. (Why? Exercise 1.4, Problem 9.) For $x(0) \in J$, we have

$$|x(1) - x^*| = |f(x(0)) - f(x^*)|.$$

By the Mean Value Theorem, there exists ξ between $x(0)$ and x^* such that

$$|f(x(0)) - f(x^*)| = |f'(\xi)|\,|x(0) - x^*|.$$

Thus

$$|f(x(0)) - x^*| \le M|x(0) - x^*|.$$

Hence

$$|x(1) - x^*| \le M|x(0) - x^*|. \tag{1.4.2}$$

Since $M < 1$, inequality (1.4.2) shows that $x(1)$ is closer to x^* than $x(0)$. Consequently, $x(1) \in J$.

By induction we conclude that

$$|x(n) - x^*| \le M^n|x(0) - x^*|.$$

For $\varepsilon > 0$ we let $\delta = \frac{\varepsilon}{2M}$. Thus $|x(0)-x^*| < \delta$ implies that $|x(n)-x^*| < \varepsilon$ for all $n \ge 0$. This conclusion suggests stability. Furthermore, $\lim_{n\to\infty} |x(n) - x^*| = 0$, and thus, $\lim_{n\to\infty} x(n) = x^*$; we conclude asymptotic stability.

The proof of part (ii) is left as Exercise 1.4, Problem 11.

Remark: In the literature of dynamical systems, the equilibrium point x^* is said to be *hyperbolic* if $|f'(x^*)| \ne 1$. So in part (i), x^* is a *hyperbolic attracting* point and in part (ii), it is a *hyperbolic repelling* point.

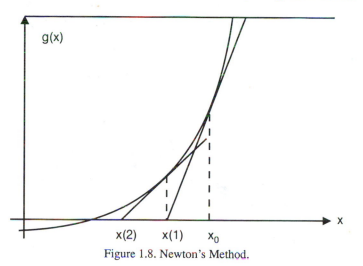

Figure 1.8. Newton's Method.

Example 1.13. The Newton–Raphson Method

The Newton–Raphson method is one of the most famous numerical methods for finding the roots of the equation $g(x) = 0$.

The Newton's algorithm for finding a zero x^* of $g(x)$ is given by the difference equation

$$x(n + 1) = x(n) - \frac{g(x(n))}{g'(x(n))}, \tag{1.4.3}$$

where $x(0) = x_0$ is your initial guess of the root x^*. Here $f(x) = x - \frac{g(x)}{g'(x)}$.

Note first that the zero x^* of $g(x)$ is also an equilibrium point of Eq. (1.4.3). To determine if Newton's algorithm provides a sequence $\{x(n)\}$ that converges to x^* we use Theorem 1.12

$$|f'(x^*)| = |1 - \frac{[g'(x^*)]^2 - g(x^*)g''(x^*)}{[g'(x^*)]^2}| = 0,$$

since $g(x^*) = 0$. By Theorem 1.12, $\lim_{n \to \infty} x(n) = x^*$ if $x(0) = x_0$ is close enough to x^* and $g'(x^*) \neq 0$.

Observe that Theorem 1.12 does not address the case where $|f'(x^*)| = 1$. Further analysis is needed here to determine the stability of the equilibrium point x^*. Our first discussion will address the case where $f'(x^*) = 1$.

Theorem 1.14. Suppose that for an equilibrium point x^* of Eq. (1.4.1), $f'(x^*) = 1$. The following statements then hold:

(i) If $f''(x^*) \neq 0$, then x^* is unstable.

(ii) If $f''(x^*) = 0$ and $f'''(x^*) > 0$, then x^* is unstable.

(iii) If $f''(x^*) = 0$ and $f'''(x^*) < 0$, then x^* is asymptotically stable.

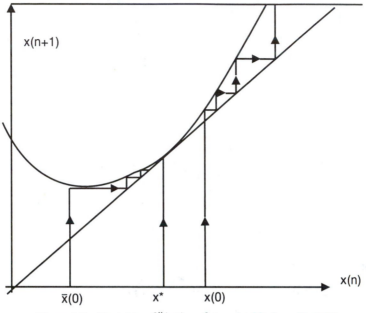

Figure 1.9a. Unstable. $f''(x*) > 0$ (semistable from the left).

Proof

(i) If $f''(x*) \neq 0$, then the curve is either concave upward where $f''(x*) > 0$ or concave downward where $f''(x*) < 0$, as shown in Figs. 1.9a–d. If $f''(x*) > 0$, then $f'(x) > 1$ for all x in a small interval $I = (x*, x* + \varepsilon)$. Using the same proof as in Theorem 1.12, it is easy to show that $x*$ is unstable. On the other hand if $f''(x*) < 0$, then $f'(x) > 1$ for all x in a small interval $I = (x* - \varepsilon, x*)$. Hence $x*$ is again unstable.

Proofs of parts (ii) and (iii) remain for the student's pleasure as Exercise 21.4, Problems 14 and 15.

We now use the preceding result to investigate the case when $f'(x*) = -1$.

Theorem 1.15. Suppose that, for the equilibrium point $x*$ of Eq. (1.1.1), $f'(x*) = -1$. The following statements then hold.

(i) If $-2f'''(x*) - 3[f''(x*)]^2 < 0$, then $x*$ is asymptotically stable.

(ii) If $-2f'''(x*) - 3[f''(x*)]^2 > 0$, then $x*$ is unstable.

Proof Contemplate the equation

$$y(n+1) = g(y(n)), \text{ where } g(y) = f^2(y). \tag{1.4.4}$$

We will make two observations about Eq. (1.4.4). First, the equilibrium point $x*$ of Eq. (1.1.1) is also an equilibrium point of Eq. (1.4.4). Second, if $x*$ is

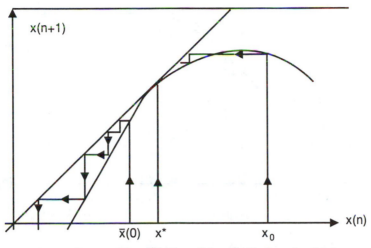

Figure 1.9b. Unstable. $f''(x*) < 0$ (semistable from the right).

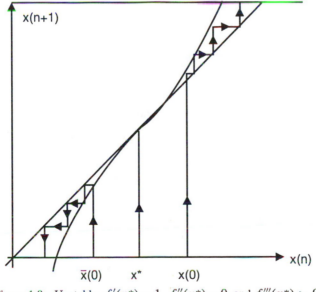

Figure 1.9c. Unstable. $f'(x*) = 1$, $f''(x*) = 0$, and $f'''(x*) > 0$.

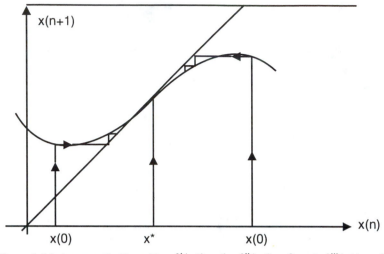

Figure 1.9d. Asymptotically stable. $f'(x^*) = 1$, $f''(x^*) = 0$, and $f'''(x^*) < 0$.

asymptotically stable (unstable) with respect to Eq. (1.4.4), then it is so with respect to Eq. (1.1.1). Why? (Exercise 1.4, Problem 13).

$$\text{Now } \frac{d}{dy} g(y) = \frac{d}{dy} f(f(y)) = f'(f(y)) f'(y).$$

Thus $\frac{d}{dy} g(x^*) = [f'(x^*)]^2 = 1$. Hence Theorem 1.14 applies to this situation. We need to evaluate $\frac{d^2}{dy^2} g(x^*)$.

$$\begin{aligned}
\frac{d^2}{dy^2} g(y) &= \frac{d^2}{dy^2} f(f(y)) \\
&= [f'(f(y)) f'(y)]' \\
&= [f'(y)]^2 f''(f(y)) + f'(f(y)) f''(y).
\end{aligned}$$

Hence

$$\frac{d^2}{dy^2} g(x^*) = 0.$$

Now Theorem 1.14 [parts (ii) and (iii)] tells us that the asymptotic stability of x^* is determined by the sign of $[g(x^*)]'''$. Using again the chain rule, one may show that

$$[g(x^*)]''' = -2 f'''(x^*) - 3[f''(x^*)]^2. \tag{1.4.5}$$

(The explicit proof with the chain rule remains as Exercise 1.4. Problem 13). This step rewards us with parts (i) and (ii), and the proof of the theorem is now complete.

Example 1.16. Consider the difference equation $x(n + 1) = x^2(n) + 3x(n)$. Find the equilibrium points and determine their stability.

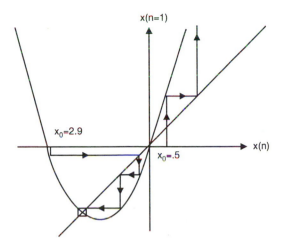

Figure 1.10. Stair step diagram for $x(n+1) = x^2(n) + 3x(n)$.

Solution The equilibrium points are 0 and -2. Now, $f'(x) = 2x+3$. Since $f'(0) = 3$, it follows from Theorem 1.12 that 0 is unstable. Now $f'(-2) = -1$, so Theorem 1.15 applies. Using Eq. (1.4.5), we obtain $-2f'''(-2) - 3[f''(-2)]^2 = -12 < 0$. Theorem 1.15 then declares that the equilibrium point -2 is asymptotically stable. Figure 1.10 illustrates the stair step diagram of the equation.

Remark: One may generalize the result in the preceding example to a general quadratic map $Q(x) = ax^2 + bx + c, a \neq 0$. It may be shown that

(i) if $f'(x^*) = -1$, then by Theorem 1.15 the equilibrium point is attracting, and

(ii) if $f'(x^*) = 1$, then by Theorem 1.14, x^* is unstable (Exercise 1.4, Problem 8)

Exercise 1.4

In Problems 1 through 7, find the equilibrium points and determine their stability using Theorems 1.12, 1.14, and 1.15.

1. $x(n+1) = \frac{1}{2}[x^3(n) + x(n)]$.

2. $x(n+1) = x^2(n) + \frac{1}{8}$.

3. $x(n+1) = \tan^{-1} x(n)$.

4. $x(n+1) = x^2(n)$.

5. $x(n+1) = x^3(n) + x(n)$.

6. $x(n+1) = \dfrac{\alpha x(n)}{1 + \beta x(n)}$, $\alpha > 1$ and $\beta > 0$.

7. $x(n + 1) = -x^3(n) - x(n)$.

8. Let $Q(x) = ax^2 + bx + c$, $a \neq 0$, and $x*$ be a fixed point of Q. Prove the following statements:

   (i) If $f'(x*) = -1$, then $x*$ is attracting.

   (ii) If $f'(x*) = 1$, then $x*$ is unstable.

9. Show that if $|f'(x*)| < 1$, there exists an interval $J = (x* - \varepsilon, x* + \varepsilon)$ such that $|f'(x)| \leq M < 1$, for all $x \in J$, and for some constant M.

10. Suppose that in Eq. (1.4.3), $g(x*) = g'(x*) = 0$ and $g''(x*) \neq 0$. Prove that $x*$ is an equilibrium point of Eq. (1.4.3). (Hint: Use L'Hopitals Rule.)

11. Prove Theorem 1.12, part (ii).

12. Prove that if $x*$ is an equilibrium point of Eq. (1.1.1), then it is an equilibrium point of Eq. (1.4.4). Show also that the converse is false.

13. Prove that if an equilibrium point $x*$ of Eq. (1.1.1) is asymptotically stable with respect to Eq. (1.4.4) (or unstable as the case may be), it is also so with respect to Eq. (1.1.1).

14. Verify Formula 1.4.5.

15. Prove Theorem 1.14, parts (ii) and (iii).

16. *Semi-Stability* Definition: An equilibrium point $x*$ of $x(n + 1) = f(x(n))$ is semi-stable (from the right) if given $\varepsilon > 0$ there exists $\delta > 0$ such that if $x(0) > x*$, $x(0) - x* < \delta$, then $x(n) - x* < \varepsilon$. Semi-stability from the left is defined similarly. If, in addition, $\lim_{n \to \infty} x(n) = x*$, whenever $x(0) - x* < \eta\{x* - x(0) < \eta\}$, then $x*$ is said to be semi-asymptotically stable from the right {or from the left, whatever the case may be}.

    Suppose that if $f'(x*) = 1$, $f''(x*) \neq 0$. Prove that $x*$ is

    (i) semi-asymptotically stable from the right if $f''(x*) < 0$,

    (ii) semi-asymptotically stable from the left if $f''(x*) > 0$.

17. Determine whether the equilibrium point $x* = 0$ is semi-asymptotically stable from the left or from the right.

    (a) $x(n + 1) = x^3(n) + x^2(n) + x(n)$.

    (b) $x(n + 1) = x^3(n) - x^2(n) + x(n)$.

1.5   Periodic Points and Cycles

The second most important notion in dynamical system is the notion of periodicity. For example, the motion of a pendulum is periodic. We have seen in Example 1.11 that if the sensitivity m_s of the suppliers to price is equal to the sensitivity of consumers to price then prices oscillate between two values only.

Definition 1.17. Let b be in the domain of f. Then

(i) b is called a *periodic* point of f or Eq. (1.4.1) if for some positive integer k, $f^k(b) = b$. Hence a point is k periodic if it is a fixed point of f^k, that is, if it is an equilibrium point of the difference equation

$$x(n + 1) = g(x(n)), \qquad\qquad (1.5.1)$$

where $g = f^k$.

The periodic orbit of b, $O^+(b) = \{b, f(b), f^2(b), \ldots, f^{k-1}(b)\}$, is often called a k cycle.

(ii) b is called eventually k periodic if for some positive integer m, $f^m(b)$ is a k periodic point. In other words, b is eventually k periodic if

$$f^{m+k}(b) = f^m(b).$$

Graphically, a k periodic point is the x coordinate of the point where the graph of f^k meets the diagonal line $y = x$. Figure 1.11a depicts the graph of f^2, where f is the logistic map, which shows that there are 4 fixed points of f^2, of which two are fixed points of f as shown in Figure 1.11b. Hence the other two fixed points of f^2 form a 2 cycle. Notice also that the point $x_0 = 0.3$ (in Fig. 1.11b) goes into a 2 cycle and thus it is an eventually 2 periodic point. Moreover, the point $x^* = 0.445$ is asymptotically stable relative to f^2 (Fig. 1.11c).

Observe also if $A = -1$ in Eq. (1.3.10), then $f^2(p_0) = -(-p_0 + B) + B = p_0$. Therefore every point is 2 periodic. (See Fig. 1.7b.) This means that in this case, if the initial price p_0 per unit of a certain commodity, then the price oscillates between p_0 and $B - p_0$.

Example 1.18. Consider again the difference equation generated by the tent function

$$T(x) = \begin{cases} 2x & \text{for } 0 \le x \le \dfrac{1}{2} \\[2mm] 2(1 - x) & \text{for } \dfrac{1}{2} < x \le 1 \end{cases}.$$

This may also be written in the compact form

$$T(x) = 1 - 2\left| x - \frac{1}{2} \right|.$$

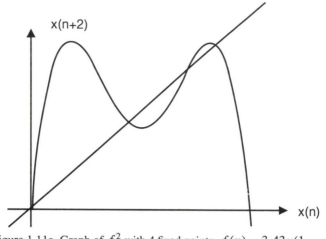

Figure 1.11a. Graph of f^2 with 4 fixed points. $f(x) = 3.43x(1 - x)$.

We first observe that the periodic points of period 2 are the fixed points of T^2. It is easy to verify that T^2 is given by

$$T^2(x) = \begin{cases} 4x & \text{for } 0 \le x < \dfrac{1}{4} \\[2ex] 2(1 - 2x) & \text{for } \dfrac{1}{4} \le x < \dfrac{1}{2} \\[2ex] 4\left(x - \dfrac{1}{2}\right) & \text{for } \dfrac{1}{2} \le x < \dfrac{3}{4} \\[2ex] 4(1 - x) & \text{for } \dfrac{3}{4} \le x \le 1 \end{cases}.$$

There are four equilibrium points (Fig. 1.12a): 0, 0.4, $\frac{2}{3}$, and 0.8, two of which 0 and $\frac{2}{3}$ are equilibrium points of T. Hence $\{0.4, 0.8\}$ is the only 2 cycle of T. Notice that, from Fig. 1.12b, that $x* = 0.8$ is not stable relative to T^2.

Figure 1.13 depicts the graph of T^3. It is easy to verify that $\left\{\frac{2}{7}, \frac{4}{7}, \frac{6}{7}\right\}$ is a 3 cycle. Now

$$T\left(\frac{2}{7}\right) = \frac{4}{7}, \qquad T\left(\frac{4}{7}\right) = \frac{6}{7}, \qquad T\left(\frac{6}{7}\right) = \frac{2}{7}.$$

Using a computer or some hand-held calculator, one may show (using the stair step diagram) that the tent map T has periodic points of all periods. This is a phenomenon shared by all equations that possess a 3 cycle and was discovered by Li and York [2] in their celebrated paper "Period Three Implies Chaos."

We now turn our attention to explore the stability of periodic points.

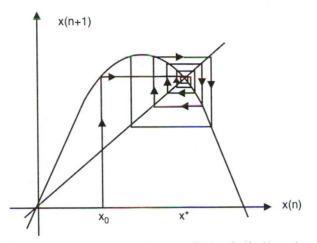

Figure 1.11b. x_0 goes into a 2-cycle. $f(x) = 3.43x(1 - x)$.

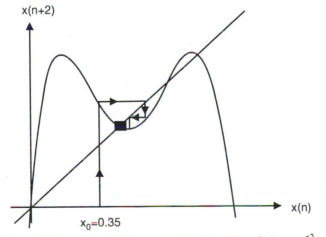

Figure 1.11c. $x^* \approx 0.445$ is asymptotically stable relative to f^2.

x(n=2)

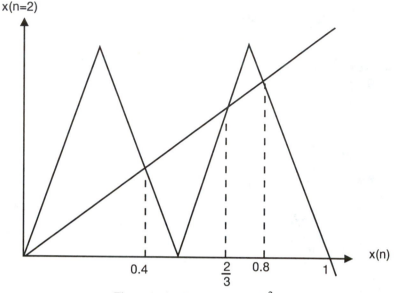

Figure 1.12a. Fixed points of T^2.

Definition 1.19. Let b be a k period point of f. Then b is

(i) stable if it is a stable fixed point of f^k,

(ii) asymptotically stable (attracting) if it is an attracting fixed point of f^k,

(iii) repelling if it is a repelling fixed point of f^k.

Notice that if b possesses a stability property then so does every point in its k cycle $\{x(0) = b, x(1) = f(b), x(2) = f^2(b), \ldots, x(k-1) = f^{k-1}(b)\}$. Hence we often speak of the stability of a k cycle or a periodic orbit. Figure 1.12b shows that the 2 cycle in the tent map is not stable since $x* = 0.8$ is not stable as a fixed point of T^2, while the 2 cycle in the logistic map is asymptotically stable. (See Fig. 1.11c.)

Since the stability of a k periodic point b of Eq. (1.1.1) reduces to the study of the stability of the point as an equilibrium point of Eq. (1.5.1), one can use all the theorems in the previous section applied on f^k. For example, Theorem 1.12 may be modified as follows.

Theorem 1.20. Let $O^+(b) = \{b = x(0), x(1), \ldots, x(k-1)\}$ be a k cycle of a continuously differentiable function f. Then the following statements hold.

(i) The k cycle $O^+(b)$ is attracting if

$$|f'(x(0))f'(x(1)), \ldots, f'(x(k-1))| < 1.$$

(ii) The k cycle $O^+(b)$ is repelling if

$$|f'(x(0))f'(x(1)), \ldots, f'(x(k-1))| > 1.$$

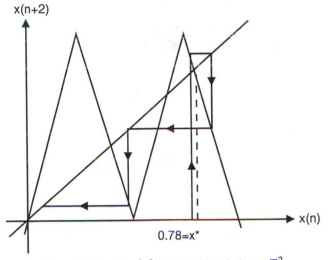

Figure 1.12b. $x^* = 0.8$ is not stable relative to T^2.

Proof We apply Theorem 1.12 on Eq. (1.5.1). Notice that by using the chain rule one may show that

$$[f^k(x(r))]' = f'(x(0))f'(x(1)), \ldots, f'(x(k-1)).$$

(See Exercise 1.5, Problem 12).

The conclusion of the theorem now follows.

Example 1.21. Consider the map $Q(x) = x^2 - 0.85$ defined on the interval $[-2,2]$. Find the 2 cycles and determine their stability.

Solution Now $Q^2(x) = (x^2 - 0.85)^2 - 0.85$. The 2 periodic points are obtained by solving the equation

$$Q^2(x) = x \text{ or } x^4 - 1.7x^2 - x - 0.1275 = 0. \tag{1.5.2}$$

This equation has four roots, two of which are fixed points of the map $Q(x)$. These two fixed points are the roots of the equation

$$x^2 - x - 0.85 = 0. \tag{1.5.3}$$

To eliminate these fixed points of $Q(x)$ from Eq. (1.5.2) we divide the left-hand side of Eq. (1.5.2) over the left-hand side of Eq. (1.5.3) to obtain the second degree equation

$$x^2 + x + 0.15 = 0 \tag{1.5.4}$$

The 2 periodic points are now obtained by solving Eq. (1.5.4) and are given by

$$a = \frac{-1 + \sqrt{0.4}}{2}, \qquad b = \frac{-1 - \sqrt{0.4}}{2}.$$

x(n+3)

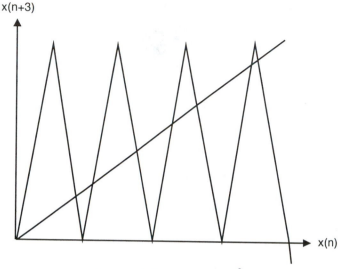

Figure 1.13. Fixed points of T^3.

To check the stability of the cycle $\{a, b\}$ we apply Theorem 1.20. Now

$$|Q'(a)Q'(b)| = |(-1 + \sqrt{0.4})(-1 - \sqrt{0.4})| = 0.6 < 1.$$

Hence by Theorem 1.20, part (i), the 2 cycle is asymptotically stable.

Exercise 1.5.

1. Suppose that the difference equation $x(n+1) = f(x(n))$ has a 2 cycle whose orbit is $\{a, b\}$. Prove that

   (i) the 2 cycle is asymptotically stable if $|f'(a)f'(b)|a < 1$, and

   (ii) the 2 cycle is unstable if $|f'(a)f'(b)| > 1$.

2. Let T be the tent map in Example 1.16. Show that $\left\{\frac{2}{9}, \frac{4}{9}, \frac{8}{9}\right\}$ is a repelling 3 cycle for T.

3. Let $f(x) = -\frac{1}{2}x^2 - x + \frac{1}{2}$. Show that 1 is an attracting 2-periodic point of f

In Problems 4 through 6 find the 2 cycle and then determine their stability.

4. $x(n+1) = 2.5x(n)[1 - x(n)]$.

5. $x(n+1) = 1 - x^2$.

6. $x(n+1) = 5 - (6/x(n))$.

7. Let $f(x) = ax^3 - bx + 1$, where $a, b \in R$. Find the values of a and b for which $\{0, 1\}$ is an attracting 2 cycle.

Consider the so-called Baker's function defined as follows.

$$B(x) = \begin{cases} 2x & \text{for } 0 \leq x \leq \dfrac{1}{2} \\[2mm] 2x - 1 & \text{for } \dfrac{1}{2} < x \leq 1 \end{cases}.$$

Problems 8, 9, and 10 are concerned with the Baker's function $B(x)$ on $[0, 1]$.

8.* (Hard) Draw the Baker's function $B(x)$. Then find the number of n-periodic points of B.

9. Sketch the graph of B^2 and then find the 2 cycles of the Baker's function B.

10. Show that if m is an odd positive integer, then $\bar{x} = k/m$ is periodic, for $k = 1, 2, \ldots, m - 1$.

11. Consider the quadratic map

$$Q(x) = ax^2 + bx + c, a \neq 0.$$

(a) If $\{d, e\}$ is a 2 cycle such that $Q'(d)Q'(e) = -1$, prove that it is asymptotically stable.

(b) If $\{d, e\}$ is a 2 cycle with $Q'(d)Q'(e) = 1$, what can you say about the stability of the cycle?

12. (This exercise generalizes the result in Problem 1.) Let $\{x(0), x(1), \ldots, x(k-1)\}$ be a k cycle of Eq. (1.2.1). Prove that

(i) if $|f'(x(0))f'(x(1)), \ldots, f'(x(k-1))| < 1$, then the k cycle is asymptotically stable

(ii) if $|f'(x(0))f'(x(1)), \ldots, f'(x(k-1))| > 1$, then the k cycle is unstable.

13. Give an example of a decreasing function that has a fixed point and a 2 cycle.

14. (i) Can a decreasing map have a k cycle for $k > 1$?

(ii) Can an increasing map have a k cycle for $k > 1$?

1.6 The Logistic Equation and Bifurcation

Let us now return to the most important example in this chapter: the logistic difference equation

$$x(n + 1) = \mu x(n)[1 - x(n)] \tag{1.6.1}$$

which arises from iterating the function

$$F_\mu(x) = \mu x(1 - x), \qquad x \in [0, 1], \qquad \mu > 0. \tag{1.6.2}$$

1.6.1 Equilibrium Points (Fixed Points of F μ)

To find the equilibrium points (fixed points of F_μ) of Eq. (1.6.1) we solve the equation

$$F_\mu(x^*) = x^*.$$

Hence the fixed points are 0, $x^* = (\mu - 1)/\mu$. Next we investigate the stability of each equilibrium point separately.

(a) The equilibrium point 0. (See Figs. 1.14a–c.) Since $F_\mu'(0) = \mu$, it follows from Theorems 1.12 and 1.14 that

(i) 0 is an attracting fixed point for $0 < \mu < 1$,

(ii) 0 is a repelling fixed point for $\mu > 1$,

(iii) 0 is an unstable fixed point for $\mu = 1$.

(b) The equilibrium point $x^* = (\mu - 1)/\mu$, $\mu \neq 1$. (See Figs. 1.15a–c.)

In order to have $x^* \in (0, 1]$ we require that $\mu > 1$. Now $F_\mu'((\mu - 1)/\mu) = 2 - \mu$. Thus using Theorems 1.12 and 1.15 we obtain the following conclusions:

(i) x^* is an attracting fixed point for $1 < \mu \leq 3$ (Fig. 1.15a).

(ii) x^* is a repelling fixed point for $\mu > 3$ (Fig. 1.15b).

1.6.2 2 Cycles

To find the 2 cycles we solve the equation $F_\mu^2(x) = x$ (or we solve $x_2 = \mu x_1(1 - x_1)$, $x_1 = \mu x_2(1 - x_2)$)

$$\mu^2 x(1 - x)[1 - \mu x(1 - x)] - x = 0. \tag{1.6.3}$$

Discarding the equilibrium points 0 and $x^* = \frac{\mu - 1}{\mu}$, then one may divide Eq. (1.6.3) by the factor $x(x - (\mu - 1)/\mu)$ to obtain the quadratic equation

$$\mu^2 x^2 - \mu(\mu + 1)x + \mu + 1 = 0.$$

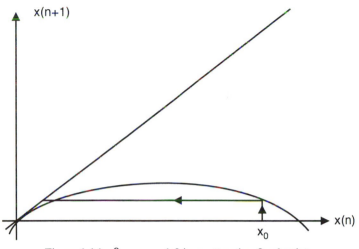

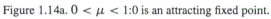

Figure 1.14a. $0 < \mu < 1$:0 is an attracting fixed point.

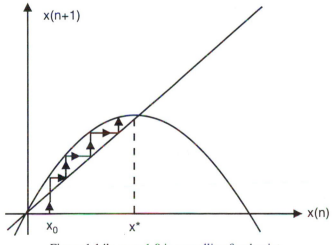

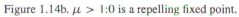

Figure 1.14b. $\mu > 1$:0 is a repelling fixed point.

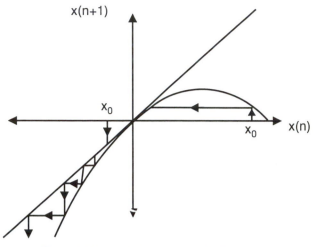

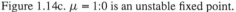

Figure 1.14c. $\mu = 1{:}0$ is an unstable fixed point.

Solving this equation produces the 2 cycle

$$x(0) = \left[(1+\mu) - \sqrt{(\mu-3)(\mu+1)}\right]/2\mu,$$

$$x(1) = \left[(1+\mu) + \sqrt{(\mu-3)(\mu+1)}\right]/2\mu. \qquad (1.6.4)$$

Clearly, there are no periodic points of period 2 for $0 < \mu \le 3$, and there is a 2 cycle for $\mu < 3$. For our reference we let $\mu_0 = 3$.

1.6.2.1 Stability of the 2 Cycle $\{x(0), x(1)\}$ for $\mu > 3$

From Theorem 1.20, this 2 cycle is attracting if

$$|F_\mu'(x(0))F_\mu'(x(1))| < 1$$

or

$$-1 < \mu^2(1-2x(0))(1-2x(1)) < 1. \qquad (1.6.5)$$

Substituting from Eq. (1.6.4) the values of $x(0)$ and $x(1)$ into Eq. (1.6.5) we obtain

$$3 < \mu < 1 + \sqrt{6} \approx 3.44949.$$

Conclusion This 2 cycle is attracting if $3 < \mu < 3.44949\ldots.$

Question What happens when $\mu = 1 + \sqrt{6}$?

Well, in this case

$$\left[F_\mu^2(x(0))\right]' = F_\mu'(x(0))F_\mu'(x(1)) = -1. \qquad (1.6.6)$$

(Verify in Exercise 1.6, Problem 7).

Hence we may use Theorem 1.15, part (ii) to conclude that the 2 cycle is also attracting. For later reference, let $\mu_1 = 1 + \sqrt{6}$. Moreover, the 2 cycle becomes unstable when $\mu > \mu_1 = 1 + \sqrt{6}$.

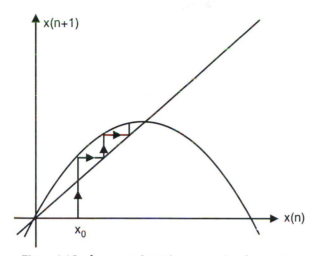

Figure 1.15a. $1 < \mu \le 3$: x^* is an attracting fixed point.

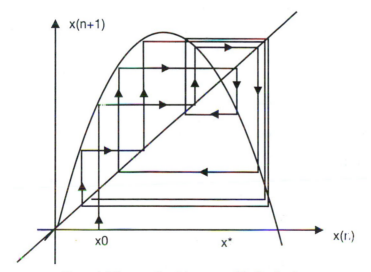

Figure 1.15b. $\mu > 3$: x^* is an unstable fixed point.

Table 1.4. Feigenbaum Table

n	μ_n	$\mu_n - \mu_{n-1}$	$\dfrac{\mu_n - \mu_{n-1}}{\mu_{n+1} - \mu_n}$
0	3	—	—
1	3.449499 ...	0.449499 ...	—
2	3.544090 ...	0.094591 ...	4.752027 ...
3	3.564407 ...	0.020313 ...	4.656673 ...
4	3.568759 ...	0.004352 ...	4.667509 ...
5	3.569692 ...	0.00093219 ...	4.668576 ...
6	3.569891 ...	0.00019964 ...	4.669354 ...

1.6.3 2^2 Cycles

To find the 4 cycles we solve $F_\mu^4(x) = x$. The computation now becomes unbearable and one should resort to a computer to do the work. It turns out that there is a 2^2 cycle when $\mu > 1 + \sqrt{6}$ which is attracting for $1 + \sqrt{6} < \mu < 3.544090 \ldots$. This 2^2 cycle becomes unstable at $\mu < \mu_2 = 3.544090 \ldots$.

When $\mu = \mu_2$, the 2^2 cycle bifurcates into a 2^3 cycle. The new 2^3 cycle is attracting for $\mu_3 < \mu \leq \mu_4$ for some number μ_4. This process of double bifurcation continues indefinitely. Thus we have a sequence $\{\mu_n\}_{n=0}^\infty$, where at μ_n there is a bifurcation from 2^{n-1} cycle to 2^n cycle. (See Figs. 1.16a–b.) The following table provides some astonishing patterns.

From Table 1.4 we bring forward the following observations.

(i) The sequence $\{\mu_n\}$ seems to converge to a number $\mu_\infty = 3.57 \ldots$.

(ii) The quotient $(\mu_n - \mu_{n-1})/(\mu_{n+1} - \mu_n)$ seems to tend to a number $\delta = 4.6692016 \ldots$. This number is called the *Feigenbaum number* after its discoverer, the physicist Mitchell Feigenbaum [2]. In fact Feigenbaum made a much more remarkable discovery: The number δ is universal and is independent of the form of the family of maps f_μ. However, the number μ_∞ depends on the family of functions under consideration.

Theorem 1.22. (Feigenbaum [1978]). For sufficiently smooth families of maps (such as F_μ) of the interval into itself, the number $\delta = 4.6692016$ does not in general depend on the family of maps.

1.6.4   The Bifurcation Diagram

Here the horizontal axis represents the μ values and the vertical axis represents higher iterates $F_\mu^n(x)$. For a fixed x_0, the diagram shows the eventual behavior of $F_\mu^n(x_0)$. The bifurcation diagram was obtained in the computer for $x_0 = \frac{1}{2}$,

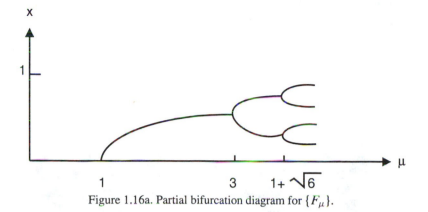

Figure 1.16a. Partial bifurcation diagram for $\{F_\mu\}$.

taking increments of $\frac{1}{500}$ for $\mu \in [0, 4]$, and plotting all points $\left(\mu, F_\mu^n\left(\frac{1}{2}\right)\right)$, for $200 \le n \le 500$.

Question What happens when $\mu > \mu_\infty$?

Answer From Fig. 1.16b, we see that for $\mu_\infty < \mu \le 4$ we have a larger number of small windows where the attracting set is an asymptotically stable cycle. The largest window appears at approximately $\mu = 3.828427\ldots$ where we have an attracting 3 cycle. Indeed, there are attracting k cycles for all positive integers k but whose windows are so small that may not be noticed without sufficient zooming. As in the situation when $\mu < \mu_\infty$, these k cycles lose stability and then double bifurcates into attracting $2^n k$ cycles. We observe that outside these windows the picture looks chaotic!

Remarks Our analysis of the logistic map F_μ may be repeated for any quadratic map $Q(x) = ax^2 + bx + c$. Indeed, the iteration of the quadratic map Q (with suitably chosen parameters) is equivalent to the iteration of the logistic map F_μ. In other words, the maps Q and F_μ possess the same type of qualitative behavior. The reader is asked in Exercise 1.6, Problem 11 to verify that one can transform the difference equation

$$y(n + 1) = y^2(n) + c \tag{1.6.7}$$

to

$$x(n + 1) = \mu x(n)[1 - x(n)] \tag{1.6.8}$$

by letting

$$y_n = -\mu x_n + \frac{\mu}{2}, \qquad c = \frac{\mu}{2} - \frac{\mu^2}{4}. \tag{1.6.9}$$

Note here that $\mu = 2$ corresponds to $c = 0$, $\mu = 3$ corresponds to $c = \frac{-3}{4}$, and $\mu = 4$ corresponds to $c = -2$. Naturally, we expect to have the same behavior of the iteration of the Eqs. (1.6.7) and (1.6.7) at these corresponding values of μ and c.

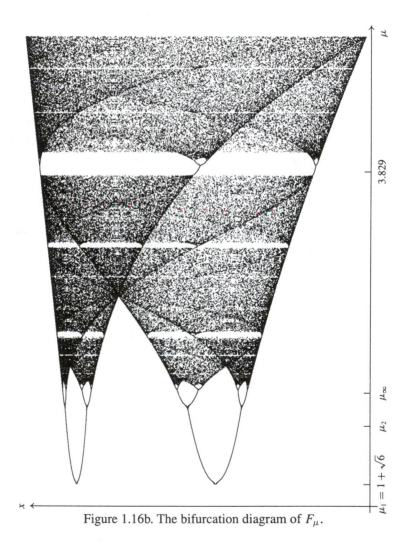

Figure 1.16b. The bifurcation diagram of F_μ.

Comments We are still plagued by numerous unanswered questions in connection with periodic orbits (cycles) of the difference equation

$$x(n + 1) = f(x(n)). \tag{1.6.10}$$

Question A Do all points converge to some asymptotically stable (attractor) periodic orbit of Eq. (1.6.8)?

The answer is definitely no.

If $f(x) = 1 - 2x^2$ in Eq. (1.6.10), then there are no asymptotically stable (attractor) periodic orbits. Can you verify this statement? If you have some difficulty here, it is not your fault. Obviously, we need some tools to help us in verifying that there are no periodic attractors.

Question B If there is a periodic attractor of Eq. (1.6.10), how many points converge to it?

Once again, we need more machinery to answer this question.

Question C Can there be several distinct periodic attractors for Eq. (1.6.10)?

This question leads us to the Li-Yorke famous result "Period 3 implies chaos." [3]. To explain this and more general results requires the introduction of the so-called "*Schwarzian derivative*" of $f(x)$.

Exercises 1.6.

All the problems here refer to the logistic difference Eq. (1.6.1).

1. Use the stair step diagram for F_4^k on $[0, 1]$ to demonstrate that F_4 has at least 2^k periodic points of period k (including periodic points of periods that are divisions of k).

2. Find the exact solution of $x(n + 1) = 4\,x(n)[1 - x(n)]$.
   [Hint: Let $x(n) = \sin^2 \theta(n)$.]

3. Let $x^* = (\mu - 1)/\mu$ be the equilibrium point of Eq. (1.6.1). Show that

   (i) for $1 < \mu \leq 3$, x^* is an attracting fixed point,

   (ii) for $\mu > 3$, x^* is a repelling fixed point.

4. Prove that $\lim_{n \to \infty} F_2^n(x) = \frac{1}{2}$ if $0 < x < 1$.

5. Let $1 < \mu \leq 2$ and $x^* = (\mu - 1)/\mu$ be the equilibrium point of Eq. (1.6.1). Show that if $x^* < x < \frac{1}{2}$, then $\lim_{n \to \infty} F_\mu^n(x) = x^*$.

6. Prove that the 2 cycle given by Eq. (1.6.4) is attracting if $3 < \mu < 1 + \sqrt{6}$.

7. Verify formula Eq. (1.6.6). Then show that the 2 cycle in Eq. (1.6.4) is attracting when $\mu = 1 + \sqrt{6}$.

8. Verify that $\mu_2 \approx 3.54$ using a calculator or a computer.

*9. (Project) Show that the map $H_\mu(x) = \sin \mu x$ leads to the same value for the Feigenbaum number δ.

(Hint: Use a calculator or a computer.)

10. Show that if $|\mu - \mu_1| < \epsilon$, then $|F\mu(x) - F\mu_1(x)| < \varepsilon$ for all $x \in [0, 1]$.

11. Show that the difference equation $y(n + 1) = y^2(n) + c$ can be transformed to the logistic equation $x(n + 1) = \mu x(n)(1 - x(n))$, with $c = \frac{\mu}{2} - \frac{\mu^2}{4}$.

(Hint: Let $y = -\mu x + \frac{1}{2}\mu$.)

12.   (a) Find the equilibrium points $y_1^* \, y_2^*$ of Eq. (1.6.7).

(b) Find the values of c where y_1^* is attracting, repelling, or unstable.

(c) Find the values of c where y_2^* is attracting, repelling, or unstable.

13. Find the value of c_0 where the Eq. (1.6.7) double bifurcates for $c > c_0$. Check your answer using Eq. (1.6.9).

*14. (Project) Use a calculator or a computer to develop a bifurcation diagram, as in Figs. 1.16a–b, for Eq. (1.6.6).

*15. (Project) Develop a bifurcation diagram for the quadratic map $Q_\lambda(x) = 1 - \lambda x^2$ on the interval $[-1, 1]$.

Appendix

There are many available computer packages that can be used to generate Cobweb diagrams and bifurcation diagrams. My favorite software is PHASER, though it does not have a program for bifurcation diagrams. Here we provide programs written specifically for the Texas Instrument calculator TI-85 (or TI-82).

Program 1
Cobweb Diagram for the Logistic Map
(written by Dr. Richard Cooper)
\START\
\COMMENT=Program file dated 06/20/94, 10:21
\NAME=COBWEB
\FILE=COBWEB.85p
Func:Fix 2:FnOff
0\->\xMin:1\->\xMax:0 \->\xScl

```
0\->\yMin:1\->\yMax:0 \->\yScl
Input "iter",N
Input "R",R
Input "START",S
y1=R*x(1-x)
y2=x
DispG
evalF(y1,x,S)\->\Y
Line(S,0,S,Y)
For(I,1,N)
Line(S,Y,Y,Y)
Y\->\S
evalF(y1,x,S)\->\Y
Line(S,S,S,Y)
End
\STOP\
```

Program 2
Bifurcation Diagram
(communicated by Dr. Denny Gulick)
```
\START\
\COMMENT=Program file dated 06/20/94, 10:21
\NAME=BIFUR
\FILE=BIFUR.85p
ClDrw
FnOff
Disp "FIRST ITER TO PLOT"
Input J
Input N
Input P
Input Q
P\->\xMin
Q\->\xMax
0\->\xScl
\(-)\2\->\yMin
2\->\yMax
.10\->\yScl
(xMax-xMin)/94\->\S
For(M,P,Q,S)
.5\->\x
For(I,1,J-1)
y1\->\x
End
For(I,J,J+N)
y1\->\x
```

```
PtOn(M,x)
End
End
\STOP\
```

References

[1] E. C. Pielou, *An Introduction to Mathematical Ecology*, Wiley Interscience, New York, 1969.

[2] M. Feigenbaum, "Quantitative Universality for a Class of Nonlinear Transformations," *J. Stat. Phys.* **19** (1978), 25–52.

[3] T.Y. Li and J.A. Yorke, "Period Three Implies Chaos," *Am. Math. Monthly* **82** (1975), 985–992.

Bibliography

R. Devaney, *A First Course in Chaotic Dynamical Systems: Theory and Experiments*, Addison–Wesley, Reading, 1992.

D. Gulick, *Encounters with Chaos*, McGraw–Hill, New York, 1992.

R. Mickens, *Difference Equations*, Van Nostrand Reinhold, New York, 1990.

J.T. Sandefur, *Discrete Dynamical Systems*, Clarendon, Oxford, 1990.

2
Linear Difference Equations of Higher Order

In this chapter we examine linear difference equations of high order, namely, those involving a single dependent variable.[1] Those equations arise in almost every field of scientific inquiry, from population dynamics (the study of a single species) to economics (the study of a single commodity) to physics (the study of the motion of a single body). We will be acquainted with some of these applications in this chapter. We start this chapter by introducing some rudiments of difference calculus which are essential in the study of linear equations.

2.1   Difference Calculus

Difference calculus is the discrete analog of the familiar differential and integral calculus. In this section, we introduce some very basic properties of two operators that are essential in the study of difference equations. These are the *difference operator* (Section 1.2.)

$$\Delta x(n) = x(n + 1) - x(n),$$

and the *shift operator*

$$Ex(n) = x(n + 1).$$

It is easy to see that

$$E^k x(n) = x(n + k).$$

---

[1]Difference equations that involve more than one dependent variable are called systems of difference equations; we will inspect those equations in Chapter 3.

However, $\Delta^k x(n)$ is not so apparent. Let I be the identity operator, i.e., $Ix = x$. Then, one may write $\Delta = E - I$ and $E = \Delta + I$.

However, $\Delta^k x(n)$ is not so apparent. Let I be the identity operator, i.e., $Ix = x$. Then, one may write $\Delta = E - I$ and $E = \Delta + 1$.

Hence,

$$\Delta^k x(n) = (E - I)^k x(n)$$

$$= \sum_{i=0}^{k} (-1)^i \binom{k}{i} E^{k-i} x(n)$$

$$= \sum_{i=0}^{k} (-1)^i \binom{k}{i} x(n + k - i). \qquad (2.1.1)$$

Similarly, one may show that

$$E^k x(n) = \sum_{i=0}^{k} \binom{k}{i} \Delta^{k-i} x(n). \qquad (2.1.2)$$

We should point out here that the operator Δ is the counterpart of the derivative operator D in calculus. Both operators E and Δ share one of the helpful features of the derivative operator D, namely, the property of *linearity*.

"Linearity" simply means that $\Delta[ax(n)+by(n)] = a\Delta x(n)+b\Delta y(n)$ and $E[ax(n)+ by(n)] = aEx(n) + bEy(n)$, for all a and $b \in R$. In Exercise 2.1, Problem 1 the reader is allowed to show that both Δ and E are linear operators.

Another interesting difference, parallel to differential calculus, is the discrete analogue of the Fundamental Theorem of Calculus[2], stated here.

Lemma 2.1. The following statements hold.

(i) $\displaystyle\sum_{k=n_0}^{n-1} \Delta x(k) = x(n) - x(n_0).$ \qquad (2.1.3)

(ii) $\displaystyle\Delta \left(\sum_{k=n_0}^{n-1} x(k) \right) = x(n).$ \qquad (2.1.4)

The proof remains as Exercise 2.1, Problem 3.   □

---

[2] The Fundamental Theorem of Calculus states that

(i) $\int_a^b df(x) = f(b) - f(a)$,

(ii) $d\left(\int_a^x f(t)dt\right) = f(x)$.

We would now like to introduce a third property that the operator Δ has in common with the derivative operator D.

Let

$$p(n) = a_0 n^k + a_1 n^{k-1} +, \ldots, +a_k$$

be a polynomial of degree k. Then

$$
\begin{aligned}
\Delta p(n) &= \left[a_0(n+1)^k + a_1(n+1)^{k-1} +, \ldots, +a_k \right] \\
&\quad - \left[a_0 n^k + a_1 n^{k-1} +, \ldots, +a_k \right] \\
&= a_0 k n^{k-1} + \text{ terms of degree lower than } (k-1).
\end{aligned}
$$

Similarly, one may show that

$$\Delta^2 p(n) = a_0 k(k-1)n^{k-2} + \text{ terms of degree lower than } (k-2).$$

Carrying out this process k times one obtains

$$\Delta^k p(n) = a_0 k! \tag{2.1.5}$$

Thus,

$$\Delta^{k+i} p(n) = 0 \text{ for } i \geq 1. \tag{2.1.6}$$

2.1.1   The Power Shift

We now discuss the action of a polynomial of degree k in the shift operator E on the term b^n, for any constant b.

Let

$$p(E) = a_0 E^k + a_1 E^{k-1} + \cdots + a_k I \tag{2.1.7}$$

be a polynomial of degree k in E.

Then

$$
\begin{aligned}
p(E)b^n &= a_0 b^{n+k} + a_1 b^{n+k-1} + \cdots + a_k b^n \\
&= (a_0 b^k + a_1 b^{k-1} + \cdots + a_k)b^n \\
&= p(b)b^n. \tag{2.1.8}
\end{aligned}
$$

A generalization of Formula (2.1.8) now follows.

Lemma 2.2. Let $p(E)$ be the polynomial in Eq. (2.1.7) and $g(n)$ be any discrete function. Then

$$p(E)(b^n g(n)) = b^n p(bE)g(n) \tag{2.1.9}$$

Proof This is left to the reader as Exercise 2.1, Problem 4.

2.1.2 Factorial Polynomials

One of the most interesting functions in difference calculus is the *factorial polynomial* $x^{(k)}$ defined as follows. Let $x \in R$. Then the kth factorial of x is given by

$$x^{(k)} = x(x-1)\cdots(x-k+1), \qquad k \in Z^{+}.$$

Thus if $x = n \in Z^{+}$ and $n \geq k$

$$n^{(k)} = \frac{n!}{(n-k)!}$$

and

$$n^{(n)} = n!$$

The function $x^{(k)}$ plays the same role here as that played by the polynomial x^{k} in differential calculus. The following Lemma 2.3 demonstrates this fact.

So far we have defined the operators Δ and E on sequences $f(n)$. One may extend the definitions of Δ and E to continuous functions $f(t)$, $t \in R$ by simply letting $\Delta f(t) = f(t+1) - f(t)$ and $Ef(t) = f(t+1)$. This extension enables us to define $\Delta f(x)$ and $Ef(x)$ where $f(x) = x^{(k)}$, by

$$\Delta x^{(k)} = (x+1)^{(k)} - x^{(k)} \text{ and } Ex^{(k)} = (x+1)^{(k)}.$$

Using this definition one may establish the following result.

Lemma 2.3. For fixed $k \in Z^{+}$ and $x \in R$, the following statements hold:

(i) $\Delta x^{(k)} = kx^{(k-1)}$. $\qquad\qquad\qquad\qquad\qquad\qquad\qquad\qquad$ (2.1.10)

(ii) $\Delta^{n}x^{(k)} = k(k-1),\ldots,(k-n+1)x^{(k-n)}$ $\qquad\qquad\qquad$ (2.1.11)

(iii) $\Delta^{k}x^{(k)} = k!$ $\qquad\qquad\qquad\qquad\qquad\qquad\qquad\qquad\qquad$ (2.1.12)

Proof (i) $\Delta x^{(k)} = (x+1)x^{k} - x^{(k)}$

$\qquad\qquad = (x+1)x(x-1),\ldots,(x-k+2) - x(x-1),$

$\qquad\qquad\quad \ldots,(x-k+2)(x-k+1)$

$\qquad\qquad = x(x-1),\ldots,(x-k+2)\cdot k$

$\qquad\qquad = kx^{(k-1)}.$

The proofs of parts (ii) and (iii) are left to the reader as Exercise 2.1, Problem 5.

If we define for $k \in Z^{+}$

$$x^{(-k)} = \frac{1}{(x+1)(x+2),\ldots,(x+k)}, \qquad (2.1.13)$$

and $x(0) = 1$, then one may extend Lemma (2.3) to hold for all $k \in Z$. In other words parts (i), (ii), and (iii) of Lemma 2.3 hold for all $k \in Z$ (Exercise 2.1, Problem 6).

The reader may wonder if the product and quotient rules of the differential calculus have discrete counterparts. The answer is affirmative as may be shown by the following two formulas where proofs are left to the reader as Exercise 2.1, Problem 7.

Product Rule:

$$\Delta[x(n)y(n)] = Ex(n)\Delta y(n) + y(n)\Delta x(n). \tag{2.1.14}$$

Quotient Rule:

$$\Delta\left[\frac{x(n)}{y(n)}\right] = \frac{y(n)\Delta x(n) - x(n)\Delta y(n)}{y(n)Ey(n)}. \tag{2.1.15}$$

2.1.3   The Antidifference Operator

The discrete analog of the indefinite integral in calculus is the antidifference operator Δ^{-1}, defined as follows. If $\Delta F(n) = 0$, then $\Delta^{-1}(0) = F(n) = c$ for some arbitrary constant c. Moreover, if $\Delta F(n) = f(n)$ then $\Delta^{-1} f(n) = F(n) + c$, for some arbitrary constant c. Hence

$$\Delta\Delta^{-1} f(n) = f(n),$$
$$\Delta^{-1}\Delta F(n) = F(n) + c,$$

and
$$\Delta\Delta^{-1} = I \text{ but } \Delta^{-1}\Delta \neq I.$$

Using Formula (2.1.4) one may readily have

$$\Delta^{-1} f(n) = \sum_{i=0}^{n-1} f(i) + c. \tag{2.1.16}$$

Formula (2.1.16) is very useful in proving that the operator Δ^{-1} is linear.

Theorem 2.4. The Operator Δ^{-1} is linear.

Proof We need to show that for $a, b \in R$, $\Delta^{-1}[ax(n) + by(n)] = a\Delta^{-1}x(n) + b\Delta^{-1}y(n)$. Now from Formula (2.1.13) we have

$$\Delta^{-1}[ax(n) + by(n)] = \sum_{i=0}^{n-1} ax(i) + by(i)$$
$$= a\sum_{i=0}^{n-1} x(i) + b\sum_{i=0}^{n-1} y(i)$$
$$= a\Delta^{-1}x(n) + b\Delta^{-1}y(n).$$

Next we derive the antidifference of some basic functions.

Lemma 2.5. The following statements hold:

(i) $\Delta^{-k}0 = c_1 n^{k-1} + c_2 n^{k-2} +, \ldots, +c_k$ \hfill (2.1.17)

(ii) $\Delta^{-k}1 = \dfrac{n_k}{k!} + c_1 n^{k-1} + c_2 n^{k-2} +, \ldots, +c_k$ (2.1.18)

(iii) $\Delta^{-1}n^{(k)} = \dfrac{n^{(k+1)}}{k+1} + c, \qquad k \neq -1.$ (2.1.19)

Proof The proofs of parts (i) and (ii) follow by applying Δ^k on the right- hand side of Formulas (2.1.17) and (2.1.18) and then applying Formulas (2.1.6) and (2.1.5), respectively. The proof of part (iii) follows from Formula (2.1.10).

Finally, we give the discrete analog of integration by parts formula, namely, the summation by parts formula.

$$\sum_{k=1}^{n-1} y(k)\Delta x(k) = x(n)y(n) - \sum_{k=1}^{n-1} x(k+1)\Delta y(k) + c.$$ (2.1.20)

To prove Formula (2.1.20) we use Formula (2.1.14) to obtain

$$y(n)\Delta x(n) = \Delta(x(n)y(n)) - x(n+1)\Delta y(n).$$

Applying Δ^{-1} on both sides and using Formula (2.1.16) we get

$$\sum_{k=0}^{n-1} y(k)\Delta x(k) = x(n)y(n) - \sum_{k=0}^{n-1} x(k+1)\Delta y(k) + c.$$ □

Exercises 2.1.

1. Show that the operators Δ and E are linear.

2. Show that $E^k x(n) = \displaystyle\sum_{i=0}^{k} \binom{k}{i} \Delta^{k-i} x(n).$

3. Verify Formulas (2.1.3) and (2.1.4).

4. Verify Formula (2.1.9).

5. Verify Formulas (2.1.11) and (2.1.12).

6. Show that Lemma 2.3 holds for $k \in Z$.

7. Verify the product and quotient rules (2.1.14) and (2.1.15).

8. (Abel's Summation Formula): Prove that

$$\sum_{k=1}^{n} x(k)y(k) = x(n+1)\sum_{k=1}^{n} y(k) - \sum_{k=1}^{n}\left(\Delta x(k)\sum_{r=1}^{k} y(r)\right).$$

9. (Newton's Theorem): If $f(n)$ is a polynomial of degree k, show that

$$f(n) = f(0) + \frac{n^{(1)}}{1!}\Delta f(0) + \frac{n^{(2)}}{2!}\Delta^2 f(0) + \cdots + \frac{n^{(k)}}{k!}\Delta^{(k)} f(0)$$

(Hint: Write $f(n) = a_0 + a_1 n^{(1)} + a_2 n^{(2)} + \cdots + a_k n^{(k)}$.)

10. (The Discrete Taylor Formula): Verify that

$$f(n) = \sum_{i=0}^{k-1}\binom{n}{i}\Delta^i f(0) + \sum_{s=0}^{n-k}\binom{n-s-1}{k-1}\Delta^k f(s).$$

11. The Stirling Numbers: The Stirling numbers of the second kind $s_i(k)$ are defined by the difference equation $s_i(m+1) = s_{i-1}(m) + i s_i(m)$ with $s_i(i) = s_1(i) = 1$ and $1 \le i \le m$, $s_1(k) = 0$ for $1 > k$. Prove that

$$x^m = \sum_{i=1}^{m} s_i(m)x^{(i)}. \tag{2.1.21}$$

(Hint: Use mathematical induction on m.)

12. Use Eq. (2.1.21) to verify Table 2.1 that gives the Stirling numbers $s_i(k)$, for $1 \le i, k \le 7$.

13. Use Table 2.1 and Formula (2.1.21) to write the following polynomials in terms of the factorial polynomials $x^{(k)}$ (e.g. $x^2 = x^{(1)} + x^{(2)}$).

$$x^3, x^4, \text{ and } x^5.$$

14. Use Problem 13 to find

   (i) $\Delta^{-1}(n^3 + 1)$.

   (ii) $\Delta^{-1}\left(\dfrac{5}{n(n+3)}\right)$.

15. Use Problem 13 to solve the difference equation $y(n+1) = y(n) + n^3$.

16. Use Problem 13 to solve the difference equation $y(n+1) = y(n) - 5n^2$.

17. [3] Consider the difference equation

$$y(n+1) = a(n)y(n) + g(n). \tag{2.1.22}$$

   (a) Put $y(n) = \left(\prod_{i=0}^{n-1} a(i)\right) u(n)$ in Eq. (2.1.22). Then show that $\Delta u(n) = g(n)/u(n)$.

---

[3] This method of solving a nonhomogeneous equation is called the method of variation of constants.

Table 2.1. Stirling Numbers $s_i(k)$.

$i \backslash k$	1	2	3	4	5	6	7
1	1	1	1	1	1	1	1
2		1	3	7	15	31	63
3			1	6	25	90	301
4				1	10	65	350
5					1	15	140
6						1	21
7							1

(b) Prove that $y(n) = \left(\prod_{i=0}^{n-1} a(i)\right) y_0 + \sum_{r=0}^{n-1} \left(\prod_{i=r+1}^{n-1} a(i)\right) g(r)$ $y_0 = y(0)$.

(Compare with Section 1.2.)

2.2 General Theory of Linear Difference Equations

The normal form of a kth order *nonhomogeneous linear* difference equation is given by

$$y(n + k) + p_1(n)\, y(n + k - 1) + \cdots + p_k(n)y(n) = g(n), \qquad (2.2.1)$$

where $p_i(n)$ and $g(n)$ are real valued functions defined for $n \geq n_0$ and $p_k(n) \neq 0$ for all $n \geq n_0$. If $g(n)$ is identically zero then Eq. (2.2.1) is said to be a homogeneous equation. Equation (2.2.1) may be written in the form

$$y(n+k) = -p_1(n)y(n+k-1) - p_2(n)y(n+k-2), \ldots, -p_k(n)y(n)+g(n). \quad (2.2.2)$$

By letting $n = 0$ in Eq. (2.2.2), we obtain $y(k)$ in terms of $y(k - 1)$, $y(k - 2), \cdots, y(0)$. Explicitly, we have

$$y(k) = -p_1(0)y(k - 1) - p_2(0)y(k - 2), \ldots - p_k(n)y(0) + g(0).$$

Once $y(k)$ is computed, we can go to the next step and evaluate $y(k + 1)$ by letting $n = 1$ in Eq. (2.2.2). This yields

$$y(k + 1) = -p_1(1)y(k) - p_2(1)y(k - 1), \ldots - p_k(1)y(1) + g(1).$$

By repeating the above process, it is possible to evaluate all values of $y(n)$ for $n \geq k$. Let us now illustrate the above procedure by an example.

Example 2.6. Consider the third order difference equation

$$y(n + 3) - \frac{n}{n + 1} y(n + 2) + ny(n + 1) - 3y(n) = n, \qquad (2.2.3)$$

where $y(1) = 0$, $y(2) = -1$, and $y(3) = 1$. Find the values of $y(4)$, $y(5)$, $y(6)$, and $y(7)$.

Solution First we rewrite Eq. (2.2.3) in the convenient form

$$y(n + 3) = \frac{n}{n + 1} y(n + 2) - ny(n + 1) + 3y(n) + n. \qquad (2.2.3)'$$

Letting $n = 1$ in Eq. (2.2.3)$'$ we have

$$y(4) = \frac{1}{2} y(3) - y(2) + 3y(1) + 1 = \frac{5}{2}.$$

For $n = 2$,

$$y(5) = \frac{2}{3} y(4) - 2y(3) + 3y(2) + 2 = -2.$$

For $n = 3$,

$$y(6) = \frac{3}{4} y(5) - 3y(4) + 3y(3) + 3 = -\frac{3}{2}.$$

For $n = 4$,

$$y(7) = \frac{4}{5} y(6) - 4y(5) + 3y(4) + 4 = 20.9.$$

Now let us go back to Eq. (2.2.1) and formally define its solution. A sequence $\{y(n)\}_{n_0}^{\infty}$ or simply $y(n)$ is said to be a *solution* of Eq. (2.2.1) if it satisfies the equation. Observe that if we specify the initial data of the equation, we are led to the corresponding initial value problem

$$y(k + 1) + p_1(n)y(n + k - 1) + \cdots + p_k(n)y(n) = g(n), \qquad (2.2.4)$$

$$y(n_0) = a_0, \; y(n_0 + 1) = a_1, \ldots, y(n_0 + k - 1) = a_{k-1}, \qquad (2.2.5)$$

where the a_i's are real numbers. In view of the above discussion, we conclude with the following result.

Theorem 2.7. The initial value problems 2.2.4 and 2.2.5 have a *unique* solution $x(n)$.

Proof The proof follows by using Eq. (2.2.2) for $n = n_0, n_0 + 1, n_0 + 2, \ldots$. Notice that any $n \geq n_0 + k$ may be written in the form $n = n_0 + k + (n - n_0 - k)$. By *uniqueness* of the solution $x(n)$ we mean that if there is another solution $\bar{y}(n)$ of the initial value problems 2.2.4 and 2.2.5, then $\bar{y}(n)$ must be identical to $y(n)$. This is again easy to see from Eq. (2.2.2). $\qquad \square$

The question still remains whether we can find a closed form solution for Eq. (2.2.1) or Eqs. (2.2.4) and (2.2.5). Well, unlike our amiable first order equations, obtaining a closed form solution of Eq. (2.2.1) is a formidable task. However, if the coefficients p_i in Eq. (2.2.1) are constants, then a solution of the equation may be easily obtained, as we see in the next section.

In this section we are going to develop the general theory of kth order linear *homogeneous* difference equations of the form

$$x(n + k) + p_1(n)x(n + k - 1) + \cdots + p_k(n)x(n) = 0. \qquad (2.2.6)$$

We start our exposition by introducing three important definitions.

Definition 2.8. The functions $f_1(n), f_2(n), \ldots, f_r(n)$ are said to be *linearly dependent* for $n \geq n_0$ if there are constants $a_1, a_2, \ldots, a_r$, not all zero, such that

$$a_1 f_1(n) + a_2 f_2(n) +, \ldots, + a_r f_r(n) = 0, \qquad n \geq n_0.$$

If $a_j \neq 0$, then we may divide Eq. (2.2.6) by a_j to obtain

$$f_j(n) = -\frac{a_1}{a_j} f_1(n) - \frac{a_2}{a_j} f_2(n), \ldots, -\frac{a_r}{a_j} f_r(n)$$

$$= -\sum_{i \neq j} \frac{a_i}{a_j} f_i(n). \qquad (2.2.7)$$

Equation (2.2.7) simply says that each f_j, with nonzero coefficient, is a *linear combination* of the other f_i's. Thus two functions $f_1(n)$ and $f_2(n)$ are linearly dependent if one is a multiple of the other, i.e., $f_1(n) = af_2(n)$, for some constant a.

The negation of linear dependence is *linear independence*. Explicitly put, the functions $f_1(n), f_2(n), \ldots, f_r(n)$ are said to be linearly *independent* for $n \geq n_0$ if whenever

$$a_1 f_1(n) + a_2 f_2(n) + \cdots + a_r f_r(n) = 0,$$

for all $n \geq n_0$, then we must have $a_1 = a_2 = \cdots = a_r = 0$. Let us illustrate this new concept by an example.

Example 2.9. Show that the functions 3^n, $n3^n$, and $n^2 3^n$ are linearly independent on $n \geq 1$.

Solution Suppose that for constants $a_1, a_2,$ and a_3 we have

$$a_1 3^n + a_2 n 3^n + a_3 n^2 3^n = 0, \qquad \text{for all } n \geq 1.$$

Then by dividing by 3^n we get

$$a_1 + a_2 n + a_3 n^2 = 0, \qquad \text{for all } n \geq 1.$$

This is impossible since a second degree equation in n possesses at most two solutions $n \geq 1$. Hence $a_1 = a_2 = a_3 = 0$ which establishes the linear independence of our functions.

Definition 2.10. A set of k linearly independent solutions of Eq. (2.2.6) is called a *fundamental set* of solutions.

As you may have noticed from Example 2.9, it is not practical to check the linear independence of a set of solutions using the definition. Fortunately, there is a simple method to check linear independence of solutions using the so called Casoratian $C(n)$, which we now define for the eager reader.

Definition 2.11. The Casoratian[4] $C(n)$ of the solutions $x_1(n), x_2(n), \ldots, x_r(n)$ is given by

$$C(n) = \det \begin{pmatrix} x_1(n) & x_2(n) & \cdots & x_r(n) \\ x_1(n+1) & x_2(n+1) & \cdots & x_r(n+1) \\ \vdots & & & \\ x_1(n+r-1) & x_2(n+r-1) & \cdots & x_r(n+r-1) \end{pmatrix}. \quad (2.2.8)$$

Example 2.12. Consider the difference equation

$$x(n+3) - 7x(n+1) + 6x(n) = 0$$

(a) Show that the sequences 1, $(-3)^n$, and 2^n are solutions of the equation.

(b) Find the Casoratian of the sequences in part (a).

Solution

(a) $x(n) = 1$ is clearly a solution since $1 - 7 + 6 = 0$. $x(n) = (-3)^n$ is a solution since

$$(-3)^{n+3} - 7(-3)^{n+1} + 6(-3)^n = (-3)^n[-27 + 21 + 6] = 0.$$

Finally $x(n) = 2^n$ is a solution since

$$(2)^{n+3} - 7(2)^{n+1} + 6(2)^n = 2^n[8 - 14 + 6] = 0.$$

(b) Now $C(n) = \det \begin{pmatrix} 1 & (-3)^n & 2^n \\ 1 & (-3)^{n+1} & 2^{n+1} \\ 1 & (-3)^{n+2} & 2^{n+2} \end{pmatrix}$

$$= 1 \begin{pmatrix} (-3)^{n+1} & (2)^{n+1} \\ (-3)^{n+2} & (2)^{n+2} \end{pmatrix} - (-3)^n \begin{pmatrix} 1 & (2)^{n+1} \\ 1 & (2)^{n+2} \end{pmatrix}$$

---

[4]This is the discrete analog of the Wronskian in differential equations.

$$+ (2)^n \begin{pmatrix} 1 & (-3)^{n+1} \\ 1 & (-3)^{n+2} \end{pmatrix}$$

$$= (2)^{n+2}(-3)^{n+1} - (2)^{n+1}(-3)^{n+2} - (-3)^n((2)^{n+2} - (2)^{n+1})$$
$$+ (2)^n((-3)^{n+2} - (-3)^{n+1})$$
$$= -12(2)^n(-3)^n - 18(2)^n(-3)^n - 4(2)^n(-3)^n$$
$$+ 2(2)^n(-3)^n + 9(2)^n(-3)^n + 3(2)^n(-3)^n$$
$$= -20(2)^n(-3)^n.$$

Next we give a formula called Abel's Formula, to compute the Casoratian $C(n)$. The significance of Abel's Formula is due to its effectiveness in the verification of linear independence of solutions.

Lemma 2.13. (Abel's Lemma). Let $x_1(n), x_2(n), \ldots, x_k(n)$ be solutions of Eq. (2.2.6) and $C(n)$ be their Casoratian. Then for $n \geq n_0$,

$$C(n) = (-1)^{k(n-n_0)} \left(\prod_{i=n_0}^{n-1} p_k(i) \right) C(n_0). \tag{2.2.9}$$

Proof We will prove the lemma for $k = 3$ since the general case may be established in a similar fashion. So let $x_1(n)$, $x_2(n)$, and $x_3(n)$ be three independent solutions of Eq. (2.2.6). Then from Formula (2.2.8) we have

$$C(n+1) = \det \begin{pmatrix} x_1(n+1) & x_2(n+1) & x_3(n+1) \\ x_1(n+2) & x_2(n+2) & x_3(n+2) \\ x_1(n+3) & x_2(n+3) & x_3(n+3) \end{pmatrix}. \tag{2.2.10}$$

From Eq. (2.2.6), we have, for $1 \leq i \leq 3$,

$$x_i(n+3) = -p_3(n)x_i(n) - [p_1(n)x_i(n+2) + p_2(n)x_i(n+1)]. \tag{2.2.11}$$

Now if we use Formula (2.2.11) to substitute for $x_1(n+3)$, $x_2(n+3)$, and $x_3(n+3)$ in the last row of Formula (2.2.10) we obtain

$$C(n+1) = \det \begin{pmatrix} x_1(n+1) & x_2(n+1) & x_3(n+1) \\ x_1(n+2) & x_2(n+2) & x_3(n+2) \\ -p_3x_1(n) & -p_3x_2(n) & -p_3x_3(n) \\ -\left(p_2x_1(n+1)\right. & -\left(p_2x_2(n+1)\right. & -\left(p_2x_3(n+1)\right. \\ \left.+p_1x_1(n+2)\right) & \left.+p_1x_2(n+2)\right) & \left.+p_1x_3(n+2)\right) \end{pmatrix}$$

$$\tag{2.2.12}$$

Using the properties of determinants, it follows from Eq. (2.2.12)

$$C(n+1) = \det \begin{pmatrix} x_1(n+1) & x_2(n+1) & x_3(n+1) \\ x_1(n+2) & x_2(n+2) & x_3(n+2) \\ -p_3(n)x_1(n) & -p_3(n)x_2(n) & -p_3(n)x_3(n) \end{pmatrix} \qquad (2.2.13)$$

$$= -p_3(n)\det \begin{pmatrix} x_1(n+1) & x_2(n+1) & x_3(n+1) \\ x_1(n+2) & x_2(n+2) & x_3(n_2) \\ x_1(n) & x_2(n) & x_3(n) \end{pmatrix}$$

$$= -p_3(n)(-1)^2 \begin{pmatrix} x_1(n) & x_2(n) & x_3(n) \\ x_1(n+2) & x_2(n+2) & x_3(n+2) \\ x_1(n+1) & x_2(n+1) & x_3(n+1) \end{pmatrix}.$$

Thus

$$C(n+1) = (-1)^3 p_3(n)C(n). \qquad (2.2.14)$$

Using Formula (1.2.3), the solution of Eq. (2.2.14) is given by

$$C(n) = \left[\prod_{i=n_0}^{n-1} (-1)^3 p_3(i) \right] C(n_0)$$

$$= (-1)^{3(n-n_0)} \prod_{i=n_0}^{n-1} p_3(i)C(n_0).$$

This completes the proof of the lemma for $k = 3$. The general case is left to the reader as Exercise 2.2, Problem 6.

We now examine and treat one of the special cases that arises as we try to apply this Casoratian. For example, if Eq. (2.2.6) has constant coefficients $p_1, p_2, \ldots, p_k$, then we have

$$C(n) = (-1)^{kn} p_k^n C(0). \qquad (2.2.15)$$

Formula (2.2.9) has the following important correspondence.

Corollary 2.14. Suppose that $p_k(n) \neq 0$ for all $n \geq n_0$. Then the Casoratian $C(n) \neq 0$ for all $n \geq n_0$ if and only if $C(n_0) \neq 0$.

Proof This Corollary follows immediately from Formula (2.2.9) (Exercise 2.2, Problem 7).

Let us have a close look at Corollary 2.14 and examine what it really says. The main point in the corollary is that either the Casoratian is identically zero (i.e., zero for all $n \geq n_0$, for some n_0) or never zero for any $n \geq n_0$. Thus to check whether $C(n) \neq 0$ for all $n \in Z^+$, we need only to check if $C(0) \neq 0$. Note that we can always choose the most suitable n_0 and compute $C(n_0)$ there.

Next we examine the relationship between the linear independence of solutions and their Casoratian. Basically, we will show that a set of k solutions is a *fundamental set* (i.e., linearly independent) if their Casoratian $C(n)$ is never zero.

To determine the preceding statement we contemplate k solutions $x_1(n), x_2(n), \ldots, x_k(n)$ of Eq. (2.2.6). Suppose that for some constants $a_1, a_2, \ldots, a_k$ and $n_0 \in Z^+$,

$$a_1 x_1(n) + a_2 x_2(n) + \cdots + a_k(n) = 0, \qquad \text{for all } n \geq n_0.$$

Then we can generate the following $k - 1$ equations

$$a_1 x_1(n + 1) + a_2 x_2(n + 1) + \cdots + a_k x_k(n + 1) = 0,$$
$$\vdots$$
$$a_1 x_1(n + k - 1) + a_2 x_2(n + k - 1) + \cdots + a_k x_k(n + k - 1) = 0.$$

This conglomeration may be transcribed as

$$X(n)\xi = 0, \tag{2.2.16}$$

where

$$X(n) = \begin{pmatrix} x_1(n) & x_2(n) & \cdots & x_k(n) \\ x_1(n + 1) & x_2(n + 1) & \cdots & x_k(n + 1) \\ \vdots & \vdots & & \vdots \\ x_1(n + k - 1) & x_2(n + k - 1) & \cdots & x_k(n + k - 1) \end{pmatrix},$$

$$\xi = \begin{pmatrix} a_1 \\ a_2 \\ \vdots \\ a_k \end{pmatrix},$$

observe that $C(n) = \det X(n)$.

Linear Algebra tells us that the vector Eq. (2.2.15) has only the trivial (or the zero) solution (i.e., $a_1 = a_2 = \cdots = a_k = 0$) if and only if the matrix $X(n)$ is nonsingular (invertible) (i.e., $\det X(n) = C(n) \neq 0$ for all $n \geq n_0$). This deduction leads us to the following conclusion.

Theorem 2.15. The set of solutions $x_1(n), x_2(n), \ldots, x_k(n)$ of Eq. (2.2.6) is a fundamental set if and only if for some $n_0 \in Z^+$, their Casoratian $C(n_0) \neq 0$.

Proof Exercise 2.2, Problem 8.

Example 2.16. Verify that $\{n, 2^n\}$ is a fundamental set of solutions of the equation

$$x(n+2) - \frac{3n-2}{n-1}x(n+1) + \frac{2n}{n-1}x(n) = 0.$$

Solution We leave it to the reader to verify that n and 2^n are solutions of the equation. Now the Casortian of the solution $n, 2^n$ is given by

$$C(n) = \det \begin{pmatrix} n & 2^n \\ n+1 & 2^{n+1} \end{pmatrix}.$$

Thus

$$C(0) = \det \begin{pmatrix} 0 & 1 \\ 1 & 2 \end{pmatrix} = -1 \neq 0.$$

Hence by Theorem 2.15, the solutions $n, 2^n$ are linearly independent and thus form a fundamental set.

Example 2.17. Consider the third order difference equation

$$x(n+3) + 3x(n+2) - 4x(n+1) - 12x(n) = 0.$$

Show that the functions $2^n, (-2)^n$, and 3^n form a fundamental set of solutions of the equation.

Solution

(i) Let us verify that 2^n is a legitimate solution by substituting $x(n) = 2^n$ into the equation.

$$2^{n+3} + (3)(2^{n+1}) - (4)(2^{n+1}) - (12)(2^n) = 2^n[8 + 12 - 8 - 12] = 0.$$

We leave it to the reader to verify that $(-2)^n$ and 3^n are solutions of the equation.

(ii) To affirm the linear independence of these solutions we construct the Casoratian

$$C(n) = \det \begin{pmatrix} 2^n & (-2)^n & 3^n \\ 2^{n+1} & (-2)^{n+1} & 3^{n+1} \\ 2^{n+2} & (-2)^{n+2} & 3^{n+2} \end{pmatrix}.$$

Thus

$$C(0) = \det \begin{pmatrix} 1 & 1 & 1 \\ 2 & -2 & 3 \\ 4 & 4 & 9 \end{pmatrix} = -20 \neq 0.$$

By Theorem 2.15, the solutions 2^n, $(-2)^n$, and 3^n are all linearly independent, and thus form a fundamental set.

We are now ready to discuss the fundamental theorem of homogeneous linear difference equations.

Theorem 2.18. (The Fundamental Theorem). If $p_k(n) \neq 0$ for all $n \geq n_0$, then Eq. (2.2.6) has a fundamental set of solutions for $n \geq n_0$.

Proof By Theorem 2.7, there are solutions $x_1(n), x_2(n), \ldots, x_k(n)$ such that $x_i(n_0 + i - 1) = 1$, $x_i(n_0) = x_i(n_0 + 1) =, \ldots, = x_i(n_0 + i - 2) = x_i(n_0 + i) = \cdots = x_i(n_0 + k - 1) = 0$, $1 \leq i \leq k$. Hence $x_1(n_0) = 1$, $x_2(n_0 + 1) = 1$, $x_3(n_0 + 2) = 1$, etc. ... It follows that $C(n_0) = \det I = 1$. This implies by Theorem 2.13 that the set $\{x_1(n), x_2(n), \ldots, x_k(n)\}$ is a fundamental set of solutions of Eq. (2.2.6).

We remark that there are infintely many fundamental sets of solutions of Eq. (2.2.6). The next result presents a method of generating fundamental sets starting from a known one.

Lemma 2.19. Let $x_1(n)$, and $x_2(n)$ be two solutions of Eq. (2.2.6). Then the following statements hold.

(i) $x(n) = x_1(n) + x_2(n)$ is a solution of Eq. (2.2.6).

(ii) $\tilde{x}(n) = ax_1(n)$ is a solution of Eq. (2.2.6) for any constant a.

Proof (Exercise 2.2, Problem 9)

From the preceding lemma we conclude the following principle.

Superposition Principle. If $x_1(n), x_2(n), \ldots, x_r(n)$ are solutions of Eq. (2.2.6), then

$$x(n) = a_1 x_1(n) + a_2 x_2(n)+, \ldots, +a_r x_r(n)$$

is also a solution of Eq. (2.2.6) (Exercise 2.2, Problem 10).

Now let $\{x_1(n), x_2(n), \ldots, x_k(n)\}$ be a fundamental set of solutions of Eq. (2.2.6) and let $x(n)$ be any given solution of Eq. (2.2.6). Then there are constants $a_1, a_2, \ldots, a_k$ such that $x(n) = \sum_{i=1}^{k} a_i x_i(n)$. To show this we use the notation (2.2.16) to write $X(n)\xi = \hat{x}(n)$, where

$$\hat{x}(n) = \begin{pmatrix} x(n) \\ x(n+1) \\ \vdots \\ x(n+k-1) \end{pmatrix}.$$

Since $X(n)$ is invertible (why?), it follow that

$$\xi = X^{-1}(n)\hat{x}(n),$$

and for $n = n_0$

$$\xi = X^{-1}(n_0)\hat{x}(n_0).$$

The above discussion leads us to define the general solution of Eq. (2.2.6).

Definition 2.20. Let $\{x_1(n), x_2(n), \ldots, x_k(n)\}$ be a fundamental set of solutions of Eq. (2.2.6).

Then the *general solution* of Eq. (2.2.6) is given by $x(n) = \sum_{i=1}^{k} a_i x_i(n)$, for arbitrary constants a_i.

It is worth noting that any solution of Eq. (2.2.6) may be obtained from the general solution by a suitable choice of the constants a_i.

The preceding results may be restated using the elegant language of Linear Algebra as follows: Let S be the set of all solutions of Eq. (2.2.6) with the operations $+, \cdot$ defined as follows:

(i) $(x + y)(n) = x(n) + y(n)$,   for $x, y \in S, n \in Z^+$,

(ii) $(ax)(n) = ax(n)$,   for $x \in S$, a constant.

Equipped with linear algebra we now summarize the results of this section in a compact form.

Theorem 2.21. The space $(S, +, \cdot)$ is a linear (vector) space of dimension k.

Proof Use Lemma 2.19. To construct a basis of S we can use the fundamental set in Theorem 2.18 (Exercise 2.2, Problem 11).

Exercises 2.2.

1. Find the Casoration of the following functions and determine whether they are linearly dependent or independent.

   (a) $5^n, 3\,5^{n+2}, e^n$.

   (b) $5^n, n\,5^n, n^2\,5^n$.

   (c) $(-2)^n, 2^n, 3$.

   (d) $0, 3^n, 7^n$.

2. Find the Casoration $C(n)$ of the solutions of the difference equations

   (a) $x(n+3) - 10x(n+2) + 31x(n+1) - 30x(n) = 0$, if $C(0) = 6$.

   (b) $x(n+3) - 3x(n+2) + 4x(n+1) - 12x(n) = 0$, if $C(0) = 26$.

3. For the following difference equations and their accompanied solutions,

   (i) determine whether these solutions are linearly independent, and

   (ii) find, if possible, using only the given solutions, the general solution.

(a) $x(n + 3) - 3x(n + 2) + 3x(n + 1) - x(n) = 0$
; $1, n, n^2$,

(b) $x(n + 2) + x(n) = 0$
; $\cos\left(\dfrac{n\pi}{2}\right)$, $\sin\left(\dfrac{n\pi}{2}\right)$,

(c) $x(n + 3) + x(n + 2) - 8x(n + 1) - 12x(n) = 0$
; $3^n, (-2)^n, (-2)^{n+3}$,

(d) $x(n + 4) - 16x(n) = 0$
; $2^n, n2^n, n^2 2^n$.

4. Verify Formula (2.2.9) for the general case.

5. Show that the Casoration $C(n)$ in Formula (2.2.3) may be given by the formula

$$
C(n) = \det
\begin{pmatrix}
x_1(n) & x_2(n) & \cdots & x_k(n) \\
\Delta x_1(n) & \Delta x^2(n) & \cdots & \Delta x_k(n) \\
\vdots & \vdots & & \vdots \\
\Delta^{k-1} x_1(n) & \Delta^{k-1} x_2(n) & \cdots & \Delta^{k-1} x_k(n)
\end{pmatrix}.
$$

6. Verify Formula (2.2.9).

7. Prove Corollary 2.14.

8. Prove Theorem 2.15.

9. Prove Lemma 2.19.

10. Prove the superposition principle: If $x_1(n), x_2(n), \ldots, x_r$ are solutions of Eq. (2.2.1), then any linear combination of them is also a solution of Eq. (2.2.1)

11. Prove Theorem 2.21.

12. Suppose that for some integer $m \geq n_0$, $p_k(m) = 0$ in Eq. (2.2.1).

(a) What is the value of the Casoratian for $n \geq m$?

(b) Does Corollary 2.14 still hold? Why?

13.* Show that the equation $\Delta^2 y(n) = p(n)y(n + 1)$ has a fundamental set of solutions whose Casoratian $C(n) = -1$.

14. Contemplate the second order difference equation $u(n+2) + p_1(n)u(n+1) + p_2u_2(n) = 0$. If $u_1(n)$ and $u_2(n)$ are solutions of the equation and $C(n)$ is their Casoratian, prove that

$$u_2(n) = u_1(n)\left[\sum_{r=0}^{n-1} C(r)/u_1(r)u_1(r+1)\right]. \qquad (2.2.17)$$

$$\left[\text{Hint: Consider } \Delta\left(\frac{u_2(n)}{u_1(n)}\right) = \frac{u_1(n)\Delta u_2(n) - u_2(n)\Delta u_1(n)}{u_1(n)u_1(n+1)}.\right]$$

15. Contemplate the second order difference equation $u(n+2) - \frac{(n+3)}{(n+2)}u(n+1) + \frac{2}{(n+2)}u(n) = 0$.

    (a) Verify that $u_1(n) = \frac{2^n}{n!}$ is a solution of the equation.
    (b) Use Formula (2.2.17) to find another solution $u_2(n)$ of the equation. [Hint: you may take $C(n_0) = 1$ or any of your favorite constants, use Formula (2.2.9) to evaluate $C(n)$.]

16. Show that $u(n) = (n+1)$ is a solution of the equation $u(n+2) - u(n+1) - 1/(n+1)u(n) = 0$ and then find a second solution of the equation by using the method of Exercise 2.2, Problem 15.

2.3 Linear Homogeneous Equations with Constant Coefficients

Consider the kth order difference equation

$$x(n+k) + p_1x(n_k - 1) + p_2x(n+k-2) + \cdots + p_kx(n) = 0, \qquad (2.3.1)$$

where the p_i's are constants, and $p_k \neq 0$. Our objective now is to find a fundamental set of solutions, and consequently, the general solution of Eq. (2.3.1). The procedure is rather simple. We suppose that solutions of Eq. (2.3.1) are in the form λ^n, where λ is a complex number. Substituting this value into Eq. (2.3.1), we obtain

$$\lambda^k + p_1\lambda^{k-1} + \cdots + p_k = 0. \qquad (2.3.2)$$

This is called the *characteristic equation* of Eq. (2.3.1) and its roots λ are called the *characteristic roots*. Notice that since $p_k \neq 0$, none of the characteristic roots is equal to zero. Why? (Exercise 2.3, Problem 19)

We have two situations to contemplate:

Case a: Suppose that the characteristic roots $\lambda_1, \lambda_2, \ldots, \lambda_k$ are distinct. We are now going to show that the set $\{\lambda_1^n, \lambda_2^n, \ldots, \lambda_k^n\}$ is a fundamental set of solutions.

To prove this by virtue of Theorem 2.15, it suffices to show that $C(0) \neq 0$, where $C(n)$ is the Casoratian of the solutions. That is,

$$
C(0) = \det
\begin{pmatrix}
1 & 1 & \cdots & 1 \\
\lambda_1 & \lambda_2 & \cdots & \lambda_k \\
\lambda_1^2 & \lambda_2^2 & \cdots & \lambda_k^2 \\
\vdots & \vdots & & \vdots \\
\lambda_1^{k-1} & \lambda_2^{k-1} & \cdots & \lambda_k^{k-1}
\end{pmatrix}.
\tag{2.3.3}
$$

This determinant is called the *Vandermonde determinant.*

It may be shown that $C(0) = \prod_{1 \leq i < j \leq k}(\lambda_j - \lambda_k)$, and the reader will prove this conclusion in Exercise 2.3, Problem 20. $C(0) \neq 0$, in this case. This fact proves that $\{\lambda_1^n, \lambda_2^n, \ldots, \lambda_k^n\}$ is a fundamental set of solutions of Eq. (2.3.1). Consequently, the general solution of Eq. (2.3.1) is

$$
x(n) = \sum_{i=1}^{k} a_i \lambda_i^n, \ a_i \text{ is a complex number.}
\tag{2.3.4}
$$

Case b: Suppose that the distinct characteristic roots are $\lambda_1, \lambda_2, \ldots, \lambda_r$ with multiplicity's $m_1, m_2, \ldots, m_r$, respectively. In this case, Eq. (2.3.1) may be written as

$$
(E - \lambda_1)^{m_1}(E - \lambda_2)^{m_2}, \ldots, (E - \lambda_r)^{m_r} x(n) = 0.
\tag{2.3.1$'$}
$$

A vital observation here is that if $\psi_1(n), \psi_2(n), \ldots, \psi_{m_i}(n)$ are solutions of

$$
(E - \lambda_i)^{m_i} x(n) = 0,
\tag{2.3.5$_i$}
$$

then they are also solutions of Eq. (2.3.1)$'$. Why? (Exercise 2.3, Problem 21)

Suppose we are able to find a fundamental set of solutions for each Eq. (2.3.5)$_i$, $1 \leq i \leq r$. It is not unreasonable to expect, then, that the union of these r fundamental sets would be a fundamental set of solutions of Eq. (2.3.1)$'$. In the following lemma we will show that this is indeed the case.

Lemma 2.22. The set $G_i = \{\lambda_i^n, n\lambda_i^n, n^2\lambda_i^n, \ldots, n^{m_i-1}\lambda_i^n\}$ is a fundamental set of solutions of Eq. (2.3.5)$_i$.

Proof First we show that $n^s \lambda_i^n$, $1 \leq s \leq m_i - 1$, is a solution of Eq. (2.3.5). From Formula (2.1.9) it follows that

$$
\begin{aligned}
(E - \lambda_i)^{m_i} \left(n^s \lambda_i^n\right) &= \lambda_i^n (\lambda_i E - \lambda_i)^{m_i} n^s \\
&= \lambda_i^{n+m_i} \Delta^{m_i}(n^s) \\
&= 0 \ (\text{using Eq. 2.1.6}).
\end{aligned}
$$

Since $\lambda_i \neq 0$, the set G_i is linearly independent if the set $\{1, n, n^2, \ldots, n^{m_i-1}\}$ is linearly independent. But the latter set is linearly independent for $n \geq n_0$ for any $n_0 > 0$ (Exercise 2.3, Problem 22). This completes the proof.

Now we are finally able to find a fundamental set of solutions

Corollary 2.23. The set $G = \bigcup_{i=1}^{r} G_i$ is a fundamental set of solutions of Eq. (2.3.1)'.

Proof (Exercise 2.3, Problem 24)

Corollary 2.24. The general solution of Eq. (2.3.1)' is given by

$$x(n) = \sum_{i=1}^{r} \lambda_i^n \left(a_{i0} + a_{i1}n + a_{i2}n^2 + \cdots a_i, m_{i-1}n^{m_i-1} \right). \tag{2.3.6}$$

Proof Use Lemma 2.22 and Corollary 2.23.

Example 2.25. Solve the equation

$$x(n+3) - 7x(n+1) + 16x(n+1) - 12x(n) = 0, \qquad x(0) = 0 \text{ and } x(1) = 1,$$

$$x(2) = 1.$$

Solution The characteristic equation is

$$r^3 - 7r^2 + 16r - 12 = 0.$$

Thus, the characteristic roots are $\lambda_1 = 2 = \lambda_2$, $\lambda_3 = 3$.

The characteristic roots give us the general solution

$$x(n) = a_0 2^n + a_1 n 2^n + b_1 3^n.$$

To find the constants a_0, a_1, and b_1, we use the initial data

$$
\begin{aligned}
x(0) &= a_0 + b_1 = 0, \\
x(1) &= 2a_0 + 2a_1 + 3b_1, \\
x(2) &= 4a_0 + 8a_1 + 9b_1 = 1.
\end{aligned}
$$

Finally, after solving the above system of equations, we obtain

$$a_0 = 3, \ a_1 = 2, \text{ and } b_1 = -3.$$

Hence the solution of the equation is given by $x(n) = 3(2^n) + 2n(2^n) - 3^{n+1}$.

Example 2.26. (Complex Characteristic Roots). Suppose that the equation $x(n+2) + p_1 x(n+1) + p_2 x(n) = 0$ has the complex roots $\lambda_1 = \alpha + i\beta$, $\lambda_2 = \alpha - i\beta$. Its general solution would then be

$$x(n) = c_1(\alpha + i\beta)^n + c_2(\alpha - i\beta)^n.$$

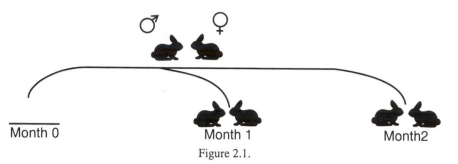

Month 1   Month2
Figure 2.1.

Recall that the point (α, β) in the complex plane corresponds to the complex number $\alpha + i\beta$. In polar coordinates,

$$\alpha = r\cos\theta, \qquad \beta = r\sin\theta, \qquad r = \sqrt{\alpha^2 + \beta^2}, \qquad \theta = \tan^{-1}\left(\frac{\beta}{\alpha}\right).$$

Hence,

$$
\begin{aligned}
x(n) &= c_1(r\cos\theta + ir\sin\theta)^n + c_2(r\cos\theta - ir\sin\theta)^n \\
&= r^n[(c_1 + c_2)\cos(n\theta) + i(c_1 - c_2)\sin(n\theta)] \text{ (See footnote 5)} \\
&= r^n[a_1\cos(n\theta) + a_2\sin(n\theta)]
\end{aligned}
\tag{2.3.7}
$$

where $a_1 = c_1 + c_2$ and $a_2 = i(c_1 - c_2)$.

Let

$$\cos\omega = \frac{a_1}{\sqrt{a_1^2 + a_2^2}}, \quad \sin\omega = \frac{a_2}{\sqrt{s_1^2 + a_2^2}}, \quad \text{and } \omega = \tan^{-1}\left(\frac{a_2}{a_1}\right).$$

Then Eq. (2.3.7) now becomes

$$
\begin{aligned}
x(n) &= r^n\sqrt{a_1^2 + a_2^2}[\cos\omega\cos(n\theta) + \sin\omega\sin(n\theta)] \\
&= r^n\sqrt{a_1^2 + a_2^2}\cos(n\theta - \omega), \\
x(n) &= Ar^n\cos(n\theta - \omega).
\end{aligned}
\tag{2.3.8}
$$

Example 2.27. The Fibonacci Sequence (The Rabbit Problem). This problem first appeared in 1202, in "Liber abaci," a book about the abacus, written by the famous Italian mathematician Leonardo di Pisa, better known as Fibonacci. The problem may be stated as follows: How many pairs of offspring does one pair of mature rabbits produce in a year, if each pair of rabbits gives birth to a new pair each month starting when it reaches its maturity age of two months? (See Fig. 2.1).

Table 2.2 shows the number of pairs of rabbits by the end of each month. The first pair has offspring at the end of the first month, and thus we have two pairs. At the end of the second month only the first pair has offspring, and thus we have three

---

[5]We used De Moivre'e Theorem: $[r(\cos\theta + i\sin\theta)]^n = r^n(\cos n\theta + i\sin n\theta)$.

Table 2.2. Rabbits' Population Size.

month	0	1	2	3	4	5	6	7	8	9	0	11	12
pairs	1	2	3	5	8	13	21	34	55	89	144	233	377

pairs. At the end of the third month, the first and second pairs will have offspring and hence we have five pairs. Continuing this procedure, we arrive at Table 2.2. If $F(n)$ is the number of pairs of rabbits at the end of n months, then the recurrence relation that represents this model is given by the second order linear difference equation

$$F(n + 2) = F(n + 1) + F(n), \qquad F(0) = 1, \qquad F(1) = 2, \qquad \text{and } 0 \leq n \leq 10.$$

This example is a special case of the Fibonacci sequence, given by

$$F(n + 2) = F(n + 1) + F(n), \qquad F(0) = 0, \qquad F(1) = 1, \qquad n \geq 0. \quad (2.3.9)$$

The first 14 terms are given by 1, 2, 3, 5, 8, 13, 21, 34, 55, 89, 144, 233, and 377 as already noted in the rabbit problem.

The characteristic equation of Eq. (2.3.9) is

$$\lambda^2 - \lambda - 1 = 0.$$

Hence the characteristic roots are $\alpha = \frac{1+\sqrt{5}}{2}$ and $\beta = \frac{1-\sqrt{5}}{2}$.

The general solution of Eq. (2.3.9) is

$$F(n) = a_1 \left(\frac{1 + \sqrt{5}}{2} \right)^n + a_2 \left(\frac{1 - \sqrt{5}}{2} \right)^n, \qquad n \geq 1. \quad (2.3.10)$$

Using the initial values $F(1) = 1$ and $F(2) = 1$, one obtains

$$a_1 = \frac{1}{\sqrt{5}}, \qquad a_2 = -\frac{1}{\sqrt{5}}.$$

Consequently,

$$F(n) = \frac{1}{\sqrt{5}} \left[\left(\frac{1 + \sqrt{5}}{2} \right)^n - \left(\frac{1 - \sqrt{5}}{2} \right)^n \right] = \frac{1}{\sqrt{5}} (\alpha^n - \beta^n). \quad (2.3.11)$$

It is interesting to note that $\lim_{n \to \infty} \frac{F(n+1)}{F(n)} = \alpha \approx 1.618$ (Exercise 2.3, Problem 15).

This number is called the *golden mean*, which supposedly represents the ratio of the sides of a rectangle that is most pleasing to the eye.

This Fibonacci sequence is very interesting to mathematicians; in fact, an entire publication, "The Fibonacci Quarterly," dwells on the intricacies of this fascinating sequence.

Exercises 2.3

1. Find difference equations whose solutions are

   (a) $2^{n-1} - 5^{n+1}$.

   (b) $3\cos\left(\dfrac{n\pi}{2}\right) - \sin\left(\dfrac{n\pi}{2}\right)$.

   (c) $(n+2)5^n \sin\left(\dfrac{n\pi}{4}\right)$.

   (d) $(c_1 + c_2 n + c_3 n^2)7^n$.

   (e) $1 + 3n - 5n^2 + 6n^3$.

2. Find the second order linear homogeneous difference equation which generates the sequence 1, 2, 5, 12, 29, 70, 169, ..., then write the solution of the obtained equation.

In each of Problems 3 through 8, write the general solution of the difference equation.

3. $x(n+2) - 16x(n) = 0$.

4. $x(n+2) + 16x(n) = 0$.

5. $(E-3)^2(E^2+4)x(n) = 0$.

6. $\Delta^3 x(n) = 0$.

7. $(E^2+2)^2 x(n) = 0$.

8. $x(n+2) - 6x(n+1) + 14x(n) = 0$.

9. Consider Example 2.24. Verify that $x_1(n) = r^n \cos n\theta$ and $x_2(n) = r^n \sin n\theta$ are two linearly independent solutions of the given equation.

10. Consider the integral defined by

$$I_k(\varphi) = \int_0^\pi \frac{\cos(k\theta) - \cos(k\varphi)}{\cos\theta - \cos\varphi} d\theta, \quad k = 0, 1, 2, \ldots, \varphi \in R.$$

   (a) Show that $I_k(\varphi)$ satisfies the difference equation

$$I_{n+2}(\varphi) - 2\cos\varphi I_{n+1}(\varphi) + I_n(\varphi) = 0, \qquad I_0(\varphi) = 0, \qquad I_1(\varphi) = \pi.$$

   (b) Solve the difference equation in part (a) to find $I_n(\varphi)$.

11. The Chebyshev Polynomials of the first and second kind are defined, respectively, as follows:

$$T_n(x) = \cos(n \cos^{-1}(x)), \qquad U_n(x) = \frac{1}{\sqrt{1-x^2}} \sin[(n+1)\cos^{-1}(x)],$$

   for $|x| < 1$.

(a) Show that $T_n(x)$ obeys the difference equation

$$T_{n+2}(x) - 2xT_{n+1}(x) + T_n(x) = 0, \qquad T_0(x) = 1, \qquad T_1(x) = x.$$

(b) Solve the difference equation in part (a) to find $T_n(x)$.

(c) Show that $U_n(x)$ satisfies the difference equation

$$U_{n+2}(x) - 2xU_{n+1}(x) + U_n(x) = 0, \qquad U_0(x) = 1, \qquad U_1(x) = 2x.$$

(d) Write down the first three terms of $T_n(x)$ and $U_n(x)$.

(e) Show that $T_n(\cos\theta) = \cos n\theta$ and that $U_n(\cos\theta) = (\sin[(n+1)\theta])$ $/\sin\theta$.

12. Show that the general solution of

$$x(n+2) - 2sx(n+1) + x(n) = 0, \qquad |s| < 1$$

is given by

$$x(n) = c_1 T_n(s) + C_2 U_{n-1}(s).$$

13. Show that the general solution of $x(n+2) + p_1 x(n+1) + p_2 x(n) = 0$, $p_2 > 0$, $p_1^2 < 4p_2$, is given by $x(n) = r^n[c_1 T_n(s) + c_2 U_{n-1}(s)]$, where $r = \sqrt{p_2}$ and $s = P_1/(2\sqrt{p_2})$.

14. The Lucas numbers L_n are defined by the difference equation

$$L_{n+2} = L_{n+1} + L_n, \, n \geq 0, \qquad L_0 = 2, \qquad L_1 = 1.$$

Solve the difference equation to find L_n.

15. Show that $\lim_{n\to\infty}(F(n+1))/F(n) = \alpha$, where $\alpha = (1 + \sqrt{5})/2$.

16.* Prove that consecutive Fibonacci numbers $F(n)$ and $F(n+1)$ are relatively prime.

17. (a) Prove that $F(n)$ is the nearest integer to $1/\sqrt{5}((1 + \sqrt{5})/2)^n$.

(b) Find $F(17)$, $F(18)$, and $F(19)$, applying part (a).

18.* Define $x = a \bmod p$ if $x = mp + a$, and $0 \leq a < p$. Let p be a prime number with $p > 5$.

(a) Show that $F(p) = 5^{(p-1)/2} \bmod p$. (Hint: $a^{p-1} = 1 \bmod p$ if a and p are relatively prime.)

(b) Show that $F(p) = \pm 1 \bmod p$.

*(c) Show that $F(p) = 1 \bmod p$ if and only if $p = 3 \bmod 5$ or $p = 4 \bmod 5$.

---

[6] Asterisk denotes difficult problems.

19. Show that if $p_k \neq 0$ in Eq. (2.3.1), then none of its characteristic roots is equal to zero.

20. Show that the Vandermonde determinant (2.3.3) is equal to $\prod_{1 \leq i \leq j \leq k}(\lambda^j - \lambda_i)$.

21. Show that if $\psi(n)$ is a solution of Eq. $(2.3.5)_i$, then it is also a solution of Eq. $(2.3.1)'$.

22. Prove that the set $\{1, n, n^2, \ldots, n^{m-1}\}$ is a linearly independent set on $n \geq n_0$, for any $n_0 > 0$.

23. Find the value of the $n \times n$ tridiagonal determinant

$$D(n) = \begin{vmatrix} b & a & 0 & - & 0 & 0 \\ a & b & a & - & 0 & 0 \\ 0 & a & b & - & 0 & 0 \\ \vdots & \vdots & & & & \vdots \\ 0 & 0 & 0 & - & b & a \\ 0 & 0 & 0 & - & a & b \end{vmatrix}.$$

(Hint: $D(n) = bD(n-1) - a^2 D(n-2)$.)

24. Prove Corollary 2.23.

25. Find the value of the $n \times n$ tridiagonal determinant

$$D(n) = \begin{vmatrix} a & b & 0 & - & 0 & 0 \\ c & a & b & - & 0 & 0 \\ 0 & c & a & - & 0 & 0 \\ \vdots & \vdots & \vdots & & \vdots & \vdots \\ 0 & 0 & 0 & - & a & b \\ 0 & 0 & 0 & - & c & a \end{vmatrix}.$$

$$g(n) \rightarrow \boxed{\text{system}} \rightarrow y(n)$$

Figure 2.2. Input-output system.

2.4 Linear Nonhomogeneous Equations: Method of Undetermined Coefficients

In the last two sections we developed the theory of linear homogeneous difference equations. Moreover, in the case of equations with constant coefficients we have shown how to construct their solutions. In this section we focus our attention to solving the kth order linear nonhomogeneous equation

$$y(n + k) + p_1(n)y(n + k - 1) + \cdots + p_k(n)y(n) = g(n), \qquad (2.4.1)$$

where $p_k(n) \neq 0$ for all $n \geq n_0$. The sequence $g(n)$ is called the *forcing term*, the *external force*, the *control*, or the *input* of the system. As we will discuss later in Chapter 6, Eq. (2.4.1) represents a physical system in which $g(n)$ is the input and $y(n)$ is the output (Fig. 2.2). Thus solving Eq. (2.4.1) amounts to determining the output $y(n)$ given the input $g(n)$. We may look at $g(n)$ as a control term that the designing engineer uses to force the system to behave in a prespecified way.

Before proceeding to present general results concerning Eq. (2.4.1) we would like to raise the following question: Do solutions of Eq. (2.4.1) form a vector space? In other words, is the sum of two solutions of Eq. (2.4.1) a solution of Eq. (2.4.1)? And is the multiple of a solution of Eq. (2.4.1) a solution of Eq. (2.4.1)? Let us answer these questions through the following example.

Example 2.28. Contemplate the equation

$$y(n + 2) - y(n + 1) - 6y(n) = 5(3^n).$$

(a) Show that $y_1(n) = n(3^{n-1})$ and $y_2(n) = (1 + n)3^{n-1}$ are solutions of the equation.

(b) Show that $y(n) = y_2(n) - y_1(n)$ is not a solution of the equation.

(c) Show that $\varphi(n) = sn(3^{n-1})$ is not a solution of the equation.

Solution

(a) The verification that y_1 and y_2 are solutions is left to the reader.

(b) $y(n) = y_2(n) - y_1(n) = 3^{n-1}$. Substituting this into the equation yields

$$3^{n+1} - 3^n - 63^{n-1} = 3^n[3 - 1 - 2] = 0 \neq 5(3^n).$$

(c) By substituting for $\varphi(n)$ into the equation we see easily that $\varphi(n)$ is not a solution.

Conclusion

(i) From the above example we conclude that, in contrast to the situation for homogeneous equation, solutions of the nonhomogeneous Eq. (2.4.1) do not form a vector space. In particular neither the sum (difference) of two solutions nor the multiple of a solution is a solution.

(ii) From part (b) in example 2.28, we found that the difference of the solutions $y_2(n)$ and $y_1(n)$ of the nonhomogeneous equation is actually a solution of the associated homogeneous equation. This is indeed true for the general nth order equation as demonstrated by the following result.

Theorem 2.29. If $y_1(n)$ and $y_2(n)$ are solutions of Eq. (2.4.1), then $x(n) = y_1(n) - y_2(n)$ is a solution of the corresponding homogeneous equation

$$x(n + k) + p_1(n)x(n + k - 1) + \cdots + p_k(n)x(n) = 0. \tag{2.4.2}$$

Proof. The reader will undertake the justification of this theorem in Exercise 2.4, Problem 12.

It is customary to refer to the general solution of the homogeneous equation (2.4.2) as the *complementary solution* of the nonhomogeneous equation (2.4.1), and will be denoted by $y_c(n)$. A solution of the nonhomogeneous equation (2.4.1) is called a *particular solution* and will be denoted by $y_p(n)$. The next result gives us an algorithm to generate all solutions of the nonhomogeneous equation (2.4.1).

Theorem 2.30. Any solution $y(n)$ of Eq. (2.4.1) may be written as

$$y(n) = y_p(n) + \sum_{i=1}^{k} a_i x_i(n),$$

where $\{x_1(n), x_2(n), \ldots, x_k(n)\}$ is a fundamental set of solutions of the homogeneous equation (2.4.2).

Proof. Observe that according to Theorem 2.29, $y(n) - y_p(n)$ is a solution of the homogeneous equation (2.4.2). Thus $y(n) - y_p(n) = \sum_{i=1}^{k} a_i x_i(n)$, for some constants a_i.

The preceding theorem leads to the definition of the *general solution* of the nonhomogeneous equation (2.4.1) as

$$y(n) = y_c(n) + y_p(n). \tag{2.4.3}$$

We now turn our attention to finding a particular solution y_p of nonhomogeneous equations with constant coefficients such as

$$y(n + k) + p_1 y(n + k - 1) + \cdots + p_k y(n) = g(n). \tag{2.4.4}$$

Because of its simplicity, we use the method of *undetermined coefficients* to compute y_p.

Basically, the method consists of making an intelligent guess as to the form of the particular solution and then substituting this function into the diifference equation. For a completely arbitrary nonhomogeneous term $g(n)$, this method is not effective. However, definite rules can be established for the determination of a particular solution by this method, if $g(n)$ is a linear combination of terms, each having one of the forms

$$a^n, \quad \sin(bn), \quad \cos(bn), \quad \text{or } n^k, \tag{2.4.5}$$

or products of these forms such as

$$a^n \sin(bn), \qquad a^n n^k, \qquad a^n n^k \cos(bn), \ldots . \tag{2.4.6}$$

Definition 2.31. A polynomial operator $N(E)$, where E is the shift operator, is said to be an *annihilator* of $g(n)$ if

$$N(E)g(n) = 0. \tag{2.4.7}$$

In other words, $N(E)$ is an annihilator of $g(n)$ if $g(n)$ is a solution of Eq. (2.4.7). For example an annihilator of $g(n) = 3^n$ is $N(E) = E - 3$, since $(E - 3)y(n) = 0$ has a solution $y(n) = 3^n$. An annihilator of $g(n) = \cos \frac{n\pi}{2}$ is $N(E) = E^2 + 1$, since $(E^2 + 1)y(n) = 0$ has a solution $y(n) = \cos \frac{n\pi}{2}$. Let us now rewrite Eq. (2.4.4) using the shift operator E as

$$p(E)y(n) = g(n), \tag{2.4.4}'$$

where $p(E) = E^{n+k} + p_1 E^{n+k-1} + p_2 E^{n+k-2} + \cdots + p_k I$.

Assume now that $N(E)$ is an annihilator of $g(n)$ in Eq. (2.4.4)'. Applying $N(E)$ on both sides of Eq. (2.4.4)' yields

$$N(E)p(E)y(n) = 0. \tag{2.4.8}$$

Let $\lambda_1, \lambda_2, \ldots, \lambda_k$ be the characteristic roots of the homogeneous equation

$$p(E)y(n) = 0 \tag{2.4.9}$$

and $\mu_1, \mu_2, \ldots, \mu_l$ be the characteristic roots of

$$N(E)y(n) = 0. \tag{2.4.10}$$

We must consider two separate cases.

Case 1. None of the λ_i's equals any of the μ_i's. In this case, write $y_p(n)$ as the general solution of Eq. (2.4.10) with undetermined constants. Substituting back this guesstimated particular solution into Eq. (2.4.4), we find the values of the constants. Table 2.3 contains several types of functions $g(n)$ and their corresponding particular solutions.

Case 2. $\lambda_i = \mu_j$ for some i, j. In this case, the set of characteristic roots of Eq. (2.4.8) is equal to the union of the sets $\{\lambda_i\}$, $\{\mu_j\}$, and consequently contains

Table 2.3. Particular Solutions $y_p(n)$

$g(n)$	$y_p(n)$
a^n	$c_1 a^n$
n^k	$c_0 + c_1 n + \cdots + c_k n^k$
$n^k a^n$	$c_0 a^n + c_1 n a^n + \cdots + c_k n^k a^n$
$\sin bn, \cos bn$	$c_1 \sin bn + c_2 \cos bn$
$a^n \sin bn, a^n \cos bn$	$(c_1 \sin bn + c_2 \cos bn)a^n$
$a^n n^k \sin bn, a^n n^k \cos bn$	$(c_0 + c_1 n + \cdots + c_k n^k)a^n \sin(bn)$
	$+(d_0 + d_1 n + \cdots d_k n^k)a^n \cos(bn)$

roots of higher multiplicity than the two individual sets of characteristic roots. To determine a particular solution $y_p(n)$, we first find the general solution of Eq. (2.4.8) and then drop all the terms that appear in $y_c(n)$. Then proceed as in Case 1 to evaluate the constants.

Example 2.32. Solve the difference equation

$$y(n+2) + y(n+1) - 12y(n) = n2^n. \tag{*}$$

Solution The characteristic roots of the homogeneous equation are $\lambda_1 = 3$ and $\lambda_2 = -4$.

Hence,

$$y_c(n) = c_1 3^n + c_2 (-4)^n.$$

Since the annihilator of $g(n) = n2^n$ is given by $N(E) = (E-2)^2$ (why?), we know that $\mu_1 = \mu_2 = 2$. This equation falls in the realm of Case 1 since $\lambda_i \neq \mu_j$, for any i, j. So we let

$$y_p(n) = a_1 2^n + a_2 n2^n.$$

Substituting this relation into the equation (*) gives

$$a_1 2^{n+2} + a_2(n+2)2^{n+2} + a_1 2^{n+1} + a_2(n+1)2^{n+1} - 12a_2 n2^n = n2^n$$

$$(10a_2 - 6a_1)2^n - 6a_2 n2^n = n2^n.$$

Hence

$$10a_2 - 6a_1 = 0 \quad \text{and} \quad -6a_2 = 1,$$

or

$$a_1 = \frac{-15}{18}, \qquad a_2 = \frac{-1}{6}.$$

The particular solution is

$$y_p(n) = \frac{-5}{18}2^n - \frac{1}{6}n2^n,$$

and the general solution is

$$y(n) = c_1 3^n + c_2(-4)^n - \frac{5}{18}2^n - \frac{1}{6}n2^n.$$

Example 2.33. Solve the diifference equation

$$(E - 3)(E + 2)y(n) = 5(3^n). \tag{2.4.11}$$

Solution The annihilator of $5(3^n)$ is $N(E) = (E - 3)$. Hence, $\mu_1 = 3$. The characteristic roots of the homogeneous equation are $\lambda_1 = 3$ and $\lambda_2 = -2$. Since $\lambda_2 = \mu_1$, we apply the procedure for Case 2.

Thus,

$$(E - 3)^2(E + 2)y(n) = 0. \tag{2.4.12}$$

Now $y_c(n) = c_1 3^n + c_2(-2)^n$.

We now know that the general solution of Eq. (2.4.12) is given by

$$\tilde{y}(n) = (a_1 + a_2n)3^n + a_3(-2)^n.$$

Omitting from $\tilde{y}(n)$ the terms 3^n and $(-2)^n$ that appeared in $y_c(n)$, we set $y_p(n) = a_2n3^n$. Substituting this $y_p(n)$ into Eq. (2.4.11) gives

$$a_2(n + 2)3^{n+2} - a_2(n + 1)3^{n-1}6a_2n3^n = 5.3^n$$

or

$$a_2 = \frac{1}{3}.$$

Thus $y_p(n) = n3^{n-1}$, and the general solution of Eq. (2.4.11) is

$$y(n) = c_1 3^n + c_2(-2)^n + n3^{n-1}.$$

Example 2.34 Solve the difference equation

$$y(n + 2) + 4y(n) = 8(2^n) \cos\left(\frac{n\pi}{2}\right). \tag{2.4.13}$$

Solution The characteristic equation of the homogeneous equation is

$$\lambda^2 + 4 = 0.$$

The characteristic roots are

$$\lambda_1 = 2i, \qquad \lambda_2 = -2i.$$

Thus $r = 2$, and $\theta = \pi/2$, and

$$y_c(n) = 2^n \left(c_1 \cos\left(\frac{n\pi}{2}\right) + c_2 \sin\left(\frac{n\pi}{2}\right)\right).$$

Notice that $g(n) = 2^n \cos\left(\frac{n\pi}{2}\right)$ appears in $y_c(n)$. Using Table 2.3, we set

$$y_p(n) = 2^n \left(an \cos\left(\frac{n\pi}{2}\right) + bn \sin\left(\frac{n\pi}{2}\right)\right). \tag{2.4.14}$$

Substituting Eq. (2.4.14) into Eq. (2.4.13) gives

$$2^{n+2}\left[a(n+2)\cos\left(\frac{n\pi}{2}+\pi\right) + b(n+2)\sin\left(\frac{n\pi}{2}+\pi\right)\right]$$
$$+(4)2^n\left[an\cos\left(\frac{n\pi}{2}\right) + bn\sin\left(\frac{n\pi}{2}\right)\right] = 8(2^n)\cos\left(\frac{n\pi}{2}\right).$$

Replacing $\cos((n\pi)/2+\pi)$ by $-\cos((n\pi)/2)$, and $\sin((n\pi)/2+\pi)$ by $-\sin((n\pi)/2)$, and then comparing the coefficients of the cosine terms lead us to $a = -1$. Then by comparing the coefficients of the sine terms, we realize that $b = 0$.

By substituting these values back into Eq. (2.4.14), we know that

$$y_p(n) = -2^{n+1}n\cos\left(\frac{n\pi}{2}\right),$$

and the general solution of Eq. (2.4.13), arrived at by adding $y_c(n)$ and $y_p(n)$, is

$$y(n) = 2^n\left(c_1\cos\frac{n\pi}{2} + c_2\sin\left(\frac{n\pi}{2}\right) - n\cos\left(\frac{n\pi}{2}\right)\right).$$

Exercise 2.4. For Problems 1 through 6, find a particular solution of the difference equation.

1. $y(n+2) - 5y(n+1) + 6y(n) = 1 + n$.

2. $y(n+2) + 8y(n+1) + 12y(n) = e^n$.

3. $y(n+2) - 5y(n+1) + 4y(n) = 4^n - n^2$.

4. $y(n+2) + 8y(n+1) + 7y(n) = ne^n$.

5. $y(n+2) - y(n) = n\cos\left(\frac{n\pi}{2}\right)$.

6. $(E^2 + 9)^2 y(n) = \sin\left(\frac{n\pi}{2}\right) - \cos\left(\frac{n\pi}{2}\right)$.

For Problems 7 through 9 find the solution of the difference equation

7. $\Delta^2 y(n) = 16$, $y(0) = 2$, $y(1) = 3$.

8. $\Delta^2 y(n) + 7y(n) = 2\sin\left(\frac{n\pi}{4}\right)$, $y(0) = 0$, $y(1) = 1$.

9. $(E - 3)(E^2 + 1)y(n) = 3^n$, $y(0) = 0$, $y(1) = 1$, $y(2) = 3$.

For Problems 10 and 11 find the general solution of the difference equation

10. $y(n + 2) - y(n) = n2^n \sin\left(\dfrac{n\pi}{2}\right)$.

11. $y(n + 2) + 8y(n + 1) + 7y(n) = n2^n$.

12. Prove Theorem 2.29.

13. Consider the difference equation $y(n+2) + p_1 y(n+1) + p_2 y(n) = g(n)$ where $p_1^2 < 4p_2$ and $0 < p_2 < 1$. Show that if y_1 and y_2 are two solutions of the equation, then $y_1(n) - y_2(n) \to 0$ as $n \to \infty$.

14. Determine the general solution of $y(n + 2) + \lambda^2 y(n) = \sum_{m=1}^{N} a_m \sin(m\pi n)$, where $\lambda > 0$ and $\lambda \neq m\pi$, $m = 1, 2, \ldots, N$.

15. Solve the difference equation

$$y(n + 2) + y(n) = \begin{cases} 1 & \text{if } 0 \leq n \leq 2 \\ -1 & \text{if } n > 2 \end{cases}.$$

with $y(0) = 0$ and $y(1) = 1$.

2.4.1   The Method of Variation of Constants (Parameters)

Contemplate the second order nonhomogeneous difference equation

$$y(n + 2) + p_1(n)y(n + 1) + p_2(n)y(n) = g(n), \tag{2.4.15}$$

and the corresponding homogeneous difference equation

$$y(n + 2) + p_1(n)y(n + 1) + p_2(n)y(n) = 0. \tag{2.4.16}$$

The method of variation of constants is commonly used to find a particular solution $y_p(n)$ of Eq. (2.4.15) when the coefficients $p_1(n)$ and $p_2(n)$ are not constants. The method assumes that a particular solution of Eq. (2.4.15) may be written in the form

$$y(n) = u_1(n)y_1(n) + u_2(n)y_2(n), \tag{2.4.17}$$

where $y_1(n)$ and $y_2(n)$ are two linearly independent solutions of the homogeneous Eq. (2.4.16), and $u_1(n)$, $u_2(n)$ are sequences to be determined later.

16.   (a) Show that

$$\begin{aligned} y(n + 1) &= u_1(n)y_1(n + 1) + u_2(n)y_2(n + 1) \\ &+ \Delta u_1(n)y_1(n + 1) + \Delta u_2(n)y_2(n + 1). \end{aligned} \tag{2.4.18}$$

(b) The method stipulates that

$$\Delta u_1(n)y_1(n+1) + \Delta u_2(n)y_2(n+1) = 0. \qquad (2.4.19)$$

Use Eqs. (2.4.18) and (2.4.19) to show that

$$\begin{aligned} y(n+2) &= u_1(n)y_1(n+2) + u_2(n)y_2(n+2) \\ &\quad + \Delta u_1(n)y_1(n+2) + \Delta u_2(n)y_2(n+2). \end{aligned}$$

(c) By substituting the above expressions for $y(n)$, $y(n+1)$, and $y(n+2)$ into Eq. (2.4.15), show that

$$\Delta u_1(n)y_1(n+2) + \Delta u_2(n)y_2(n+2) = g(n). \qquad (2.4.20)$$

(d) Using Expressions (2.4.19) and (2.4.20), show that

$$\Delta u_1(n) = \frac{-g(n)y_2(n+1)}{C(n+1)} \quad \text{and} \quad u_1(n) = \sum_{r=0}^{n-1} \frac{-g(r)y_2(r+1)}{C(r+1)} y_1(n),$$
$$(2.4.21)$$

$$\Delta u_2(n) = \frac{g(n)y_1(n+1)}{C(n+1)} \quad \text{and} \quad u_2(n) = \sum_{r=0}^{n-1} \frac{g(r)y_1(r+1)}{C(r+1)} y_2(n),$$
$$(2.4.22)$$

where $C(n)$ is the Casoratian of $y_1(n)$ and $y_2(n)$.

17. Use Formulas (2.4.21) and (2.4.22) to solve the equation

$$y(n+2) - 7y(n+1) + 6y(n) = n.$$

18. Use the variation of constants method to solve the initial value problem

$$y(n+2) - 5y(n+1) + 6y(n) = 2^n, \qquad y(1) = y(2) = 0.$$

19. Use 16_d to show that the unique solution of Eq. (2.4.15) with $y(0) = y(1) = 0$ is given by

$$y(n) = \sum_{r=0}^{n-1} \frac{y_1(r+1)y_2(n) - y_2(r+1)y_1(n)}{C(r+1)}.$$

20. Consider the equation

$$x(n+1) = ax(n) + f(n). \qquad (2.4.23)$$

(a) Show that

$$x(n) = a^n \left[x(0) + \frac{f(0)}{a} + \frac{f(1)}{a^2} +, \ldots, + \frac{f(n-1)}{a^n} \right] \qquad (2.4.24)$$

is a solution of Eq. (2.4.23).

(b) Show that if $|a| < 1$ and $\{f(n)\}$ is a bounded sequence, i.e., $|f(n)| \leq M$, for some $M > 0$, $n \in Z^+$, then all solutions of Eq. (2.4.23) are bounded.

(c) Suppose that $a > 1$ and $\{f(n)\}$ is bounded on Z^+. Show that if we choose

$$x(0) = -\left(\frac{f(0)}{a} + \frac{f(1)}{a^2} +, \ldots, + \frac{f(n)}{a^{n+1}} +, \ldots\right) = -\sum_{i=0}^{\infty} \frac{f(i)}{a^{i+1}}$$

(2.4.25)

then the solution $x(n)$ given by Eq. (2.4.24) is bounded on Z^+. Give an explicit expression for $x(n)$ in this case.

(d) Under the assumptions of part (c), show that for any choice of $x(0)$, excepting that value given by Eq. (2.4.25), the solution of Eq. (2.4.23) is unbounded.

2.5   Limiting Behavior of Solutions

To simplify our exposition we restrict our discussion to the second order difference equation

$$y(n+2) + p_1 y(n+1) + p_2 y(n) = 0. \qquad (2.5.1)$$

Suppose that λ_1 and λ_2 are the characteristic roots of the equation. Then we have the following three cases:

(a) λ_1 and λ_2 are distinct real roots. Then $y_1(n) = \lambda_1^n$ and $y_2(n) = \lambda_2^n$ are two linearly independent solutions of Eq. (2.5.1). If $|\lambda_1| > |\lambda_2|$, then we call $y_1(n)$ the *dominant solution*, and λ_1 the *dominant characteristic root*. Otherwise, $y_2(n)$ is the dominant solution, and λ_2 is the dominant characteristic root. We will now show that the limiting behavior of the general solution $y(n) = a_1 \lambda_1^n + a_2 \lambda_2^n$ is determined by the behavior of the dominant solution. So assume, without loss of generality, that $|\lambda_1| > |\lambda_2|$. Then

$$y(n) = \lambda_1^n \left[a_1 + a_2 \left(\frac{\lambda_2}{\lambda_1}\right)^n\right].$$

Since

$$\left|\frac{\lambda_2}{\lambda_1}\right| < 1, \left(\frac{\lambda_2}{\lambda_1}\right) \to 0$$

as $n \to \infty$.

Consequently, $\lim_{n \to \infty} y(n) = \lim_{n \to \infty} a_1 \lambda_1^n$. There are six different situations that may arise here depending on the value of λ_1. (See Fig. 2.3)

1. $\lambda_1 > 1$: the sequence $\{a_1 \lambda_1^n\}$ diverges to ∞ (unstable system).

2. $\lambda_1 = 1$: the sequence $\{a_1 \lambda_1^n\}$ is a constant sequence.

3. $0 \leq \lambda_1 < 1$: the sequence $\{a_1 \lambda_1^n\}$ is monotonically decreasing to zero (stable system).

4. $-1 < \lambda_1 \leq 0$: the sequence $\{a_1 \lambda_1^n\}$ is oscillating around zero (i.e., alternating in sign) and converging to zero (stable system).

5. $\lambda_1 = -1$: the sequence $\{a_1 \lambda_1^n\}$ is oscillating between two values a_1 and $-a_1$.

6. $\lambda_1 < -1$: the sequence $\{a_1 \lambda_1^n\}$ is oscillating but increasing in magnitude (unstable system).

(b) $\lambda_1 = \lambda_2 = \lambda$.

The general solution of Eq. (2.5.1) is given by $y(n) = (a_1 + a_2 n)\lambda^n$. Clearly if $|\lambda| \geq 1$, the solution $y(n)$ diverges either monotonically if $\lambda \geq 1$ or oscillatory if $\lambda \leq -1$. However if $|\lambda| < 1$, then the solution converges to zero since $\lim_{n \to \infty} n\lambda^n = 0$ (why?).

(c) Complex roots: $\lambda_1 = \alpha + i\beta$ and $\lambda_2 = \alpha - i\beta$, where $\beta \neq 0$.

As we have seen in Section 2.3, Formula (2.3.8) the solution of Eq. (2.5.1) is given by $y(n) = ar^n \cos(n\theta - \omega)$ where

$$r = \sqrt{\alpha^2 + \beta^2} \theta = \tan^{-1}\left(\frac{\beta}{\alpha}\right).$$

The solution $y(n)$ clearly oscillates since the cosine function oscillates. However, $y(n)$ oscillates in three different ways depending on the location of the conjugate characteristic roots as may be seen in Fig. 2.4.

1. $r > 1$: here λ_1 and $\lambda_2 = \bar{\lambda}_1$ are outside the unit circle. Hence $y(n)$ is oscillating but increasing in magnitude (unstable system).

2. $r = 1$: here λ_1 and $\lambda_2 = \bar{\lambda}_1$ lie on the unit circle. In this case $y(n)$ is oscillating but constant in magnitude.

3. $r < 1$: here λ_1 and $\lambda_2 = \bar{\lambda}_1$ lie inside the unit disk. The solution $y(n)$ oscillates but converges to zero as $n \to \infty$ (stable system).

Finally we summarize the above discussion in the following theorem.

Theorem 2.35. The following statements hold.

(i) All solutions of Eq. (2.5.1) oscillate (about zero) if and only if the equation has no positive real characteristic roots.

(ii) All solutions of Eq. (2.5.1) converge to zero (i.e., the zero solution is asymptotically stable) if and only if max $\{|\lambda_1|, |\lambda_2|\} < 1$.

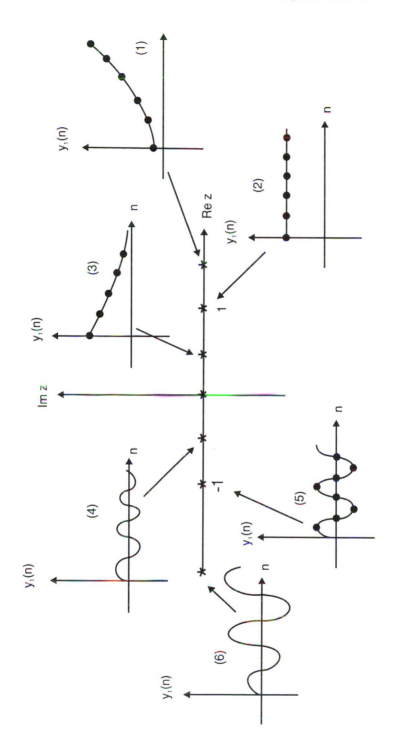

Figure 2.3 $(n, y(n))$ diagrams for real roots.

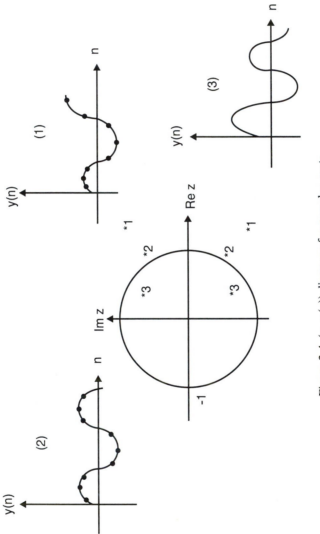

Figure 2.4. $(n, y(n))$ diagrams for complex roots.

Next we consider nonhomogeneous difference equations in which the input is constant, that is equations of the form

$$y(n + 2) + p_1 y(n + 1) + p_2 y(n) = M, \tag{2.5.2}$$

where M is a nonzero constant input or forcing term. Unlike Eq. (2.5.1), the zero sequence $y(n) = 0$ for all $n \in Z^+$ is not a solution of Eq. (2.5.2). Instead we have the equilibrium point or solution $y(n) = y^*$. From Eq. (2.5.2) we have

$$y^* + p_1 y^* + p_2 y^* = M$$

or

$$y^* = \frac{M}{1 + p_1 + p_2}. \tag{2.5.3}$$

Thus $y_p(n) = y^*$ is a particular solution of Eq. (2.5.2). Consequently, the general solution of Eq. (2.5.2) is given by

$$y(n) = y^* + y_c(n). \tag{2.5.4}$$

It is clear that $y(n) \to y^*$ if and only if $y_c(n) \to 0$ as $n \to \infty$. Furthermore, $y(n)$ oscillates[6] about y^* if and only if $y_c(n)$ oscillates about zero. These observations are summarized in the following theorem.

Theorem 2.36. The following statements hold.

(i) All solutions of the nonhomogeneous equation (2.5.2) oscillate about the equilibrium solution y^* if and only if none of the characteristic roots of the homogeneous equation (2.5.1) is a positive real number.

(ii) All solutions of Eq. (2.5.2) converge to y^* as $n \to \infty$ if and only if max $\{|\lambda_1|, |\lambda_2|\} < 1$, where λ_1 and λ_2 are the characteristic roots of the homogeneous equation (2.5.1).

Theorems 2.35 and 2.36 give necessary and sufficient conditions under which a second order difference equation is asymptotically stable. In many applications, however, one needs to have explicit criteria for stability based on the values of the coefficients p_1 and p_2 of Eq. (2.5.2) or (2.5.1). The following result provides us with such needed criteria.

Theorem 2.37 The conditions

$$1 + p_1 + p_2 > 0, \qquad 1 - p_1 + p_2 > 0, \qquad 1 - p_2 > 0 \tag{2.5.5}$$

are necessary and sufficient for the equilibrium point (solution) of Eqs. (2.5.1) and (2.5.2) to be asymptotically stable (i.e., all solutions converge to y^*).

---

[6]We say $y(n)$ oscillates about y^* if $y(n) - y^*$ alternates sign, i.e., if $y(n) > y^*$, then $y(n + 1) < y^*$.

88   2. Linear Difference Equations of Higher Order

Proof Assume that the equilibrium point of Eq. (2.5.1) or (2.5.2) is asymptotically stable. In virtue of Theorems 2.35 and 2.36, the roots λ_1, λ_2 of the characteristic equation $\lambda^2 + p_1\lambda + p_2 = 0$ lie inside the unit disk, i.e., $|\lambda_1| < 1$ and $|\lambda_2| < 1$. By the quadratic formula, we have

$$\lambda_1 = \frac{-p_1 + \sqrt{p_1^2 - 4p_2}}{2} \quad \text{and} \quad \lambda_2 = \frac{-p_1 - \sqrt{p_1^2 - 4p_2}}{2}. \tag{2.5.6}$$

Then we have two cases to consider.

Case 1. λ_1, λ_2 are real roots, i.e., $p_1^2 - 4p_2 \geq 0$. From Formula (2.5.6) we have

$$-2 < -p_1 + \sqrt{p_1^2 - 4p_2} < 2$$

or

$$-2 + p_1 < \sqrt{p_1^2 - 4p_2} < 2 + p_1. \tag{2.5.7}$$

Similarly, one obtains

$$-2 + p_1 < -\sqrt{p_1^2 4 p_2} < 2 + p_1. \tag{2.5.8}$$

Squaring the second inequality in Expression (2.5.7) yields $p_1 + p_2 + 1 > 0$. Similarly, if we square the first inequality in Expression (2.5.8) we obtain $1 - p_1 + p_2 > 0$.

Case 2. λ_1 and λ_2 are complex conjugates, i.e., $p_1^2 - 4p_2 < 0$. In this case we have

$$\lambda_1 = \frac{-p_1}{2} + \frac{i}{2}\sqrt{4p_2 - p_1^2}, \qquad \lambda_2 = \frac{-p_1}{2} - \frac{i}{2}\sqrt{4p_2 - p_1^2}.$$

Since $-1 < \lambda_1 < 1$ and $-1 < \lambda_2 < 1$, we have

$$\lambda_1\lambda_2 = p_2 < 1$$

or

$$1 - p_2 > 0.$$

This completes the proof of the necessary conditions. The converse is left to the reader as Exercise 2.5, Problem 8.

Example 2.38. Find conditions under which the solutions of the equation

$$y(n + 2) - \alpha(1 + \beta)y(n + 1) + \alpha\beta y(n) = 1, \qquad \alpha, \qquad \beta > 0,$$

(a) converge to the equilibrium point y^*, and

(b) oscillate about y^*.

Solution Let us first find the equilibrium point y^*. Be letting $y(n) = y^*$ in the equation we obtain

$$y^* = \frac{1}{1 - \alpha}, \qquad \alpha \neq 1.$$

(a) Applying Condition (2.5.5) on our equation yields

$$\alpha < 1, \qquad 1 + \alpha + 2\alpha\beta > 0, \qquad \text{and } \alpha\beta < 1.$$

Clearly the second inequality $1 + \alpha + 2\alpha\beta > 0$ is always satisfied since α, β are both positive numbers.

(b) The solutions are oscillatory about y^* if either λ_1, λ_2 are negative real numbers or complex conjugates. In the first case we have

$$\alpha^2(1 + \beta)^2 > 4\alpha\beta \text{ or } \alpha > \frac{4\beta}{(1 + \beta)^2}$$

and

$$\alpha(1 + \beta) < 0 \text{ which is impossible.}$$

Thus if $\alpha > 4\beta/(1 + \beta)^2$ we have no oscillatory solutions.

now λ_1 and λ_2 are complex conjugates if

$$\alpha^2(1 + \beta)^2 < 4\alpha\beta \text{ or } \alpha < \frac{4\beta}{(1 + \beta)^2}.$$

Hence all solutions are oscillatory if

$$\alpha < \frac{4\beta}{(1 + \beta)^2}.$$

For the treatment of the general kth order scalar difference equations, the reader is referred to Chapter 4 on stability and Chapter 8 on oscillation.

Exercise 2.5. In Problems 1 through 4

(a) Determine the stability of the equilibrium point by using Theorem 2.35 or Theorem 2.36.

(b) Determine the oscillatory behavior of the solutions of the equation.

1. $y(n + 2) - 2y(n + 1) + 2y(n) = 0$.

2. $y(n + 2) + \frac{1}{4}y(n) = \frac{5}{4}$.

3. $y(n + 2) + y(n + 1) + \frac{1}{2}y(n) = -5$.

4. $y(n + 2) - 5y(n + 1) + 6y(n) = 0$.

5. Determine the stability of the equilibrium point of the equations in problems 1 through 4 by using Theorem 2.37.

6. Show that the stability conditions (2.5.5) for the equation $y(n + 2) - \alpha y(n + 1) + \beta y(n) = 0$, where α, β are constants, may be written as

$$-1 - \beta < \alpha < 1 + \beta, \qquad \beta < 1.$$

7. Contemplate the equation $y(n + 2) - p_1 y(n + 1) - p_2 y(n) = 0$. Show that if $|p_1| + |p_2| < 1$, then all solutions of the equation converge to zero.

8. Prove that Conditions (2.5.5) imply that all solutions of Eq. (2.5.2) converge to the equilibrium point y^*.

9. Determine conditions under which all solutions of the difference equation in problem 7 oscillate.

10. Determine conditions under which all solutions of the difference equation in problem 6 oscillate.

11. Suppose that p is a real number. Prove that every solution of the difference equation $y(n + 2) - y(n + 1) + py(n) = 0$ oscillates if and only if $p > \frac{1}{4}$.

*12. Prove that a necessary and sufficient condition for the asymptotic stability of the zero solution of the equation

$$y(n + 2) + p_1 y(n + 1) + p_2 y(n) = 0$$

is

$$|p_1| < 1 + p_2 < 2.$$

13. Determine the limiting behavior of solutions of the equation

$$y(n + 2) = \alpha c + \alpha \beta (y(n + 1) - y(n))$$

if

(i) $\alpha\beta = 1$,
(ii) $\alpha\beta = 2$,
(iii) $\alpha\beta = \dfrac{1}{2}$

provided that $\alpha, \beta,$ and c are positive constants.

14. If $p_1 > 0$ and $p_2 > 0$, show that all solutions of the equation

$$y(n + 2) + p_1 y(n + 1) + p_2 y(n) = 0$$

are oscillatory.

15. Determine the limiting behavior of solutions of the equation

$$y(n + 2) - \frac{\beta}{\alpha} y(n + 1) + \frac{\beta}{\alpha} y(n) = 0,$$

where α and β are constants if

(i) $\beta > 4\alpha$,
(ii) $\beta > 4\alpha$.

2.6 Nonlinear Equations Transformable to Linear Equations

In general, most nonlinear difference equations cannot be solved explicitly. However, a few types of nonlinear equations can be solved, usually by transforming them into linear equations. In this section we discuss some tricks of the trade.

Type I. Equations of Riccati type:

$$x(n+1)x(n) + p(n)x(n+1) + q(n)x(n) = 0. \tag{2.6.1}$$

To solve the Riccati equation, we let

$$z(n) = 1/x(n)$$

in Eq. (2.6.1) to give us

$$q(n)z(n+1) + p(n)z(n) + 1 = 0. \tag{2.6.2}$$

The nonhomogeneous equation requires a different transformation.

$$y(n+1)y(n) + p(n)y(n+1) + q(n)y(n) = g(n). \tag{2.6.3}$$

If we let $y(n) = (z(n+1)/z(n)) - p(n)$ in Eq. (2.6.3) we obtain

$$z(n+2) + (q(n) - p(n+1))z(n+1) - (g(n) + p(n)q(n))z(n) = 0.$$

Example 2.39 (Pielou Logistic Equation). The most popular continuous model of the growth of a population is the well known Verhulst–Pearl equation given by

$$x'(t) = x(t)[a - bx(t)], \qquad a, b > 0, \tag{2.6.4}$$

where $x(t)$ is the size of the population at time t, a is the rate of the growth of the population if the resources were unlimited and the individuals did not affect one another, and $-bx^2(t)$ represents the negative effect on the growth due to crowdedness and limited resources. The solution of Eq. (2.6.4) is given by

$$x(t) = \frac{a/b}{1 + (e^{-at}/cb)}.$$

Now

$$x(t+1) = \frac{a/b}{1 + \left(e^{-a(t+1)}/cb\right)}$$

$$= \frac{e^a(a/b)}{1 + (e^{-at}/cb) + (e^a - 1)}.$$

Dividing by $\left[1 + (e^{-at}/cb)\right]$, we obtain

$$x(t+1) = e^a x(t) / \left[1 + \frac{b}{a}(e^a - 1)x(t)\right]$$

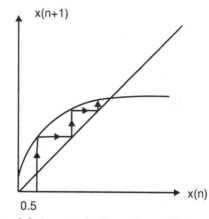

Figure 2.5. Asymptotically stable equilibrium points.

or

$$x(n + 1) = \alpha x(n)/[1 + \beta x(n)], \qquad (2.6.5)$$

where $\alpha = e^a$ and $\beta = \frac{b}{a}(e^a - 1)$.

This equation is titled *Pielou logistic equation.*

Equation (2.6.5) is of the Riccati type and may be solved by leting $x(n) = 1/z(n)$. This gives us the equation

$$z(n + 1) = \frac{1}{\alpha}z(n) + \frac{\beta}{\alpha},$$

whose solution is given by

$$z(n) = \begin{cases} \left[c - \dfrac{\beta}{\alpha - 1}\right]\alpha^n + (\beta/\alpha - 1) & \text{if } \alpha \neq 1 \\[3mm] c - \beta n & \text{if } \alpha = 1 \end{cases}.$$

Thus

$$x(n) = \begin{cases} \alpha^n(\alpha - 1)/[\beta\alpha^n + c(\alpha - 1) - \beta] & \text{if } \alpha \neq 1 \\[3mm] \dfrac{1}{c - \beta n} & \text{if } \alpha = 1 \end{cases}.$$

Hence

$$\lim_{n \to \infty} x(n) = \begin{cases} (\alpha - 1)/\beta & \text{if } \alpha \neq 1 \\[2mm] 0 & \text{if } \alpha = 1 \end{cases}.$$

This conclusion shows the equilibrium point $(\alpha - 1)/\beta$ is globally asymptotically stable if $\alpha \neq 1$. Figure 2.5 illustrates this for $\alpha = 3$, $\beta = 1$, and $x(0) = 0.5$.

Type II. *Equations of general Riccati type*:

$$x(n+1) = \frac{a(n)x(n) + b(n)}{c(n)x(n) + d(n)} \tag{2.6.6}$$

such that $c(n) \neq 0$, $a(n)d(n) - b(n)c(n) \neq 0$ for all $n \geq 0$.

To solve this equation we let

$$c(n)x(n) + d(n) = \frac{y(n+1)}{y(n)}. \tag{2.6.7}$$

Then by substituting

$$x(n) = \frac{y(n+1)}{c(n)y(n)} - \frac{d(n)}{c(n)}$$

into Eq. (2.6.6) we obtain

$$\frac{y(n+2)}{c(n+1)y(n+1)} - \frac{d(n+1)}{c(n+1)} = \frac{a(n)\left[\dfrac{y(n+1)}{c(n)y(n)} - \dfrac{d(n)}{c(n)}\right] + b(n)}{\dfrac{y(n+1)}{y(n)}}.$$

This equation simplifies to

$$y(n+2) + p_1(n)y(n+1) + p_2(n)y(n), \quad y(0) = 1, \qquad y(1) = c(0)x(0) + d(0) \tag{2.6.8}$$

where

$$p_1(n) = -\frac{c(n)d(n+1) + a(n)c(n+1)}{c(n)},$$

$$p_2(n) - (a(n)d(n) - b(n)c(n))\frac{c(n+1)}{c(n)}.$$

Example 2.40. Solve the difference equation

$$x(n+1) = \frac{2x(n) + 3}{3x(n) + 2}.$$

Solution Here $a = 2$, $b = 3$, $c = 3$, and $d = 2$. Hence $ad - bc \neq 0$. Using the transformation

$$3x(n) + 2 = \frac{y(n+1)}{y(n)}, \tag{2.6.9}$$

we obtain as in Eq. (2.6.8)

$$y(n+2) - 4y(n+1) - 5y(n) = 0, \qquad y(0) = 1, \qquad y(1) = 3x(0) + 2,$$

with characteristic roots $\lambda_1 = 5$, $\lambda_2 = -1$.

Hence

$$y(n) = c_1 5^n + c_2(-1)^n. \tag{2.6.10}$$

From Formula (2.6.9) we have

$$
\begin{aligned}
x(n) &= \frac{1}{3}\frac{y(n+1)}{y(n)} - \frac{2}{3} \\
&= \frac{1}{3}\frac{c_1 5^{n+1} + c_2(-1)^{n+1}}{c_1 5^n + c_2(-1)^n} - \frac{2}{3} \\
&= \frac{(c_1 5^n - c_2(-1)^n)}{(c_1 5^n + c_2(-1)^n)} \\
&= \frac{5^n - c(-1)^n}{5^n + c(-1)^n}
\end{aligned}
$$

where

$$
c = \frac{c_1}{c_2}.
$$

Type III. Homogeneous difference equations of the type

$$
f\left(\frac{x(n+1)}{x(n)}, n\right) = 0.
$$

Use the transformation $z(n) = \frac{x(n+1)}{x(n)}$ to convert them to a linear equation in $z(n)$, thus allowing us to solve them.

Example 2.41. Solve the difference equation

$$
x^2(n+1) - 3x(n+1)x(n) + 2x^2(n) = 0. \tag{2.6.11}
$$

Solution By dividing over $x^2(n)$, Eq. (2.6.6) becomes

$$
\left[\frac{x(n+1)}{x(n)}\right]^2 - 3\left[\frac{x(n+1)}{x(n)}\right] + 2 = 0 \tag{2.6.12}
$$

which is of Type II.

Letting $z(n) = \frac{x(n+1)}{x(n)}$ in Eq. (2.6.12) creates

$$
z^2(n) - 3z(n) + 2 = 0.
$$

We can factor this down to

$$
[z(n) - 2][z(n) - 1] = 0,
$$

and thus, either $z(n) = 2$ or $z(n) = 1$.

This leads to

$$
x(n+1) = 2x(n) \text{ or } x(n+1) = x(n).
$$

Consequently, $x(n) = 2^n x(0)$ or $x(n) = x(0)$.

These are two different solutions of Eq. (2.6.11) with the same initial value $x(0)$. Does this violate Theorem 2.6? (Exercise 2.6, Problem 15)

Type IV. Consider the difference equation of the form

$$(y(n + k))^{r_1} (y(n + k - 1))^{r_2}, \ldots, (y(n))^{r_{k+1}} = g(n). \tag{2.6.13}$$

Let $z(n) = \ln y(n)$, and rearrange to obtain

$$r_1 z(n + k) + r_2 z(n + k - 1) +, \ldots, + r_{k+1} z(n) = \ln g(n). \tag{2.6.14}$$

Example 2.42. Solve the difference equation

$$x(n + 2) = \frac{x^2(n + 1)}{x^2(n)}. \tag{2.6.15}$$

Solution Let $z(n) = \ln x(n)$ in Eq. (2.6.15). Then as in Eq. (2.6.12) we obtain

$$z(n + 2) - 2z(n + 1) + 2z(n) = 0.$$

The characteristic roots are $\lambda_1 = 1 + i$, $\lambda_2 = 1 = -i$.

Thus,

$$z(n) = (2)^{n/2} \left[c_1 \cos \left(\frac{n\pi}{4} \right) + c_2 \sin \left(\frac{n\pi}{4} \right) \right].$$

Therefore

$$x(n) = \mathrm{Exp} \left[(2)^{n/2} \left\{ c_1 \cos \left(\frac{n\pi}{4} \right) + c_2 \sin \left(\frac{n\pi}{4} \right) \right\} \right].$$

Exercises 2.6.

1. Find the general solution of the difference equation

$$y^2(n + 1) - 2y(n + 1)y(n) - 3y^2(n) = 0.$$

2. Solve the difference equation

$$y^2(n + 1) - (2 + n)y(n + 1)y(n) - 2ny^2(n) = 0.$$

3. Solve $y(n + 1)y(n) - y(n + 1) + y(n) = 0$.

4. Solve $y(n + 1)y(n) - \dfrac{2}{3} y(n + 1) + \dfrac{1}{6} y(n) = \dfrac{5}{18}$.

5. Solve $y(n + 1) = 5 - \dfrac{6}{y(n)}$.

6. Solve $x(n + 1) = \dfrac{x(n) + a}{x(n) + 1}$, $1 \neq a > 0$.

$$\left[\text{Hint: let } x(n) = \sqrt{a} + \tfrac{1}{y(n)}. \right]$$

7. Solve $x(n + 1) = x^2(n)$.

8. Solve the logistic difference equation

$$x(n + 1) = 2x(n)(1 - x(n))$$

$\left[\text{Hint: let } x(n) = \frac{1}{2}(1 - y(n)), \text{ and then use Problem 7.}\right]$

9. Solve the logistic equation

$$x(n + 1) = 4x(n)[1 - x(n)].$$

$\left[\text{Hint: let } x(n) = \sin^2 \theta(n).\right]$

10. Solve $x(n + 1) = \frac{1}{2}\left(x(n) - \frac{a}{x(n)}\right)$, $a > 0$.

$\left[\text{Hint: let } x(n) = \sqrt{a} \cot y(n).\right]$

11. Solve $y(n + 2) = y^3(n + 1)/y^2(n)$.

12. Solve $x(n + 1) = \dfrac{2x + 4}{x - 1}$.

13. Solve $y(n + 1) = \dfrac{2 - y^2(n)}{2(1 - y(n))}$.

    [Hint: let $y(n) = 1 - x(n)$ and then use Problem 10.]

14. Solve $x(n + 1) = \dfrac{2x}{x + 3}$.

15. Solve $y(n + 1) = 2y(n)\sqrt{1 - y^2(n)}$.

    [Hint: let $y(n) = \sin x(n)$.]

16. The "regular falsi" method for finding the roots of $f(x) = 0$ is given by

$$x(n + 1) = \frac{x(n - 1)f(x(n)) - x(n)f(x(n - 1))}{f(x(n)) - f(x(n - 1))}.$$

    (a) Show that for $f(x) = x^2$, this difference equation becomes

$$x(n + 1) = \frac{x(n - 1)x(n)}{x(n - 1) + x(n)}.$$

    (b) Let $x(1) = 1$, $x(2) = 1$ for the equation in part (a). Show that the solution of the equation is $x(n) = 1/F(n)$, where $F(n)$ is the nth Fibonacci number.

2.7 Applications

2.7.1 *Propagation of Annual Plants* [1]

Our objective here is to develop a mathematical model that describes the number of plants in any desired generation. It is known that plants produce seeds at the end of their growth season (say August), after which they die. Furthermore, only a fraction of these seeds survive the winter and those who survive germinate at the beginning of the season (say May) giving rise to the new generation of plants.

Let

γ = number of seeds produced per plant in August,

α = fraction of one-year-old seeds that germinate in May,

β = fraction of two-year-old seeds that germinate in May,

σ = fraction of seeds that survive a given winter.

If $p(n)$ denotes the number of plants in generation n, than

$$p(n) = \left(\begin{matrix} \text{plants from} \\ \text{one-year-old seeds} \end{matrix} \right) + \left(\begin{matrix} \text{plants from} \\ \text{two-year-old seeds} \end{matrix} \right),$$
$$p(n) = \alpha s_1(n) + \beta s_2(n), \tag{2.7.1}$$

where $s_1(n)$ $\{s_2(n)\}$ is the number of one-year {two -year} old seeds in April (before germination). Observe that seeds left after germination may be written as

$$\text{seeds left} = \left(\begin{matrix} \text{fraction} \\ \text{not germinated} \end{matrix} \right) \times \left(\begin{matrix} \text{original number} \\ \text{of seeds in April} \end{matrix} \right).$$

This gives rise to two equations

$$\tilde{s}_1(n) = (1 - \alpha)s_1(n), \tag{2.7.2}$$

$$\tilde{s}_2(n) = (1 - \beta)s_2(n), \tag{2.7.3}$$

where $\tilde{s}_1(n)\{\tilde{s}_2(n)\}$ is the number of one-year {two-year}-old seeds left in May after some have germinated. New seeds $s_0(n)$ (0-year-old) are produced in August (Fig. 2.6) at the rate of γ per plant,

$$s_0(n) = \gamma p(n). \tag{2.7.4}$$

After winter, seeds $s_0(n)$ that were new in generation n will be one year old in the next generation $n + 1$ and a fraction $\sigma s_0(n)$ of them will survive. Hence

$$s_1(n + 1) = \sigma s_0(n)$$

or by using Formula (2.7.4) we have

$$s_1(n + 1) = \sigma \gamma p(n). \tag{2.7.5}$$

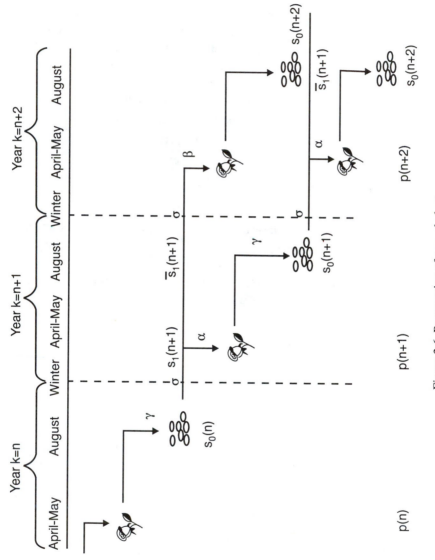

Figure 2.6. Propagation of annual plants.

And similarly,
$$s_2(n + 1) = \sigma \tilde{s}_1(n)$$

which yields by Formula (2.7.2)
$$s_2(n + 1) = \sigma(1 - \alpha)s_1(n),$$
$$s_2(n + 1) = \sigma^2 \gamma(1 - \alpha)p(n - 1) \text{(by Eq. 2.7.5)}. \tag{2.7.6}$$

Substituting for $s_1(n + 1)$, $s_2(n + 1)$ in expressions (2.7.5) and (2.7.6) into Formula (2.7.1) gives
$$p(n + 1) = \alpha \gamma \sigma p(n) + \beta \gamma \sigma^2 (1 - \alpha)p(n - 1)$$
or
$$p(n + 2) = \alpha \gamma \sigma p(n + 1) + \beta \gamma \sigma^2 (1 - \alpha)p(n). \tag{2.7.7}$$

The characteristic equation of the equation is given by
$$\gamma^2 - \alpha \gamma \sigma \lambda - \beta \gamma \sigma^2 (1 - \alpha) = 0$$

with characteristic roots
$$\lambda_1 = \frac{\alpha \gamma \sigma}{2}\left[1 + \sqrt{1 + \frac{4\beta}{\gamma \alpha^2}(1 - \alpha)}\right],$$
$$\lambda_2 = \frac{\alpha \gamma \sigma}{2}\left[1 - \sqrt{1 + \frac{4\beta}{\gamma \alpha^2}(1 - \alpha)}\right].$$

Observe that λ_1 and λ_2 are real roots since $1 - \alpha > 0$. Furthermore, $\lambda_1 > 0$ and $\lambda_2 < 0$. To insure propagation (i.e., $p(n)$ increases indefinitely as $n \to \infty$) we need to have $\lambda_1 > 1$. We are not going to do the same with λ_2 since it is negative and leads to undesired fluctuation (oscillation) in the size of the plants population. Hence
$$\frac{\alpha \gamma \sigma}{2}\left[1 + \sqrt{1 + \frac{4\beta}{\gamma \alpha^2}(1 - \alpha)}\right] > 1$$
or
$$\frac{\alpha \gamma \sigma}{2}\sqrt{1 + \frac{4\beta(1 - \alpha)}{\gamma \alpha^2}} > 1 - \frac{\alpha \gamma \sigma}{2}.$$

Squaring both sides and simplifying yields
$$\gamma > \frac{1}{\alpha \sigma + \beta \sigma^2 (1 - \alpha)}. \tag{2.7.8}$$

If $\beta = 0$, that is if no two-year-old seeds germinate in May, then condition (2.7.8) becomes
$$\gamma > \frac{1}{\alpha \sigma}. \tag{2.7.9}$$

Condition (2.7.9) says that plant propagation occurs if the product of the number of seeds produced per plant in August, the fraction of one-year-old seeds that germinate in May, and the fraction of seeds that survive a given winter, exceeds 1.

2.7.2   Gambler's Ruin

A gambler plays a sequence of games against an adversary in which the probability that the gambler wins \$1.00 in any given game is a known value q, and the probability of his losing \$1.00 is $1 - q$, where $0 \le q \le 1$. He quits gambling if he either loses all his money or if he reaches his goal of acquiring N dollars. If the gambler runs out of money first, we say that the gambler has been ruined. Let $p(n)$ denote the probability that the gambler will be ruined if he possesses n dollars. He may be ruined in two ways. First, by winning the next game; the probability of this event is q; then his fortune will be $n + 1$ and the probability of being ruined will become $p(n + 1)$. Secondly, by losing the next game, the probability of this event is $1 - q$ and the probability of being ruined is $p(n - 1)$. Hence applying the theorem of total probabilities we have

$$p(n) = qp(n + 1) + (1 - q)p(n - 1).$$

Replacing n by $n + 1$ we get

$$p(n + 2) - \frac{1}{q}p(n + 1) + \frac{(1 - q)}{q}p(n) = 0, n = 0, 1, \ldots, N \qquad (2.7.10)$$

with $p(0) = 1$ and $p(N) = 0$. The characteristic equation is given by

$$\lambda^2 - \frac{1}{q}\lambda + \frac{1 - q}{q} = 0$$

and the characteristic roots are given by

$$\lambda_1 = \frac{1}{2q} + \frac{1 - 2q}{2q} = \frac{1 - q}{q},$$

$$\lambda_2 = \frac{1}{2q} - \frac{1 - 2q}{2q} = 1.$$

Hence the general solution may be written as

$$p(n) = c_1 + c_2 \left(\frac{1 - q}{q}\right)^n, \qquad \text{if } q \ne \frac{1}{2}.$$

Now using the initial conditions $p(0) = 1$, $P(N) = 0$ we obtain

$$c_1 + c_2 = 1,$$

$$c_1 + c_2 \left(\frac{1 - q}{q}\right)^N = 0$$

which gives

$$c_1 = \frac{-\left(\dfrac{1 - q}{q}\right)^N}{1 - \left(\dfrac{1 - q}{q}\right)^N}, \qquad c_2 = \frac{1}{1 - \left(\dfrac{1 - q}{q}\right)^N}.$$

Thus

$$p(n) = \frac{\left(\dfrac{1-q}{q}\right)^n - \left(\dfrac{1-q}{q}\right)^N}{1 - \left(\dfrac{1-q}{q}\right)^N}. \tag{2.7.11}$$

The special case when $q = 1/2$ must be treated separately, since in this case we have repeated roots $\lambda_1 = \lambda_2 = 1$. This is certainly the case when we have a fair game. The general solution in this case may be given by

$$p(n) = a_1 + a_2 n$$

which when using the initial conditions yields

$$p(n) = 1 - \frac{n}{N} = \frac{N - n}{N}. \tag{2.7.12}$$

For example, if you start with \$4, the probability that you win a dollar is 0.3 and you will quit if you run out of money or have a total of \$10. Then $n = 4$, $q = 0.3$, and $N = 10$, and the probability of being ruined is given by

$$p(4) = \frac{\left(\dfrac{7}{3}\right)^4 - \left(\dfrac{7}{3}\right)^{10}}{1 - \left(\dfrac{7}{3}\right)^{10}} = 0.994.$$

On the other hand if $q = 0.5$, $N = \$100.00$, and $n = 20$, then from Formula (2.7.12) we have

$$p(40) = 1 - \frac{20}{100} = 0.8.$$

Observe that if $q \leq 0.5$ and $N \to \infty$, $p(n)$ tends to 1 in both Formulas 2.7.11 and 2.7.12, and the gambler's ruin is certain.

The probability that the gambler wins is given by

$$\tilde{p}(n) = 1 - p(n) = \begin{cases} \dfrac{1 - \left(\dfrac{1-q}{q}\right)^n}{1 - \left(\dfrac{1-q}{q}\right)^N}, & \text{if } q \neq 0.5 \\[4mm] \dfrac{n}{N}, & \text{if } q = 0.5 \end{cases}. \tag{2.7.13}$$

2.7.3   National Income [2, 3]

In a capitalist country the national income $y(n)$ in a given period n may be written as

$$Y(n) = C(n) + I(n) + G(n) \tag{2.7.14}$$

where

$C(n)$ = Consumer expenditure for purchase of consumer goods,
$I(n)$ = Induced private investment for buying capital equipment, and
$G(n)$ = Government expenditure,

where n is usually measured in years.

We now make some assumptions that are widely accepted by economists (see for example Samuelson [3]).

(a) Consumer expenditure $C(n)$ is proportional to the national income $Y(n-1)$ in the preceding year $n-1$, that is

$$C(n) = \alpha Y(n-1) \tag{2.7.15}$$

where $\alpha > 0$ is commonly called the *marginal propensity to consume*.

(b) Induced private investment $I(n)$ is proportional to the increase in consumption $C(n) - C(n-1)$, that is

$$I(n) = \beta[C(n) - C(n-1)] \tag{2.7.16}$$

where $\beta > 0$ is called the *relation*.

(c) Finally, the government expenditure $G(n)$ is constant over the years and we may choose our units such that

$$G(n) = 1. \tag{2.7.17}$$

Employing Formulas (2.7.15), (2.7.16), and (2.7.17) in Formula (2.7.14) produces the second order difference equation

$$Y(n+2) - \alpha(1+\beta)Y(n+1) + \alpha\beta Y(n) = 1, \qquad n \in Z^+. \tag{2.7.18}$$

Observe that this is the same equation we have already studied, in detail, in Example 2.38. As we have seen there, the equilibrium state of the national income $Y^* = 1/(1-\alpha)$ is asymptotically stable (or just stable in the theory of economics) if and only if the following conditions hold:

$$\alpha < 1, \qquad 1 + \alpha + 2\alpha\beta > 0, \qquad \text{and } \alpha\beta < 1. \tag{2.7.19}$$

Furthermore, the national income $Y(n)$ fluctuates (oscillates) around the equilibrium state Y^* if and only if

$$\alpha < \frac{4\beta}{(1+\beta)^2}. \tag{2.7.20}$$

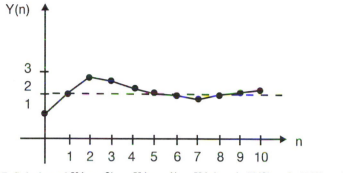

Figure 2.7. Solution of $Y(n+2) - Y(n+1) + Y(n) = 1, Y(0) = 1, Y(1) = 2$.

Now consider a concrete example where $\alpha = \frac{1}{2}, \beta = 1$. Then $Y^* = 2$, i.e., $Y^* =$ twice the government expenditure. Then clearly conditions (2.7.19) and (2.7.12) are satisfied. Hence the national income $Y(n)$ always converges in an oscillatory fashion, to $Y^* = 2$, regardless of what the initial national income $Y(0)$ and $Y(1)$ are. (See Figure 2.7.)

The actual solution may be given by

$$Y(n) = A \left(\frac{1}{\sqrt{2}} \right)^n \cos \left(\frac{n\pi}{4} - \omega \right) + 2.$$

Figure 2.8 depicts the solution $Y(n)$ if $y(0) = 1$ and $Y(1) = 2$. Here we find $A = -\sqrt{2}$ and $\omega = \pi/4$ and consequently, the solution is

$$Y(n) = - \left(\frac{1}{\sqrt{2}} \right)^{n-1} \cos \left[\frac{(n+1)}{4} \pi \right] + 2.$$

Finally, Fig. 2.8 depicts the parameter diagram $(\beta - \alpha)$ which shows regions of stability and regions of instability.

2.7.4 The Transmission of Information [4]

Suppose that a signaling system has two signals s_1 and s_2 such as dots and dashes in telegraphy. Messages are transmitted by first encoding them into a string or sequence of these two signals. Suppose that s_1 requires exactly n_1 units of time and s_2 exactly n_2 units of time to be transmitted. Let $M(n)$ be the number of possible message sequences of duration n. Now a signal of duration time n either ends with an s_1 signal or with an s_2 signal.

If the message ends with s_1, the last signal must start at $n - n_1$ (since s_1 takes n_1 units of time). Hence there are $M(n - n_1)$ possible messages to which the last s_1 may be appended. Hence there are $M(n - n_2)$ messages of duration n that ends with s_1. By a similar argument, one may conclude that there are $M(n - n_1)$ messages of duration n that ends with s_2. Consequently the total number of messages $x(n)$

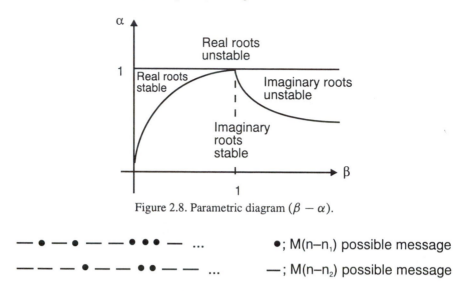

Figure 2.8. Parametric diagram $(\beta - \alpha)$.

$\qquad \bullet - \bullet - - - \bullet \bullet \bullet - \ldots$ $\bullet$; $M(n-n_1)$ possible message

$- - - \bullet - - \bullet \bullet - - \ldots$ $-$; $M(n-n_2)$ possible message

Figure 2.9. Two signals, one ends with s_1 and the other with s_2.

of duration n may be given by

$$M(n) = M(n - n_1) + M(n - n_2).$$

If $n_1 \geq n_2$, then the above equation may be written in the familiar form of an n_1th order equation

$$M(n + n_1) - M(n + n_1 - n_2) - M(n) = 0. \qquad (2.7.21)$$

On the other hand, if $n_1 \leq n_2$, then we obtain the n_2th order equation

$$M(n + n_2) - M(n + n_2 - n_1) - M(n) = 0. \qquad (2.7.22)$$

An interesting special case is when $n_1 = 1$ and $n_2 = 2$. In this case we have

$$M(n + 2) - M(n + 1) - M(n) = 0$$

or

$$M(n + 2) = M(n + 1) + M(n)$$

which is nothing but our Fibonacci sequence $\{0, 1, 1, 2, 3, 5, 8, \ldots\}$ that we encountered in Example 2.27. The general solution (see Formula 2.3.10) is given by

$$M(n) = a_1 \left(\frac{1 + \sqrt{5}}{2} \right)^n + a_2 \left(\frac{1 - \sqrt{5}}{2} \right)^n, \qquad n = 0, 1, 2, \ldots . \qquad (2.7.23)$$

To find a_1 and a_2 we need to specify $M(0)$ and $M(1)$. Here a sensible assumption is to let $M(0) = 0$ and $M(1)$. Using this initial date in Eq. (2.7.23) yields

$$a_1 = \frac{1}{\sqrt{5}}, \qquad a_2 = -\frac{1}{\sqrt{5}},$$

and the solution of our problem now becomes

$$M(n) = \frac{1}{\sqrt{5}}\left(\frac{1+\sqrt{5}}{2}\right)^n - \frac{1}{\sqrt{5}}\left(\frac{1-\sqrt{5}}{2}\right)^n. \qquad (2.7.24)$$

In information theory, [4] the capacity C of the channel is defined as

$$C = \lim_{n\to\infty} \frac{\log_2 M(n)}{n} \qquad (2.7.25)$$

where $\log_2$ denotes the logarithm base 2.

From Eq. (2.7.24) we have

$$C = \lim_{n\to\infty} \frac{\log_2 \frac{1}{\sqrt{5}}}{n} + \lim_{n\to\infty} \frac{1}{n}\log_2\left[\left(\frac{1+\sqrt{5}}{2}\right)^n - \left(\frac{1-\sqrt{5}}{2}\right)^n\right]. \qquad (2.7.26)$$

Since $\left(\frac{1-\sqrt{5}}{2}\right) \approx 0.6 < 1$, $\qquad \left(\frac{1-\sqrt{5}}{2}\right)^n \to 0$ as $n \to \infty$.

Observe also that the first term in the right-hand side of Eq. (2.7.2) goes to zero as $n \to \infty$.

Thus

$$C = \lim_{n\to\infty} \frac{1}{n}\log_2\left(\frac{1+\sqrt{5}}{2}\right)^n,$$

$$C = \log_2\left(\frac{1+\sqrt{5}}{2}\right)^n \approx 0.7. \qquad (2.7.27)$$

Exercise 2.7.

1. The model for annual plants was given by Eq. (2.7.7) in terms of the plant population $p(n)$.

   (a) Write the model in terms of $s_1(n)$.

   (b) Let $\alpha = \beta = 0.01$ and $\sigma = 1$. How big should γ be to ensure that the plant population increases in size?

2. An alternative formulation for the annual plant model in which we define the beginning of a generation at the time when seeds are produced. Figure 2.10 shows the new method.

   Write the difference equation in $p(n)$ that represents this model. Then find conditions on γ under which plant propagation occurs.

3. A planted seed produces a flower with one seed at the end of the first year and a flower with two seeds at the end of two years and each year thereafter. Suppose that each seed is planted as soon as it is produced.

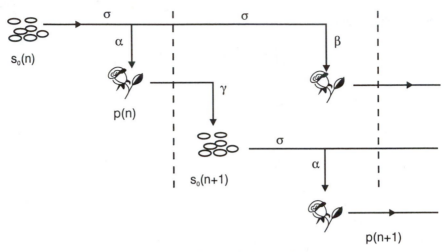

Figure 2.10. Annual plant model.

(a) Write the difference equation that describes the number of flowers $F(n)$ at the end of the nth year.

(b) Compute the number of flowers at the end of 3, 4, and 5 years.

4. Find the number of flowers at the end of n years in problem 3 if a plant seed produces a flower with four seeds at the end of the first year and a flower with ten seeds at the end of two years and each year thereafter.

(a) Show that this system can be given by

$$s^0(n + 1) = \gamma(\beta s^1(n) + \alpha s^0(n)),$$
$$s^1(n + 1) = \sigma(1 - \alpha)s^0(n).$$

(b) Show that $p(n + 2) - \alpha\gamma\sigma p(n + 1) - \beta\gamma\sigma(1 - \alpha)p(n) = 0$.

(c) Show that the adult plant population is preserved (i.e., $\lambda_1 \geq 1$) if and only if

$$\gamma \geq \frac{1}{\alpha\sigma + \beta\sigma^2(1 - \alpha)}.$$

5. Suppose that the probability of winning any particular bet is 0.49. If you start with $50 and will quit when you have $100, what is the probability of ruin (i.e., losing all your money),

(i) if you make $1 bets?

(ii) if you make $10 bets?
   (Hint: Let $10 equal 1 unit, $n = 5$, $N = 10$.)

(iii) if you make $50 bets?

6. John has m chips and Robert has $(N - m)$ chips. Suppose that John has a probability p of winning each game where one chip is bet on in each play. If $G(m)$ is the expected value of the number of games that will be played before either John or Roberto is ruined,

   (a) Show that $G(m)$ satisfies the second order equation

   $$G(m + 2) + pG(m + 1) + (1 - p)G(m) = 0. \qquad (*)$$

   (b) What are the values of $G(0)$ and $G(N)$?

   (c) Solve the difference equation (*) with the initial conditions in part (b).

7. Suppose that in a game we have the following situation: on each play, the probability that you will win \$2 is 0.1, the probability that you will win \$1 is 0.3, and the probability that you will lose \$1 is 0.6. If you quit when either you are broke or when you have at least N dollars, write a third order difference equation that describes the probability $p(n)$ of eventually going broke if you have n dollars. Then find the solution of the equation.

8. Suppose that Becky plays a roulette wheel that has 37 divisions: 18 are red, 18 are black, and one is green. Becky can bet on either the red or black, and she wins a sum equal to her bet if the outcome is a division of that color; otherwise, she loses the bet. If the bank has one million dollars and she has \$5000, what is the probability that Becky can break the bank, assuming that she bets \$100 on either red or black for each spin of the wheel?

9. In the national income model [Eq. (2.7.18)], assume that the government expenditure $G(n)$ is proportional to the national income $Y(n-2)$ two periods past, i.e., $G(n) = \gamma Y(n - 2), 0 < \gamma < 1$. Derive the difference equation for the national income $Y(n)$. Find the conditions for stability and oscillations of solutions.

10. Determine the behavior (stability, oscillations) of solutions of Eq. (2.7.18) for the cases

    (a) $\alpha = \dfrac{4\beta}{(1 + \beta)^2}$,

    (b) $\alpha > \dfrac{4\beta}{(1 + \beta)^2}$.

11. Modify the national income model such that, instead of the government having fixed expenditures, it increases its expenditures by 5% each time period, that is

    $$G(n) = (1.05)^n.$$

    (a) Write down the second order difference equation that describes this model.

(b) Find the equilibrium value.

(c) If $\alpha = 0.5$, $\beta = 1$, find the general solution of the equation.

12. Suppose that in the national income we make the following assumptions:

(i) $Y(n) = C(n) + I(n)$

i.e., there is no government expenditure.

(ii) $C(n) + a_1 Y(n-1) + a_2 Y(n-2) + K$

i.e., consumption in any period is a linear combination of the incomes of the two preceding periods, where c_1, c_2, and K are constants.

(iii) $I(n+1) = I(n) + h$

i.e., investment increases a fixed amount $h > 0$ each period.

(a) Write down the second order difference equation that model the national income $Y(n)$.

(b) Find the general solution if $a_1 = \dfrac{1}{2}$, $a_2 = \dfrac{1}{4}$.

(c) Show that $Y(n)$ is asymptotic to the equilibrium $Y^* = \alpha + \beta n$.

13. (Inventory Analysis). Let $S(n)$ be the number of units of consumers' goods produced for sale in period n, and $T(n)$ be the number of units of consumer goods produced for inventories in period n. Assume that there is a constant noninduced net investment V_0 in each period. Then the total income $Y(n)$ produced in time n is given by $Y(n) = T(n) + S(n) + V_0$.

(a) Develop a difference equation that models the total income $Y(n)$, under the assumptions

(i) $S(n) = \beta Y(n-1)$,

(ii) $T(n) = \beta Y(n-1) - \beta Y(n-2)$.

(b) Obtain conditions under which

(i) solutions converge to the equilibrium, and

(ii) solutions are oscillatory.

(c) Interpret your results in problem (b). [Hint: use problem 11(a)]

14. Let $I(n)$ denote the level of inventories at the close of period n.

(a) Show that $I(n) = I(n-1) + S(n) + T(n) - \beta Y(n)$.

(b) Assuming that $S(n) = 0$ (passive inventory adjustment), show that

$$I(n) - I(n-1) = (1 - \beta)Y(n) - V_0.$$

(c) Suppose as in problem (b) that $s(n) = 0$. Show that

$$I(n+2) - (\beta+1)I(n+1) + \beta I(n) = 0. \text{ [Hint: use problem 11 part (a)]}$$

(d) With $\beta \neq 1$, show that

$$I(n) = \left(I(0) - \frac{c}{1-\beta} \right) \beta^n + \frac{c}{1-\beta},$$

where $(E - \beta)I(n) = c$.

15. Consider Eq. (2.7.21) with $n_1 = n_2 = 2$ (i.e., both signals s_1 and s_2 take 2 units of time for transmission).

   (a) Solve the obtained difference equation with the initial conditions $M(2) = M(3) = 2$.

   (b) Find the channel capacity c.

16. Consider Eq. (2.7.21) with $n_1 = n_2 = 1$ (i.e., both signals take 1 unit of time for transmission).

   (a) Solve the obtained difference equation.

   (b) Find the channel capacity c.

17. Two billiard balls of equal mass m lie on a billiards table such that the line between their centers is perpendicular to two sides of the table. One of the balls is hit so that it travels with constant speed s toward the other ball. Let $u(n)$ and $v(n)$ be the speeds of the balls before the nth impact occurs, where $v(n)$ refers to the ball that is hit initially.

   (a) Show that by the conservation of momentum we have

$$mu(n) + mv(n) = mu(n+1) - mv(m-1).$$

   (b) Use the Newton's Collision rule to conclude that

$$u(n+1) + v(n+1) = -\alpha(u(n) - v(n)).$$

   (c) Use the results in parts (a) and (b) to determine the speeds $u(n)$ and $v(n)$ of the balls before the nth impact.

18. (Euler's method for solving a second order differential equations.) Recall from Example 1.9 that one may approximate $x'(t)$ by $(x(n+1) - x(n))/h$, where h is the step size of the approximation and $x(n) = x(t_0 + nh)$.

   (a) Show that $x''(t)$ may be approximated by $\dfrac{x(n+2) - 2x(n+1) + x(n)}{h}$.

   (b) Write down the corresponding difference equation of the differential equation

$$x''(t) = f(x(t), x'(t)).$$

19. Use Euler's method described in Problem 18 to write the corresponding difference equation of

$$x''(t) - 4x(t) = 0, \qquad x(0) = 0, x'(0) = 1$$

Solve both differential and difference equations and compare the results.

20. (The Midpoint Method). The midpoint method stipulates that one may approximate $x'(t)$ by $(x(n + 1) - x(n - 1))/h$, where h is the step size of the approximation and $t = t_0 + nh$.

    (a) Use the method to write the corresponding difference equation of the differential equation $x'(t) = g(t, x(t))$.

    (b) Use the method to write the corresponding difference equation of $x'(t) = 0.7x^2 + 0.7, x(0) = 1, t \in [0, 1]$. Then solve the obtained difference equation.

    (c) Compare your findings in part (b) with the results in Example 1.9. Determine which of the two methods, Euler or the Midpoint, is more accurate.

References

[1] L. Edelstein-Keshet, *Mathematical Models in Biology*, Random House, New York, 1988.

[2] S. Goldberg, *Introduction to Difference Equations*, Dover, New York, 1986.

[3] P.A. Samuelson, "Interactions Between the Multiplier Analysis and the Principle of Acceleration," *Rev. Econom. Stat.*, **21** (1939), 75–78; reprinted in Readings in Business Cycle Theory, Blakiston Co., Philadelphia, 1944.

[4] C.E. Shannon and W. Weaver, *The Mathematical Theory of Communication*, University of Illinois, Urbana, 1949, pp. 7–8.

Bibliography

R.P. Agarwal, *Difference Equations and Inequalities*, Marcel Dekker, New York, 1992.

W.G. Kelley and A.C. Peterson, *Difference Equations*, Academic, New York, 1991.

V. Lakshmikantham and D. Trigiante, *Theory of Difference Equations: Numerical Methods and Applications*, Academic, New York, 1988.

R. Mickens, *Difference Equations*, Van Nostrand, Reinhold, New York, 1990.

K.S. Miller, *Linear Difference Equations*, W.A. Benjamin, New York, 1968.

3

Systems of Difference Equations

In the last chapter, we concerned ourselves with linear difference equations, namely, those equations with only one independent and one dependent variable. Since not every situation that we will encounter will be this facile, we must be prepared to deal with systems of more than one dependent variable.

Thus, in this chapter, we deal with those equations of two or more dependent variables, known as the first-order difference equations. These equations naturally apply to various fields of scientific endeavor, like biology (the study of competitive species in population dynamics), physics (the study of the motions of interacting bodies), the study of control systems, neurology, and electricity. Furthermore, we will also transform those high order linear difference equations that we investigated in Chapter 2 into systems of first order equations. This transformation will probably prove to be of little practical use in the realm of boundary value problems and oscillations, but will be substantiated as an immensely helpful tool in the study of stability theory later on [see 1,2,3].

3.1 Autonomous (Time-Invariant) Systems

In this section, we are interested in finding solutions of the following system of k linear equations:

$$x_1(n + 1) = a_{11}x_1(n) + a_{12}x_2(n) + \cdots + a_{1k}x_k(n)$$
$$x_2(n + 1) = a_{21}x_1(n) + a_{22}x_2(n) + \cdots + a_{2k}x_k(n)$$

$$\vdots \qquad \vdots \qquad\qquad \vdots \qquad\qquad \vdots$$

$$x_k(n + 1) = a_{k1}x_1(n) + a_{k2}x_2(n) + \cdots + a_{kk}x_k(n).$$

This system may be written in the vector form

$$x(n + 1) = Ax(n), \qquad\qquad (3.1.1)$$

where $x(n) = (x_1(n), x_2(n), \ldots, x_k(n))^T$ (see footnote 1) $\in R^k$, and $A = (a_{ij})$ is a $k \times k$ real nonsingular matrix. System (3.1.1) is considered autonomous or time-invariant, since the values of A are all constants. Nonautonomous or time-variant systems will be considered later, in Section 3.3.

If for some $n_0 \geq 0$, $x(n_0) = x_0$ is specified, then System (3.1.1) is called an initial value problem. Furthermore, by simple iteration (or by direct substitution into the equation), one may show that the solution is given by

$$x(n, n_0, x_0) = A^{n-n_0}x_0, \qquad\qquad (3.1.2)$$

where $A^0 = I$, the $k \times k$ identity matrix. Notice that $x(n_0, n_0, x_0) = x_0$. If $n_0 = 0$, then the solution in Formula (3.1.2) may be written as $x(n, x_0)$, or simply $x(n)$. We now show that we may assume that $n_0 = 0$ without loss of generality.

Let $y(n - n_0) = x(n)$. Then Eq. (3.1.1) becomes

$$y(n + 1) = Ay(n) \qquad\qquad (3.1.3)$$

with $y(0) = x(n_0)$ and

$$y(n) = A^n y(0). \qquad\qquad (3.1.4)$$

3.1.1   The Discrete Analogue of the Putzer Algorithm

In differential equations, the Putzer Algorithm is used to compute e^{At}. Here, we introduce an analogous algorithm to compute A^n. First, let us review the rudiments of matrix theory that are vital in the development of this algorithm. In what follows C denotes the set of complex numbers.

Recall that for a real $k \times k$ matrix $A = (a_{ij})$, an eigenvalue of A is a real or complex number λ such that $A\xi = \lambda\xi$ for some nonzero $\xi \in C^k$. Equivalently, this relation may be written as

$$(A - \lambda I)\xi = 0. \qquad\qquad (3.1.5)$$

Equation (3.1.5) has a nonzero solution x if and only if

$$det(A - \lambda I) = 0$$

or

$$\lambda^k + a_1\lambda^{k-1} + a_2\lambda^{k-2} + \cdots + a_{k-1}\lambda + a_k = 0. \qquad\qquad (3.1.6)$$

---

[1] T indicates the transpose of a vector.

Equation (3.1.6) is called *the characteristic equation* of A whose roots λ are called the *eigenvalues* of A. If $\lambda_1, \lambda_2, \ldots, \lambda_k$ are the eigenvalues of A (some of them may be repeated), then one may write Eq. (3.1.6) as

$$p(\lambda) = \prod_{j=1}^{k}(\lambda - \lambda_j) = 0. \qquad (3.1.7)$$

We are now ready to state the Cayley–Hamilton theorem, one of the fundamental results of Matrix Theory.

Theorem 3.1. Every matrix satisfies its characteristic equation. That is to say,

$$p(A) = \prod_{j=1}^{k}(A - \lambda_j I) = 0, \qquad (3.1.8)$$

or

$$A^k + a_1 A^{k-1} + a_2 A^{k-2} + \cdots + a_k I = 0. \qquad (3.1.8)'$$

3.1.2   The Development of the Algorithm for A^n

Let A be a $k \times k$ real matrix. We look for a representation of A^n in the form

$$A^n = \sum_{j=1}^{s} u_j(n)M(j - 1), \qquad (3.1.9)$$

where $u_j(n)$'s are scalar functions to be determined later, and

$$M(j) = (A - \lambda_j I)M(j - 1), \; M(0) = I \qquad (3.1.10)$$

or

$$M(j + 1) = (A - \lambda_{j+1} I)M(j), \; M(0) = I.$$

By iteration, one may show that

$$M(n) = (A - \lambda_n I)(A - \lambda_{n-1} I) \ldots (A - \lambda_1 I)$$

or in a compact form

$$M(n) = \prod_{j=1}^{n}(A - \lambda_j I). \qquad (3.1.11)$$

Notice that by the Cayley–Hamilton Theorem we have

$$M(k) = \prod_{j=1}^{k}(A - \lambda_j I) = 0.$$

And consequently, $M(n) = 0$ for all $n \geq k$. In light of this observation, we may rewrite Formula (3.1.9) as

$$A^n = \sum_{j=1}^{k} u_j(n)M(j - 1). \qquad (3.1.12)$$

If we let $n = 0$ in Formula (3.1.12) we obtain

$$A^0 = I = u_1(0)I + u_2(0)M(1) + \cdots + u_k(0)M(k-1). \qquad (3.1.13)$$

Now Eq. (3.1.13) is satisfied if

$$u_1(0) = 1 \text{ and } u_2(0) = u_3(0) = \cdots = u_k(0) = 0. \qquad (3.1.14)$$

From Formula (3.1.12) we have

$$\sum_{j=1}^{k} u_j(n+1)M(j-1) \quad = \quad AA^n$$

$$= \quad A\left[\sum_{j=1}^{k} u_j(n)M(j-1)\right]$$

$$= \quad \sum_{j=1}^{k} u_j(n)AM(j-1).$$

Substituting for $AM(j-1)$ from Eq. (3.1.10) yields

$$\sum_{j=1}^{k} u_j(n+1)M(j-1) = \sum_{j=1}^{k} u_j(n)[M(j) + \lambda_j M(j-1)]. \qquad (3.1.15)$$

Comparing the coefficients of $M(j)$, $1 \leq j \leq k$, in Eq. (3.1.15), and applying Condition (3.1.14), we obtain

$$\left. \begin{aligned} u_1(n+1) &= \lambda_1 u_1(n), & u_1(0) &= 1 \\ u_j(n+1) &= \lambda_j u_j(n) + u_{j-1}(n), & u_j(0) &= 0, \ j = 2, 3, \ldots, k \end{aligned} \right\} \qquad (3.1.16)$$

The solutions of Eqs. (3.1.16) are given by

$$u_1(n) = \lambda_1^n, \qquad u_j(n) = \sum_{i=0}^{n-1} \lambda_j^{n-1-i} u_{j-1}(i), \qquad j = 2, 3, \ldots, k. \qquad (3.1.17)$$

Equations (3.1.11) and (3.1.17) together constitute an algorithm for computing A^n which henceforth will be called the *Putzer algorithm*.

Example 3.2. Find the solution of the difference system $x(n+1) = Ax(n)$, where

$$A = \begin{pmatrix} 4 & 1 & 2 \\ 0 & 2 & -4 \\ 0 & 1 & 6 \end{pmatrix}.$$

Solution: The eigenvalues of A may be obtained by solving the characteristic equation $\det(A - \lambda I) = 0$. Now

$$\det \begin{pmatrix} 4-\lambda & 1 & 2 \\ 0 & 2-\lambda & -4 \\ 0 & 1 & 6 \end{pmatrix} = (4-\lambda)(\lambda-4)^2 = 0.$$

Hence, the eigenvales of A are $\lambda_1 = \lambda_2 = \lambda_3 = 4$.

$$\text{So } M(0) = I, \qquad M(1) = A - 4I = \begin{pmatrix} 0 & 1 & 2 \\ 0 & -2 & -4 \\ 0 & 1 & 2 \end{pmatrix}.$$

$$M(2) = (A - 4I)M(1) = \begin{pmatrix} 0 & 0 & 0 \\ 0 & 0 & 0 \\ 0 & 0 & 0 \end{pmatrix}.$$

Now,

$$u_1(n) = 4^n,$$

$$u_2(n) = \sum_{i=0}^{n-1}(4^{n-1-i})(4^i) = n(4^{n-1}),$$

$$u_3(n) = \sum_{i=0}^{n-1} 4^{n-1-i}(i\,4^{i-1})$$

$$= 4^{n-2}\sum_{i=0}^{n-1} i$$

$$= \frac{n(n-1)}{2}\, 4^{n-2}.$$

Using Eq. (3.1.12), we have

$$A^n = 4^n \begin{pmatrix} 1 & 0 & 0 \\ 0 & 1 & 0 \\ 0 & 0 & 1 \end{pmatrix} + n4^{n-1} \begin{pmatrix} 0 & 1 & 2 \\ 0 & -2 & -4 \\ 0 & 1 & 2 \end{pmatrix}$$

$$+ \frac{n(n-1)}{2}\, 4^{n-2} \begin{pmatrix} 0 & 0 & 0 \\ 0 & 0 & 0 \\ 0 & 0 & 0 \end{pmatrix}$$

$$= \begin{pmatrix} 4^n & n4^{n-1} & 2n4^{n-1} \\ 0 & 4^n - 2n4^{n-1} & -n4^n \\ 0 & n4^{n-1} & 4^n + 2n4^{n-1} \end{pmatrix}.$$

The solution of the difference equation is given by

$$x(n) = A^n x(0) = \begin{pmatrix} 4^n x_1(0) + n4^{n-1}x_2(0) + 2n4^{n-1}x_3(0) \\ (4^n - 2n4^{n-1})x_2(0) - n4^n x_3(0) \\ n4^{n-1}x_2(0) + (4^n + 2n4^{n-1})x_3(0) \end{pmatrix}$$

where $x(0) = (x_1(0), x_2(0), x_3(0))^T$.

Exercise 3.1.

In Problems 1 through 4, use the discrete Putzer algorithm to evaluate A^n.

1. $A = \begin{bmatrix} 1 & 1 \\ -2 & 4 \end{bmatrix}$.

2. $A = \begin{bmatrix} -1 & 2 \\ 3 & 0 \end{bmatrix}$.

3. $A = \begin{bmatrix} 1 & 2 & -1 \\ 0 & 1 & 0 \\ 4 & -4 & 5 \end{bmatrix}$.

4. $A = \begin{bmatrix} 2 & 1 & 0 \\ 0 & 2 & 1 \\ 0 & 0 & 2 \end{bmatrix}$.

5. Solve the system
$$x_1(n + 1) = -x_1(n) + x_2(n), \qquad x_1(0) = 1.$$
$$x_2(n + 1) = 2x_2(n), \qquad\qquad x_2(0) = 2.$$

6. Solve the system
$$x_1(n + 1) = x_2(n),$$
$$x_2(n + 1) = x_3(n),$$
$$x_3(n + 1) = 2x_1(n) - x_2(n) + x_3(n).$$

7. Solve the system
$$x(n + 1) = \begin{bmatrix} 1 & -2 & -2 \\ 0 & 0 & -1 \\ 0 & 2 & 3 \end{bmatrix} x(n), \qquad x(0) = \begin{pmatrix} 1 \\ 1 \\ 0 \end{pmatrix}.$$

8. Solve the system
$$x(n + 1) = \begin{pmatrix} 1 & 3 & 0 & 0 \\ 0 & 2 & 1 & -1 \\ 0 & 0 & 2 & 0 \\ 0 & 0 & 0 & 3 \end{pmatrix} x(n).$$

9. Verify that the matrix $A = \begin{pmatrix} 2 & -1 \\ 1 & 3 \end{pmatrix}$ satisfies its characteristic equation (Cayley–Hamilton Theorem).

10. Let $\rho(A) = \max \{|\lambda| : \lambda \text{ is an eigenvalue of } A\}$. Suppose that $\rho(A) = \rho_0 < \beta$.

(a) Show that $|u_j(n)| \leq \dfrac{\beta^n}{(\beta - \rho_0)}, j = 1, 2, \ldots, k$ (Hint: Use Eq. 3.1.14.)

(b) Show that if $\rho_0 < 1$, then $u_j(n) \to 0$ as $n \to \infty$. Concluding with $A^n \to 0$ as $n \to \infty$.

(c) If $\alpha < \min \{|\lambda| : \lambda \text{ is an eigenvalue of } A\}$, establish a lower bound for $|u_j(n)|$.

11. If a $k \times k$ matrix A has distinct eigenvalues $\lambda_1, \lambda_2, \ldots, \lambda_k$, then one may compute A^n, $n \geq k$ using the following method. Let $p(\lambda)$ be the characteristic polynomial of A. Divide λ^n by $p(\lambda)$ to obtain $\lambda^n = p(\lambda)q(\lambda) + r)(\lambda)$, where the remainder $r(\lambda)$ is a polynomial of degree at most $(k-1)$. Thus one may write $A^n = p(A)q(A) + r(A)$.

(a) Show that $A^n = r(A) = a_0 I + a_1 A + a_2 A^2 + \cdots + a_{k-1} A^{k-1}$.

(b) Show that $\lambda_1^n = r(\lambda_1)$, $\lambda_2^n = r(\lambda_2)$, $\ldots$, $\lambda_k^n = r(\lambda_k)$.

(c) Use part (b) to find $a_0, a_1, \ldots, a_{k-1}$.

12. Extend the method of problem 11 to the case of repeated roots. (Hint: If $\lambda_1 = \lambda_2 = \lambda$ and $\lambda^n = a_0 + a_1 \lambda + a_2 \lambda^2 + \cdots + a_{k-1} r^{k-1}$, differentiate to get another equation $n\lambda^{n-1} = a_1 + 2a_2 \lambda + \cdots (k-1)a_{k-1}\lambda^{k-2}$.)

13. Apply the method of problem 11 to find A^n for

(i) $A = \begin{pmatrix} 1 & 1 \\ -2 & 4 \end{pmatrix}$,

(ii) $A = \begin{pmatrix} 1 & 2 & -1 \\ 1 & 0 & 1 \\ 4 & -4 & 5 \end{pmatrix}$.

14. Apply the method of problem 13 to find A^n for

$$A = \begin{pmatrix} 4 & 1 & 2 \\ 0 & 2 & -4 \\ 0 & 1 & 6 \end{pmatrix}.$$

15^2. Consider the right triangle in Fig. 3.1 where $p(0) = (0, 0)$, $p(1) = (1/2, 1/2)$, and $p(2) = (1/2, 0)$. For $p(n) = (x(n), y(n))$ with $n \geq 3$, we have

$$x(n+3) = 1/2(x(n) + x(n+1)),$$

$$y(n+3) = 1/2(y(n) + y(n+1)).$$

(a) Write each equation as a system $z(n+1) = Az(n)$.

(b) Find $\lim_{n \to \infty} p(n)$.

---

[2]Proposed by C.V. Eynden and solved by Trinity University Problem Solving Group (1994).

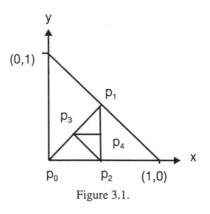

Figure 3.1.

3.2   The Basic Theory

Now contemplate the system

$$x(n + 1) = A(n)x(n), \tag{3.2.1}$$

where $A(n) = (a_{ij}(n))$ is a $k \times k$ nonsingular matrix function. This is a *homogeneous linear difference system* which is *nonautonomous* or *time-variant*.

The corresponding nonhomogeneous system is given by

$$y(n + 1) = A(n)y(n) + g(n), \tag{3.2.2}$$

where $g(n) \in R^k$.

We now establish the existence and uniqueness of solutions of Eq. (3.2.1).

Theorem 3.3. For each $x_0 \in R^k$ and $n_0 \in Z^+$, there exists a unique solution $x(n, n_0, x_0)$ of Eq. (3.2.1) with $x(n_0, n_0, x_0) = x_0$.

Proof From Eq. (3.2.1),

$$x(n_0 + 1, n_0, x_0) = A(n_0)x(n_0) = A(n_0)x_0,$$

$$x(n_0 + 2, n_0, x_0) = A(n_0 + 1)x(n_0 + 1) = A(n_0 + 1)A(n_0)x_0.$$

Inductively, one may conclude that

$$x(n, n_0, x_0) = \left[\prod_{i=n_0}^{n-1} A(i) \right] x_0, \tag{3.2.3}$$

where

$$\prod_{i=n_0}^{n-1} A(i) = \left\{ \begin{array}{ll} A(n-1)A(n-2)\ldots A(n_0) & \text{if } n > n_0 \\ I & \text{if } n = n_0 \end{array} \right\}.$$

Formula (3.2.3) gives the unique solution with the desired properties.

We will now develop the notion of a fundamental matrix; a central building block in the theory of linear systems.

Definition 3.4. The solutions $x_1(n)$, $x_2(n)$, ..., $x_k(n)$ of Eq. (3.2.1) are said to be linearly independent for $n \geq n_0 \geq 0$ if, whenever $c_1 x_1(n) + c_2 x_2(n) + \cdots + c_k x_k(n) = 0$ for all $n \geq n_0$, then $c_i = 0$, $1 \leq i \leq k$.

Let $\Phi(n)$ be a $k \times k$ matrix whose columns are solutions of Eq. (3.2.1). We write

$$\Phi(n) = [x_1(n), x_2(n), \ldots, x_k(n)].$$

Now

$$
\begin{aligned}
\Phi(n+1) &= [A(n)x_1(n), A(n)x_2(n), \ldots, A(n)x_k(n)] \\
&= A(n)[x_1(n), x_2(n), \ldots, x_k(n)] \\
&= A(n)\Phi(n).
\end{aligned}
$$

Hence, $\Phi(n)$ satisfies the matrix difference equation

$$\Phi(n+1) = A(n)\Phi(n). \tag{3.2.4}$$

Furthermore, the solutions $x_1(n)$, $x_2(n)$, ..., $x_k(n)$ are linearly independent for $n \geq n_0$, if and only if the matrix $\Phi(n)$ is nonsingular ($\det \Phi(n) \neq 0$) for all $n \geq n_0$. (Why?) This actually leads to the next definition.

Definition 3.5. If $\Phi(n)$ is a matrix that is nonsingular for all $n \geq n_0$ and satisfies Eq. (3.2.4), then it is said to be a *fundamental matrix* for system equation (3.2.1).

Note that if $\Phi(n)$ is a fundamental matrix, and C is any nonsingular matrix, then $\Phi(n)C$ is also a fundamental matrix (Exercise 3.2, Problem 6). Thus there are infinitely many fundamental matrices for a given system. However, there is one fundamental matrix that we already know, namely,

$$\Phi(n) = \prod_{i=n_0}^{n-1} A(i), \text{ with } \Phi(n_0) = I \qquad \text{(Exercise 3.2, Problem 5).}$$

In the autonomous case when A is a constant matrix, $\Phi(n) = A^{n-n_0}$, and if $n_0 = 0$, $\Phi(n) = A^n$. Consequently, it would be much more suitable to use the Putzer Algorithm to compute the fundamental matrix for an autonomous system.

Theorem 3.6. There is a unique solution $\Psi(n)$ of the matrix Eq. (3.2.4) with $\Psi(n_0) = I$.

Proof One may think of the matrix difference equation (3.2.4) as a system of k^2 first order difference equations. Thus, to complete the point, we may apply the "existence and uniqueness" Theorem 3.3, to obtain a k^2-vector solution v such that $v(n_0) = (1, 0, \ldots, 1, 0, \ldots)^T$, where 1's appear at the first, $k+1$, $2k+2$, ... slots and 0's everywhere else. The vector v is then converted to the $k \times k$ matrix $\Psi(n)$

by grouping the components into sets of k elements in which each set will be a column. Clearly $\Psi(n_0) = I$.

We may add here, that starting with any fundamental matrix $\Phi(n)$, then the fundamental matrix $\Phi(n)\Phi^{-1}(n_0)$ is such a matrix. This special fundamental matrix is denoted by $\Phi(n, n_0)$ and is referred to as the *state transition* matrix.

One may in general write $\Phi(n,m) = \Phi(n)\Phi^{-1}(m)$ for any two positive integers n,m with $n \geq m$. The fundamental matrix $\Phi(n,m)$ has some agreeable properties that we ought to list here. Observe first that $\Phi(n,m)$ is a solution of the matrix difference equation $\Phi(n + 1,m) = A(n)\Phi(n,m)$ (Exercise 3.2, Problem 2). The reader is asked to prove the following statements.

(i) $\Phi^{-1}(n,m) = \Phi(m,n)$ (Exercise 3.2, Problem 3),

(ii) $\Phi(n,m) = \Phi(n,r)\Phi(r,m)$ (Exercise 3.2, Problem 3),

(iii) $\Phi(n,m) = \prod_{i=m}^{n-1} A(i)$ (Exercise 3.2, Problem 3).

Corollary 3.7. The unique solution of $x(n, n_0, x_0)$ of Eq. (3.2.1) with $x(n, n_0, x_0) = x_0$ is given by

$$x(n, n_0, x_0) = \Phi(n, n_0)x_0. \tag{3.2.5}$$

Checking the linear independence of a fundamental matrix $\Phi(n)$ for $n \geq n_0$ is a formidable task. We will instead show that it suffices to establish linear independence at n_0.

Lemma 3.8. (Abel's Formula). For any $n \geq n_0 \geq 0$,

$$\det\Phi(n) = \left(\prod_{i=n_0}^{n-1}[\det A(i)]\right)\det\Phi(n_0). \tag{3.2.6}$$

Proof Taking the determinant of both sides of Eq. (3.2.4) we obtain the scalar difference equation

$$\det\Phi(n + 1) = \det A(n)\,\det\Phi(n)$$

whose solution is given by Eq. (3.2.6).

Corollary 3.9. If in Eq. (3.2.1), A is a constant matrix, then

$$\det\Phi(n) = [\det A]^n\,\det\Phi(n_0) \tag{3.2.7}$$

Proof The proof allows from Formula (3.2.5).

Corollary 3.10. The fundamental matrix $\Phi(n)$ is nonsingular for all $n \geq n_0$ if and only if $\Phi(n_0)$ is nonsingular.

Proof This follows from Formula (3.2.6), having noted that $\det A(i) \neq 0$, for $i \geq n_0$.

Corollary 3.11. The solutions $x_1(n), x_2(n), \ldots, x_k(n)$ of Eq. (3.2.1) are linearly independent for $n \geq n_0$ if and only if $\Phi(n_0)$ is nonsingular.

Proof This follows immediately from Corollary 3.10.

The following theorem establishes the existence of k linearly independent solutions of Eq. (3.2.1).

Theorem 3.12. There are k linearly independent solutions of System (3.2.1) for $n \geq n_0$.

Proof For each $i = 1, 2, \ldots, k$, let $e_i = (0, 0, \ldots, 0, \ldots, 0)^T$ be the standard unit vector in R^k where all the components are zero except the ith component, which is equal to 1. By Theorem 3.3, for each e_i, $1 \leq i \leq k$, there exists a solution $x(n, n_0, e_i)$ of Eq. (3.2.1) with $x(n_0, n_0, e_i) = e_i$. To prove that the set $\{x(n, n_0, e_i) | 1 \leq i \leq k\}$ is lineraly independent, according to Corollary 3.8, it suffices to show that $\Phi(n_0)$ is nonsingular. But, this fact is obvious since $\Phi(n_0) = I$. The proof of the theorem is now complete.

Corollary 3.13. The set S of all the solutions of Eq. (3.2.1) is a linear space of dimension k.

Proof **(Exercise 3.2, Problem 9).**

One important consequence of Corollary 3.13 is that if $x_1(n), x_2(n), \ldots, x_r(n)$ are solutions of Eq. (3.2.1), then $x(n) = \sum_{i=1}^{r} c_i x_i(n)$ is also a solution of Eq. (3.2.1). This fact leads us up to the definition of the general solution of Eq. (3.2.1).

Definition 3.14. Assuming that $\{x_i(n) | 1 \leq i \leq k\}$ is any linearly independent set of solutions of Eq. (3.2.1), the general solution of Eq. (3.2.1) is defined to be

$$x(n) = \sum_{i=1}^{k} c_i x_i(n), \tag{3.2.8}$$

where $c_i \in R$ and at least one $c_i \neq 0$.

Formula (3.2.8) may be written as

$$x(n) = \Phi(n)c, \tag{3.2.9}$$

where $\Phi(n) = (x_1(n), x_2(n), \ldots, x_k(n))$ is a fundamental matrix, and $c = (c_1, c_2, \ldots, c_k)^T \in R^k$.

Let us now focus our attention on the nonhomogeneous system (3.2.2). We define a particular solution $y_p(n)$ of Eq. (3.2.2) as any k-vector function that satisfies the nonhomogeneous difference system. The following result gives us a mechanism to find the general solution of System (3.2.2).

Theorem 3.15. Any solution $y(n)$ of Eq. (3.2.2) can be written as

$$y(n) = \Phi(n)c + y_p(n) \qquad (3.2.10)$$

for an appropriate choice of the constant vector c.

Proof (**Exercise 3.2, Problem 8**).

We now give a formula to evaluate $y_p(n)$.

Lemma 3.16. A particular solution of Eq. (3.2.2) may be given by

$$y_p(n) = \sum_{r=n_0}^{n-1} \Phi(n, r+1)g(r)$$

with $y_p(n_0) = 0$.

Proof

$$
\begin{aligned}
y_p(n+1) &= \sum_{r=n_0}^{n} \Phi(n+1, r+1)g(r) \\
&= \sum_{r=n_0}^{n-1} A(n)\Phi(n, r+1)g(r) + \Phi(n+1, n+1)g(n) \\
&= A(n)y(n) + g(n).
\end{aligned}
$$

Hence, $y_p(n)$ is a solution of Eq. (3.2.2). Furthermore, $y_p(n_0) = 0$.

Theorem 3.17 (Variation of Constants Formula). The unique solution of the initial value problem

$$y(n+1) = A(n)y(n) + g(n), \; y(n_0) = y_0 \qquad (3.2.11)$$

is given by

$$y(n, n_0, y_0) = \Phi(n, n_0)y_0 + \sum_{r=n_0}^{n-1} \Phi(n, r+1)g(r) \qquad (3.2.12)'$$

or more explicitly by

$$y(n, n_0, y_0) = \left(\prod_{i=n_0}^{n-1} A(i) \right) y_0 + \sum_{r=n_0}^{n-1} \left(\prod_{i=r+1}^{n-1} A(i) \right) g(r) \qquad (3.2.12)'$$

Proof This theorem follows immediately from Theorem (3.15) and Lemma (3.16).

Corollary 3.18. For autonomous systems when A is a constant matrix, the solution of Eq. (3.2.11) is given by

$$y(n, n_0, y_0) = A^{n-n_0}y_0 + \sum_{r=n_0}^{n-1} A^{n-r-1}g(r). \qquad (3.2.13)$$

Example 3.19. Solve the system $y(n + 1) = Ay(n) + g(n)$, where

$$A = \begin{pmatrix} 2 & 1 \\ 0 & 2 \end{pmatrix}, \quad g(n) = \begin{pmatrix} n \\ 1 \end{pmatrix}, \quad y(0) = \begin{pmatrix} 1 \\ 0 \end{pmatrix}.$$

Solution: Using the Putzer algorithm, one may show that

$$A^n = \begin{pmatrix} 2^n & n2^{n-1} \\ 0 & 2^n \end{pmatrix}.$$

Hence,

$$
\begin{aligned}
y(n) &= \begin{pmatrix} 2^n & n2^{n-1} \\ 0 & 2^n \end{pmatrix}\begin{pmatrix} 1 \\ 0 \end{pmatrix} \\
&\quad + \sum_{r=0}^{n-1}\begin{pmatrix} 2^{n-r-1} & (n-r-1)2^{n-r-2} \\ 0 & 2^{n-r-1} \end{pmatrix}\begin{pmatrix} r \\ 1 \end{pmatrix} \\
&= \begin{pmatrix} 2^n \\ 0 \end{pmatrix} + \sum_{r=0}^{n-1}\begin{pmatrix} r2^{n-r-1} + (n-r-1)2^{n-r-2} \\ 2^{n-r-1} \end{pmatrix} \\
&= \begin{pmatrix} 2^n \\ 0 \end{pmatrix} + 2^n\begin{pmatrix} \frac{1}{4}\sum_{r=1}^{n-1} r\left(\frac{1}{2}\right)^r + \frac{n-1}{4}\sum_{r=0}^{n-1}\left(\frac{1}{2}\right)^r \\ \frac{1}{2}\sum_{r=0}^{n-1}\left(\frac{1}{2}\right)^r \end{pmatrix} \quad \text{(See footnote 3)}\\
&= \begin{pmatrix} 2^n \\ 0 \end{pmatrix} + 2^n\begin{pmatrix} \frac{1}{2}\left[1-\left(\frac{1}{2}\right)^n\right] - (n)\left(\frac{1}{2}\right)^{n+2} \\ +\frac{n-1}{2}\left[1-\left(\frac{1}{2}\right)^n\right]1 - \left(\frac{1}{2}\right)^n \end{pmatrix} \\
&= \begin{pmatrix} 2^n \\ 0 \end{pmatrix} + 2^n\begin{pmatrix} -\frac{n}{4}\left(\frac{1}{2}\right)^n + \frac{n}{2} - \frac{n}{2}\left(\frac{1}{2}\right) \\ 1 - \left(\frac{1}{2}\right)^n \end{pmatrix} \\
&= \begin{pmatrix} 2^n \\ 0 \end{pmatrix} + \begin{pmatrix} n2^{n-1} - \frac{3}{4}n \\ 1 - \left(\frac{1}{2}\right)^n \end{pmatrix} \\
&= \begin{pmatrix} 2^n + n2^{n-1} - \frac{3}{4}n \\ 1 - \left(\frac{1}{2}\right)^n \end{pmatrix}.
\end{aligned}
$$

---

3 $\sum_{r=1}^{n-1} ra^r = \dfrac{a(1-a^n) - na^{n+1}(1-a)}{(1-a)^2}.$

We now revisit scalar equations of order k and demonstrate how to transform them into a k- dimensional system of first order equations. Consider again the equation

$$y(n + k) + p_1(n)y(n + k - 1) + \cdots + p_k(n)y(n) = g(n). \qquad (3.2.14)$$

This relation may be written as a system of first order equations of dimension k. We let

$$
\begin{aligned}
z_1(n) &= y(n), \\
z_2(n) &= y(n + 1) = z_1(n + 1), \\
z_3(n) &= y(n + 2) = z_2(n + 1), \ldots, z_k(n) = y(n + k - 1) = z_{k-1}(n + 1).
\end{aligned}
$$

Let $z(n) = (z_1(n), z_2(n), \ldots, z_k(n))$.

Hence,

$$z_1(n + 1) = z_2(n)$$

$$z_2(n + 1) = z_3(n)$$

$$\vdots \qquad\qquad \vdots$$

$$
\begin{aligned}
z_{k-1}(n + 1) &= z_k(n) \\
z_k(n + 1) &= -p_k(n)z_1(n) - p_{k-1}(n)z_2(n), \ldots, \\
&\quad -p_1(n)z_k(n) + g(n).
\end{aligned}
$$

In vector notation, we transcribe this system as

$$z(n + 1) = A(n)z(n) + h(n), \qquad (3.2.15)$$

where

$$
A(n) = \begin{pmatrix}
0 & 1 & 0 & \cdots & 0 \\
0 & 0 & 1 & \cdots & 0 \\
0 & 0 & 0 & \cdots & 0 \\
\vdots & \vdots & \vdots & & \vdots \\
0 & 0 & 0 & & 1 \\
-p_k(n) & -p_{k-1}(n) & -p_{k-2}(n) & \cdots & -p_1(n)
\end{pmatrix}, \qquad (3.2.16)
$$

and

$$
h(n) = \begin{pmatrix}
0 \\
0 \\
0 \\
\vdots \\
g(n)
\end{pmatrix}.
$$

If $g(n) = 0$, we arrive at the homogeneous system

$$z(n + 1) = A(n)z(n). \tag{3.2.17}$$

The matrix $A(n)$ is called the *companion matrix* of Eq. (3.2.14).

Consider now the kth order homogeneous equation with constant coefficients

$$x(n + k) + p_1 x(n + k - 1) + p_2 x(n + k - 2) + \cdots + p_k(n)x(n) = 0$$

which is equivalent to the system where A is the companion matrix defined in Formula (3.2.16) with all p_i's constant

$$z(n + 1) = Az(n). \tag{3.2.18}$$

We first observed that the Casoration of Eq. (2.3.1) is denoted by $C(n) = \det \Phi(n)$, where $\Phi(n)$ is a fundamental matrix of Eq. (3.2.18). Why? (Exercise 3.2, Problem 16). The characteristic equation of A is given by

$$\lambda^k + p_1 \lambda^{k-1} + p_2 \lambda^{k-2} + \cdots + p_{k-1}\lambda + p_k = 0, \quad \text{(Exercise 3.2, Problem 15)}$$

which correlates with Eq. (2.3.2). Hence, the eigenvalues of A are the roots of the characteristic equation of Eq. (2.3.1).

Exercise 3.2.

1. Let $\Phi_1(n)$ and $\Phi_2(n)$ be two fundamental matrices of System (3.2.1). Prove that $\Phi_1(n)\Phi_1^{-1}(n_0) = \Phi_2(n)\Phi_2^{-1}(n_0)$ for any $n_0 \geq 0$.

2. Let $\Phi(n,m)$ be a fundamental matrix of Eq. (3.2.1). Show that

   (i) $\Phi(n,m)$ is a solution of $\Phi(n + 1,m) = A(n)\Phi(n,m)$,

   (ii) $\Phi(n,m)$ is a solution of $\Phi(n,m + 1) = \Phi(n,m)A^{-1}(m)$.

3. Let $\Phi(n,m)$ be a fundamental matrix of Eq. (3.2.1). Show that

   (a) $\Phi(n,m) = \Phi(n - m) = A^{n-m}$ if $A(n) \equiv A$ is a constant matrix.

   (b) $\Phi(n,m) = \Phi(n,r)\Phi(r,m)$.

   (c) $\Phi^{-1}(n,m) = \Phi(m,n)$.

   (d) $\Phi(n,m) = \sum_{i=m}^{n-1} A(i)$.

4. Let $\Phi(n)$ be a fundamental matrix of Eq. (3.2.1). Show that each column of $\Phi(n)$ is a solution of (3.2.1).

5. Show that $\Phi(n) = \prod_{i=n_0}^{n-1} A(i)$ is a fundamental matrix of Eq. (3.2.1).

6. Show that if $\Phi(n)$ is a fundamental matrix of Eq. (3.2.1), any C is any nonsingular matrix, then $\Phi(n)C$ is also a fundamental matrix of Eq. (3.2.1).

7. Show that if $\Phi_1(n)$, $\Phi_2(n)$ are two fundamental matrices of Eq. (3.2.1), then there exists a nonsingular matrix C such that $\Phi_2(n) = \Phi_1(n)C$.

8. Prove that any solution of Eq. (3.2.2) may be written as $y(n) = \Phi(n)c + y_p(n)$, where $y_p(n)$ is a particular solution of Eq. (3.2.2).

9. Show that the set S of all the solutions of Eq. (3.2.1) forms a linear space of dimension k.

10. Solve the system
$$y_1(n + 1) = y_2(n),$$
$$y_2(n + 1) = y_3(n) + 2,$$
$$y_3(n + 1) = y_1(n) + 2y_3(n) + n^2.$$

11. Solve the system
$$y_1(n + 1) = 2y_1(n) + 3y_2(n) + 1,$$
$$y_2(n + 1) = y_1(n) + 4y_2(n),$$
$$y_1(0) = 0, \ y_2(0) = -1.$$

12. Solve the system $y(n + 1) = Ay(n) + g(n)$ if

$$A = \begin{pmatrix} 2 & 2 & -2 \\ 0 & 3 & 1 \\ 0 & 1 & 3 \end{pmatrix}, \qquad g(n) = \begin{pmatrix} 1 \\ n \\ n^2 \end{pmatrix}.$$

13. For system equation (3.2.17), show that
$$\det A(n) = (-1)^k p_k(n).$$

14. If $\Phi(n)$ is a fundamental matrix of Eq. (3.2.17), prove that
$$\det \Phi(n) = (-1)^{k(n-n_0)} \left(\prod_{i=n_0}^{n-1} p(i) \right) \Phi(n_0).$$

15. Prove by induction that the characteristic equation of

$$A = \begin{pmatrix} 0 & 1 & 0 & \cdots & 0 \\ 0 & 0 & 1 & \cdots & 0 \\ \vdots & \vdots & \vdots & & \\ 0 & 0 & 0 & \cdots & 1 \\ -p_k & -p_{k-1} & -p_{k-2} & \cdots & -p_1 \end{pmatrix}$$

is
$$\lambda^k + p_1\lambda^{k-1} + p_2\lambda^{k-2} + \cdots + p_{k-1}\lambda + p_k = 0.$$

16. Let $C(n)$ be the casoration of Eq. (3.2.14) with $g(n) = 0$. Prove that there exists a fundamental matrix $\Phi(n)$ of Eq. (3.2.17) such that $C(n) = \det \Phi(n)$.

Use the methods of systems to solve the difference equation for Problems 17 through 21.

17. $x(n + 2) + 8x(n + 1) + 12x(n) = 0.$

18. $x(n + 2) - 16x(n) = 0.$

19. $y(n + 2) - 5y(n + 1) + 4y(n) = 4^n.$

20. $\Delta^2 y(n) = 16.$

21. $\Delta^2 x(n) + \Delta x(n) - x(n) = 0.$

3.3   The Jordan Form: Autonomous (Time-Invariant) Systems Revisited

The Jordan form of a matrix is vital for both theoretical and computation purposes in autonomous systems. In this section, we will briefly describe the Jordan form and derive a new method for computing fundamental matrices.

We say that the two $k \times k$ matrices A and B, are *similar* if there exists a nonsingular matrix P such that $P^{-1}AP = B$. It may be shown in this case that A and B have the same eigenvalues, and, in fact, the eager student will show this supposition in Exercise 3.3, Problem 15. If a matrix A is similar to a diagonal matrix $D = \text{diag}(\lambda_1, \lambda_2, \ldots, \lambda_k)$, then A is said to be *diagonalizable*. Notice here that the diagonal elements of D, namely, $\lambda_1, \lambda_2, \ldots, \lambda_k$, are the eigenvalues of A. We remark here that only special types of matrices are diagonalizable. For those particular diagonalizable matrices, computing A^n is simple. For if

$$P^{-1}AP = D = \text{diag}[\lambda_1, \lambda_2, \ldots, \lambda_k],$$

then

$$A = PDP^{-1},$$

and consequently,

$$A^n = (PDP^{-1})^n = PD^n P^{-1}.$$

Explicitly,

$$A^n = P \begin{bmatrix} \lambda_1^n & & & 0 \\ & \lambda_2^n & & \\ & & \ddots & \\ 0 & & & \lambda_k^n \end{bmatrix} P^{-1}. \tag{3.3.1}$$

If we are interested in finding another (but simpler) fundamental matrix of the equation

$$x(n + 1) = Ax(n), \tag{3.3.2}$$

then we let

$$\Phi(n) = A^n P = P \begin{bmatrix} \lambda_1^n & & & 0 \\ & \lambda_2^n & & \\ & & \ddots & \\ 0 & & & \lambda_k^n \end{bmatrix} \tag{3.3.3}$$

From Formula (3.3.3), we have $\Phi(0) = P$ and consequently,

$$A^n = \Phi(n)\Phi^{-1}(0). \tag{3.3.4}$$

Now, Formula (3.3.3) is useful only if one can pinpoint the matrix P. Fortunately, this is an easy task. We will now reveal how to compute P.

Let $P = (\xi_1, \xi_2, \ldots, \xi_k)$, where ξ_i is the ith column of P. Since $P^{-1}AP = D$, then $AP = PD$. This implies that $A\xi_i = \lambda_i\xi_i, i = 1, 2, \ldots, k$ (Exercise 3.3, Problem 15). Thus, $\xi_i, 1 \le i \le k$, is the eigenvector of A corresponding to the eigenvalue λ_i, and hence the ith column of P is the eigenvector of A corresponding to the ith eigenvalue λ_i of A. Since P is nonsingular, its columns (and hence the eigenvectors $\xi_1, \xi_2, \ldots, \xi_k$ of A) are linearly independent. Reversing the above steps, one may show that the converse of the above statement is true. Namely, if there are k linearly independent eigenvectors of a $k \times k$ matrix A, then it is diagonalizable. The following theorem summarizes the above discussion.

Theorem 3.22. A $k \times k$ matrix is diagonalizable if and only if it has k linearly independent eigenvectors.

Let us revert back to Formula (3.3.3), which gives us a computational method to find a fundamental matrix $\Phi(n)$. Let $\lambda_1, \lambda_2, \ldots, \lambda_k$ be the eigenvalues of A and $\xi_1, \xi_2, \ldots, \xi_k$ be the corresponding linearly independent eigenvectors of A. Then from Formula (3.3.3) we have

$$\begin{aligned} \Phi(n) &= [\xi_1, \xi_2, \ldots, \xi_k] \begin{bmatrix} \lambda_1^n & & & 0 \\ & \lambda_2^n & & \\ & & \ddots & \\ 0 & & & \lambda_k^n \end{bmatrix} \\ &= [\lambda_1^n\xi_1, \lambda_2^n\xi_2, \ldots, \lambda_k^n\xi_k]. \end{aligned} \tag{3.3.5}$$

Notice that since columns of $\Phi(n)$ are solutions of Eq. (3.3.2) it follows that for each i, $1 \le i \le k$, $x(n) = \lambda_i^n\xi_i$ is a solution of Eq. (3.3.2).

Hence, the general solution of Eq. (3.3.2) may be given by

$$x(n) = c_1\lambda_1^n\xi_1 + c_2\lambda_2^n\xi_2 + \cdots + c_k\lambda_k^n\xi_k. \tag{3.3.6}$$

The following example illustrates the above method.

Example 3.23. Find the general solution of $x(n + 1) = A(n)$, where

$$A = \begin{pmatrix} 2 & 2 & 1 \\ 1 & 3 & 1 \\ 1 & 2 & 2 \end{pmatrix}.$$

Solution: The eigenvalues of A may be obtained by solving the characteristic equation

$$\det(A - \lambda I) = \det \begin{pmatrix} 2 - \lambda & 2 & 1 \\ 1 & 3 - \lambda & 1 \\ 1 & 2 & 2 - \lambda \end{pmatrix} = 0.$$

This determinant produces $(\lambda - 1)^2(\lambda - 5) = 0$. Thus, $\lambda_1 = 5$, and $\lambda_2 = \lambda_3 = 1$. To find the corresponding eigenvectors, we solve the equation $(A - \lambda I)x = 0$. Hence, for $\lambda_1 = 5$,

$$\begin{pmatrix} -3 & 2 & 1 \\ 1 & -2 & 1 \\ 1 & 2 & -3 \end{pmatrix} \begin{pmatrix} x_1 \\ x_2 \\ x_3 \end{pmatrix} = \begin{pmatrix} 0 \\ 0 \\ 0 \end{pmatrix}.$$

Solving this system gives us the first eigenvector

$$\xi_1 = \begin{pmatrix} 1 \\ 1 \\ 1 \end{pmatrix}.$$

For $\lambda_2 = \lambda_3 = 1$, we have

$$\begin{pmatrix} 1 & 2 & 1 \\ 1 & 2 & 1 \\ 1 & 2 & 1 \end{pmatrix} \begin{pmatrix} x_1 \\ x_2 \\ x_3 \end{pmatrix} = \begin{pmatrix} 0 \\ 0 \\ 0 \end{pmatrix}.$$

Consequently, $x_1 + 2x_2 + x_3 = 0$ is the only equation obtained from this algebraic system. To solve the system, then two of the three unknown terms x_1, x_2, and x_3 must be arbitrarily chosen. So if we let $x_1 = 1$ and $x_2 = 0$, then $x_3 = -1$ and we obtain the eigenvector

$$\xi_2 = \begin{pmatrix} 1 \\ 0 \\ -1 \end{pmatrix}.$$

On the other hand, if we let $x_1 = 0$ and $x_2 = 1$, then $x_3 = -2$ and we obtain the third eigenvector

$$\xi_3 = \begin{pmatrix} 0 \\ 1 \\ -2 \end{pmatrix}.$$

Obviously, there are infinitely many choices for ξ_2, ξ_3. Using Formula (3.3.6), the general solution is

$$x(n) = c_1 5^n \begin{pmatrix} 1 \\ 1 \\ 1 \end{pmatrix} + c_2 \begin{pmatrix} 1 \\ 0 \\ -1 \end{pmatrix} + c_3 \begin{pmatrix} 0 \\ 1 \\ -2 \end{pmatrix}$$

or

$$x(n) = \begin{pmatrix} c_1 5^n + c_2 \\ c_1 5^n + c_3 \\ c_1 5^n - c_2 - 2c_3 \end{pmatrix}. \tag{3.3.7}$$

Suppose that in the above problem we are given an initial value

$$x(0) = \begin{pmatrix} 0 \\ 1 \\ 0 \end{pmatrix},$$

and must find the solution $x(n)$ with this initial value. One way of doing this is by letting $n = 0$ in the solution given by Formula (3.3.7) and evaluate the constants c_1, c_2, and c_3.

Thus

$$c_1 + c_2 = 0,$$

$$c_1 + c_3 = 1,$$

$$c_1 - c_2 - 2c_3 = 0.$$

Solving this system gives $c_1 = \frac{1}{2}$, $c_2 = -\frac{1}{2}$, and $c_3 = \frac{1}{2}$, leading us to the solution

$$x(n) = \begin{pmatrix} \dfrac{1}{2}5^n - \dfrac{1}{2} \\[2mm] \dfrac{1}{2}5^n + \dfrac{1}{2} \\[2mm] \dfrac{1}{2}5^n - \dfrac{1}{2} \end{pmatrix}.$$

We now introduce yet another method to find the solution. Let

$$x(n) = \Phi(n)\Phi^{-1}(0)x(0),$$

where

$$\begin{aligned} \Phi(n) &= (\lambda_1^n \xi_1, \lambda_2^n \xi_2, \lambda_3^n \xi_3) \\[2mm] &= \begin{pmatrix} 5^n & 1 & 0 \\ 5^n & 0 & 1 \\ 5^n & -1 & -2 \end{pmatrix} \end{aligned}$$

and

$$\Phi(0) = \begin{pmatrix} 1 & 1 & 0 \\ 1 & 0 & 1 \\ 1 & -1 & -2 \end{pmatrix}.$$

Thus,

$$\Phi^{-1}(0) = \begin{pmatrix} \dfrac{1}{4} & \dfrac{1}{2} & \dfrac{1}{4} \\[2mm] \dfrac{3}{4} & -\dfrac{1}{2} & -\dfrac{1}{4} \\[2mm] -\dfrac{1}{4} & \dfrac{1}{2} & -\dfrac{1}{4} \end{pmatrix}.$$

This gives

$$x(n) = \begin{pmatrix} 5^n & 1 & 0 \\ 5^n & 0 & 1 \\ 5^n & -1 & -2 \end{pmatrix} \begin{pmatrix} \dfrac{1}{4} & \dfrac{1}{2} & \dfrac{1}{4} \\ \dfrac{3}{4} & -\dfrac{1}{2} & -\dfrac{1}{4} \\ -\dfrac{1}{4} & \dfrac{1}{2} & -\dfrac{1}{4} \end{pmatrix} \begin{pmatrix} 0 \\ 1 \\ 0 \end{pmatrix}$$

$$= \begin{pmatrix} \dfrac{1}{2}5^n - \dfrac{1}{2} \\ \dfrac{1}{2}5^n + \dfrac{1}{2} \\ \dfrac{1}{2}5^n - \dfrac{1}{2} \end{pmatrix}.$$

In the next example we will examine the case when the matrix A has complex eigenvalues. Notice that if A is a real matrix (which we are assuming here) and if $\lambda = \alpha + i\beta$ is an eigenvalue of A, then $\bar{\lambda} = \alpha - i\beta$ is also an eigenvalue of A. Moreover, if ξ is the eigenvector of A corresponding to the eigenvalue $\lambda = \alpha + i\beta$, then $\bar{\xi}$ is the eigenvector of A corresponding to the eigenvalue $\bar{\lambda} = \alpha - i\beta$. Taking advantage of these observations, one may be able to considerably simplify the computation involved in finding a fundamental matrix of system equation (3.3.2).

Suppose that $\xi = \xi_1 + i\xi_2$. A solution of System (3.3.2) may then be given by $x(n) = (\alpha + i\beta)^n(\xi_1 + i\xi_2)$. Also, if

$$r = \sqrt{\alpha^2 + \beta^2},$$

then

$$\theta = \tan^{-1}\left(\frac{\beta}{\alpha}\right).$$

This solution may now be written as

$$
\begin{aligned}
x(n) &= [r(\cos\theta + i\sin\theta)]^n(\xi_1 + i\xi_2) \\
&= r^n(\cos n\theta + i\sin n\theta)(\xi_1 + i\xi_2) \\
&= r^n[(\cos n\theta)\xi_1 - (\sin n\theta)\xi_2] + ir^n[(\cos n\theta)\xi_2 + (\sin n\theta)\xi_1] \\
&= u(n) + i\,v(n),
\end{aligned}
$$

where $u(n) = r^n[(\cos n\theta)\xi_1 - (\sin n\theta)\xi_2]$ and $v(n) = r^n[(\cos n\theta)\xi_2 + (\sin n\theta)\xi_1]$. One might show (Exercise 3.3, Problem 7) that $u(n)$ and $v(n)$ are linearly independent solutions of System (3.3.2). Hence, we do not need to consider the solution generated by $\bar{\lambda}$ and $\bar{\xi}$.

Example 3.24. Find a general solution of the system $x(n + 1) = A\,x(n)$, where

$$A = \begin{pmatrix} 1 & -5 \\ 1 & -1 \end{pmatrix}.$$

Solution The eigenvales of A are $\lambda_1 = 2i, \lambda_2 = -2i$, and the corresponding eigenvectors are

$$\xi_1 = \begin{pmatrix} \frac{1}{5} - \frac{2}{5}i \\ 1 \end{pmatrix}, \xi_2 = \begin{pmatrix} \frac{1}{5} + \frac{2}{5}i \\ 1 \end{pmatrix}.$$

Hence,

$$x(n) = (2i)^n \begin{pmatrix} \frac{1}{5} - \frac{2}{5}i \\ 1 \end{pmatrix},$$

$$i = \xi\frac{\pi}{2} + i \sin \frac{\pi}{2},$$

$$i^n = \xi\frac{n\pi}{2} + i \sin \frac{n\pi}{2}$$

is a solution.

This solution may be written as

$$x(n) = 2^n \left(\cos\left(\frac{n\pi}{2}\right) + i \sin\left(\frac{n\pi}{2}\right)\right) \begin{pmatrix} \frac{1}{5} - \frac{2}{5}i \\ 1 \end{pmatrix}$$

$$= 2^n \begin{pmatrix} \frac{1}{5} \cos\left(\frac{n\pi}{2}\right) + \frac{2}{5} \sin \frac{n\pi}{2} \\ \cos\left(\frac{n\pi}{2}\right) \end{pmatrix} + i2^n$$

$$\begin{pmatrix} \frac{-2}{5} \cos\left(\frac{n\pi}{2}\right) + \frac{1}{5} \sin\left(\frac{n\pi}{2}\right) \\ \sin\left(\frac{n\pi}{2}\right) \end{pmatrix}.$$

Thus,

$$u(n) = 2^n \begin{pmatrix} \frac{1}{5} \cos\left(\frac{n\pi}{2}\right) + \frac{2}{5} \sin\left(\frac{n\pi}{2}\right) \\ \cos\left(\frac{n\pi}{2}\right) \end{pmatrix}$$

and

$$v(n) = 2^n \begin{pmatrix} \frac{-2}{5} \cos\left(\frac{n\pi}{2}\right) + \frac{1}{5} \sin\left(\frac{n\pi}{2}\right) \\ \sin\left(\frac{n\pi}{2}\right) \end{pmatrix}$$

are two linearly independent solutions. A general solution may be given as

$$x(n) = c_1 2^n \begin{pmatrix} \frac{1}{5} \cos\left(\frac{n\pi}{2}\right) + \frac{2}{5} \sin\left(\frac{n\pi}{2}\right) \\ \cos\left(\frac{n\pi}{2}\right) \end{pmatrix} + c_2 2^n$$

$$\begin{pmatrix} \frac{-2}{5} \cos\left(\frac{n\pi}{2}\right) + \frac{1}{5} \sin\left(\frac{n\pi}{2}\right) \\ \sin\left(\frac{n\pi}{2}\right) \end{pmatrix}$$

$$= 2^n \left[\begin{array}{l} \left(\frac{1}{5}c_1 - \frac{2}{5}c_2\right) \cos\left(\frac{n\pi}{2}\right) + \left(\frac{2}{5}c_1 + \frac{1}{5}c_2\right) \sin\left(\frac{n\pi}{2}\right) \\ c_1 \cos\left(\frac{n\pi}{2}\right) + c_2 \sin\left(\frac{n\pi}{2}\right) \end{array} \right].$$

So far, we have discussed the solution of System (3.3.2) if the matrix A is diagonalizable. We remark here that a sufficient condition for a $k \times k$ matrix A to be diagonalizable would be that it has k distinct eigenvalues (Exercise 3.3, Problem 20). If the matrix A has repeated roots, then it is diagonalizable if it is *normal*, that is to say, if $A^T A = A A^T$. (For a proof see Ref. 5.) Examples of normal matrices are

(i) symmetric matrices ($A^T = A$),

(ii) skew symmetric matrices ($A^T = -A$),

(iii) unitary matrices ($A^T A = A A^T = I$).

We now turn our attention to the general case where the matrix A is not diagonalizable. This happens when A has repeated eigenvalues and one is not able to generate k linearly independent eigenvectors. For example, the following matrices are not diagonalizable:

$$\begin{bmatrix} 2 & 1 \\ 0 & 2 \end{bmatrix}, \begin{bmatrix} 2 & 0 & 0 \\ 0 & 2 & 1 \\ 0 & 0 & 2 \end{bmatrix}, \begin{bmatrix} 2 & 0 & 0 & 0 \\ 0 & 3 & 0 & 0 \\ 0 & 0 & 4 & 1 \\ 0 & 0 & 0 & 4 \end{bmatrix}.$$

If a $k \times k$ matrix A is not diagonalizable, then it is akin to the so-called *Jordan form*, i.e., $P^{-1}AP = J$, where

$$J = \mathrm{diag}(J_1, J_2, \ldots, J_r), \qquad 1 \le r \le k \qquad (3.3.8)$$

and

$$J_i = \begin{pmatrix} \lambda_i & 1 & 0 & \cdots & 0 \\ 0 & \lambda_i & 1 & & 0 \\ & & \ddots & & \\ 0 & 0 & \ddots & \ddots & \vdots \\ \vdots & \vdots & & \ddots & 1 \\ 0 & 0 & & & \lambda_i \end{pmatrix}. \qquad (3.3.9)$$

The matrix J_i is called a *Jordan block*.

These remarks are formalized in the following theorem.

Theorem 3.25 (The Jordan Canonical Form). Any $k \times k$ matrix A is similar to a Jordan form given by the Formula (3.3.8), where each J_i is a $s_i \times s_i$ matrix of the form (3.3.9), and $\sum_{i=1}^{r} s_i = k$.

The number of Jordan blocks corresponding to one eigenvalue λ is called the *geometric multiplicity* of λ and this number in turn equals the number of linearly independent eigenvectors corresponding to λ.

The *algebraic multiplicity* of an eigenvalue λ is the number of times it is repeated. If the algebraic multiplicity of λ is 1 (i.e., λ is not repeated), then we refer to λ as *simple*. If the geometric multiplicity of λ is equal to its algebraic multiplicity (i.e., only 1×1 Jordan blocks correspond to λ), then it is called *semi-simple*. For example, the matrix

$$
\begin{bmatrix}
3 & 0 & 0 & 0 & 0 \\
0 & 2 & 0 & 0 & 0 \\
0 & 0 & 2 & 0 & 0 \\
0 & 0 & 0 & 5 & 1 \\
0 & 0 & 0 & 0 & 5
\end{bmatrix}
$$

has one simple eigenvalue 3, one semisimple eigenvalue 2, and one eigenvalue 5, neither simple nor semi-simple.

To illustrate the theorem, we list below the possible Jordan forms of a 3×3 matrix with an eigenvalue $\lambda = 5$, of multiplicity 3. In the matrix, different Jordan blocks are indicated by lines.

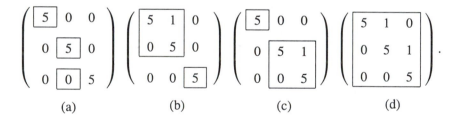

      (a)   (b)   (c)   (d)

Recall that s_i is the order of the ith Jordan block and r is the number of Jordan blocks in a Jordan form. In (a), the matrix is diagonalizable and we have three Jordan blocks of order 1. Thus, $s_1 = s_2 = s_3 = 1, r = 3$ and the geometric multiplicity of λ is 3.

In (b), there are two Jordan blocks with $s_1 = 2, s_2 = 1, r = 2$ and the geometric multiplicity of λ is 2.

In (c), there are also two Jordan blocks with $s_1 = 1, s_2 = 2, r = 2$ and the geometric multiplicity of λ is 2. In (d), there is only one Jordan block with $s_1 = 3, r = 1$ and the geometric multiplicity of λ is 1. The linearly independent eigenvectors corresponding to $\lambda = 5$ in (a), (b), (c), (d) are, respectively,

$$\underbrace{\begin{pmatrix} 1 \\ 0 \\ 0 \end{pmatrix}, \begin{pmatrix} 0 \\ 1 \\ 0 \end{pmatrix}, \begin{pmatrix} 0 \\ 0 \\ 1 \end{pmatrix}}_{\text{(a)}} \quad \underbrace{\begin{pmatrix} 1 \\ 0 \\ 0 \end{pmatrix}, \begin{pmatrix} 0 \\ 0 \\ 1 \end{pmatrix}}_{\text{(b)}} \quad \underbrace{\begin{pmatrix} 1 \\ 0 \\ 0 \end{pmatrix}, \begin{pmatrix} 0 \\ 1 \\ 0 \end{pmatrix}}_{\text{(c)}} \quad \underbrace{\begin{pmatrix} 1 \\ 0 \\ 0 \end{pmatrix}}_{\text{(d)}}.$$

Note that a matrix of the form

$$\begin{pmatrix} \lambda & 1 & 0 & \cdots & 0 \\ 0 & \lambda & 1 & \cdots & 0 \\ \vdots & \vdots & \vdots & & \vdots \\ 0 & 0 & 0 & \cdots & 1 \\ 0 & 0 & 0 & \cdots & \lambda \end{pmatrix}$$

has only one eigenvector, namely, the unit vector $e_1 = (1, 0, \ldots, 0)^T$. This shows us that the linearly independent eigenvectors of the Jordan form J given by Formula (3.3.8) are

$$e_1, e_{s_1+1}, e_{s_1+s_2+1}, \ldots, e_{s_1+s_2+\cdots+s_{r-1}+1}.$$

Now, since $P^{-1}AP = J$, then

$$AP = PJ. \tag{3.3.10}$$

Let $P = (\xi_1, \xi_2, \ldots, \xi_k)$. Equating the first s_1 columns of both sides in Formula (3.3.10) we obtain

$$A\xi_1 = \lambda_1\xi_1, \ldots, A\xi_i = \lambda_1\xi_i + \xi_{i-1}, \qquad i = 2, 3, \ldots, s_1. \tag{3.3.11}$$

Clearly ξ_1 is the only eigenvector of A in the *Jordan Chain* $\xi_1, \xi_2, \ldots, \xi_{s_1}$. The other vectors $\xi_2, \xi_3, \ldots, \xi_{s_1}$ are called *generalized eigenvectors* of A, and may be obtained by using the difference equation

$$(A - \lambda_1 I)\xi_i = \xi_{i-1}, \qquad i = 2, 3, \ldots, s_1. \tag{3.3.12}$$

Repeating this process for the remainder of the Jordan blocks, one may find the generalized eigenvectors corresponding to the mth Jordan block using the difference equation

$$(A - \lambda_m I)\xi_{m_i} = \xi_{m_i-1}, \qquad i = 2, 3, \ldots, s_m. \tag{3.3.13}$$

Now we know that $A^n = (PJP^{-1})^n = PJ^nP^{-1}$ where

$$J^n = \begin{bmatrix} J_1^n & & & 0 \\ & J_2^n & & \\ & & \ddots & \\ 0 & & & J_k^n \end{bmatrix}.$$

Notice that for any J_i, $i = 1, 2, \ldots, r$, we have $J_i = \lambda_i I + N_i$ where

$$N_i = \begin{pmatrix} 0 & 1 & 0 & \cdots & 0 \\ 0 & 0 & 1 & & 0 \\ \vdots & \vdots & & & 1 \\ 0 & 0 & & \cdots & 0 \end{pmatrix}$$

is an $s_i \times s_i$ nilpotent matrix (i.e., $N_i^k = 0$ for all $k \geq s_i$). Hence,

$$J_i^n = (\lambda_i I + N_i)^n = \lambda_i^n I + \binom{n}{1} \lambda_i^{n-1} N_i \tag{3.1}$$

$$+ \binom{n}{2} \lambda_i^{n-2} N_i^2 + \cdots + \binom{n}{s_i - 1} \lambda_i^{n-s_i+1} N_i^{s_i-1} \tag{3.2}$$

$$= \begin{pmatrix} \lambda_i^n & \binom{n}{1}\lambda_i^{n-1} & \binom{n}{2}\lambda_i^{n-2} & \cdots & \binom{n}{s_i-1}\lambda_i^{n-s_i+1} \\ 0 & \lambda_i^n & \binom{n}{1}\lambda_i^{n-1} & \cdots & \binom{n}{s_i-2}\lambda_i^{n-s_i+2} \\ \vdots & \vdots & & \ddots & \vdots \\ & & & & \binom{n}{1}\lambda_i^{n-1} \\ 0 & 0 & & \cdots & \lambda_i^n \end{pmatrix}. \tag{3.3.14}$$

The lines inside J_i^n indicate that the entries in each diagonal are identical.
We can now substantiate that the general solution of System (3.3.2) is

$$\begin{aligned} x(n) &= A^n c \\ &= P J^n P^{-1} c \end{aligned}$$

or

$$x(n) = P J^n \hat{c}, \tag{3.3.15}$$

where

$$\hat{c} = P^{-1} c.$$

Hence, a fundamental matrix of System (3.3.2) may be given by $\Phi(n) = P J^n$. Also, the state transition matrix may be given by $\Phi(n, n_0) = P J^{n-n_0} P^{-1}$.

The following Corollary arises directly from an immediate consequence of Formula (3.3.14).

Corollary 3.26. Assuming that A is any $k \times k$ matrix, then $\lim_{n \to \infty} A^n = 0$ if and only if $|\lambda| < 1$ for all eigenvalues λ of A.

Proof (**Exercise 3.3, Problem 21**).

The importance of the preceding corollary lies in the fact that if $\lim_{n \to \infty} A^n = 0$, then $\lim_{n \to \infty} x^n = \lim_{n \to \infty} A^n x(0) = 0$. This fact reminds us that if $|\lambda| < 1$ for all eigenvalues of A, then all solutions $x(n)$ of Eq. (3.3.1) tend toward the zero vector as $n \to \infty$.

Example 3.27. Find the general solution of $x(n + 1) = A\, x(n)$ with

$$A = \begin{pmatrix} 4 & 1 & 2 \\ 0 & 2 & -4 \\ 0 & 1 & 6 \end{pmatrix}.$$

Solution Note that this example uses conclusions from Example 3.2. The eigenvalues are $\lambda_1 = \lambda_2 = \lambda_3 = 4$. To find the eigenvectors, we solve the equation $(A - \lambda I)\xi = 0$, or

$$\begin{pmatrix} 0 & 1 & 2 \\ 0 & -2 & -4 \\ 0 & 1 & 2 \end{pmatrix} \begin{pmatrix} d_1 \\ d_2 \\ d_3 \end{pmatrix} = \begin{pmatrix} 0 \\ 0 \\ 0 \end{pmatrix}.$$

Hence,

$$d_2 + 2d_3 = 0,$$

$$-2d_2 - 4d_3 = 0,$$

$$d_2 + 2d_3 = 0.$$

These equations imply that $d_2 = -2\, d_3$, thus generating two eigenvectors,

$$\xi_1 = \begin{pmatrix} 0 \\ -2 \\ 1 \end{pmatrix} \text{ and } \xi_2 = \begin{pmatrix} 1 \\ -2 \\ 1 \end{pmatrix}.$$

We must now find one generalized eigenvector ξ_3. Applying Formula (3.3.11) let us test $(A - 4I)\xi_3 = \xi_1$.

$$\begin{pmatrix} 0 & 1 & 2 \\ 0 & -2 & -4 \\ 0 & 1 & 2 \end{pmatrix} \begin{pmatrix} a_1 \\ a_2 \\ a_3 \end{pmatrix} = \begin{pmatrix} 0 \\ -2 \\ 1 \end{pmatrix}.$$

This system is an inconsistent system which has no solution. The second attempt will use

$$(A - 4I)\xi_3 = \xi_2$$

or

$$\begin{pmatrix} 0 & 1 & 2 \\ 0 & -2 & -4 \\ 0 & 1 & 2 \end{pmatrix} \begin{pmatrix} a_1 \\ a_2 \\ a_3 \end{pmatrix} = \begin{pmatrix} 1 \\ -2 \\ 1 \end{pmatrix}.$$

Hence, $a_2 + 2a_3 = 1$. One may now set

$$\xi_3 = \begin{pmatrix} 0 \\ -1 \\ 1 \end{pmatrix}.$$

Thus,

$$P = \begin{pmatrix} 0 & 1 & 0 \\ -2 & -2 & -1 \\ 1 & 1 & 1 \end{pmatrix},$$

$$J = \begin{pmatrix} 4 & 0 & 0 \\ 0 & 4 & 1 \\ 0 & 0 & 4 \end{pmatrix},$$

and

$$J^n = \begin{pmatrix} 4^n & 0 & 0 \\ 0 & 4^n & n4^{n-1} \\ 0 & 0 & 4^n \end{pmatrix}.$$

Hence,

$$x(n) = PJ^n c = \begin{pmatrix} 0 & 4^n & n4^{n-1} \\ -2 \, 4^n & -2 \, 4^n & -2 \, n4^{n-1} \\ -4^n & 4^n & n4^{n-1} + 4^n \end{pmatrix} \begin{pmatrix} c_1 \\ c_2 \\ c_3 \end{pmatrix}.$$

Example 3.28. Solve the system

$$x(n+1) = Ax(n), \quad x(0) = \begin{pmatrix} 1 \\ 1 \\ 1 \end{pmatrix},$$

where

$$A = \begin{pmatrix} 3/2 & 1/2 & 1/2 \\ 1/2 & 5/2 & -1/2 \\ 0 & 1 & 2 \end{pmatrix}.$$

Solution: The eigenvalues of A are $\lambda_1 = \lambda_2 = \lambda_3 = 2$. We have a sole eigenvector,

$$\xi_1 - \begin{pmatrix} 1 \\ 0 \\ 1 \end{pmatrix}.$$

We now need to compose two generalized eigenvectors, using Eq. (3.3.13).

$$(A - 2I)\xi_2 = \xi_1 \text{ gives } \xi_2 = \begin{pmatrix} 1 \\ 1 \\ 2 \end{pmatrix}$$

and

$$(A - 2I)\xi_3 = \xi_2 \text{ gives } \xi_3 = \begin{pmatrix} 1 \\ 2 \\ 1 \end{pmatrix}.$$

So

$$P = \begin{pmatrix} 1 & 1 & 1 \\ 0 & 1 & 2 \\ 1 & 2 & 1 \end{pmatrix}, \qquad J = \begin{pmatrix} 2 & 1 & 0 \\ 0 & 2 & 1 \\ 0 & 0 & 2 \end{pmatrix},$$

$$J^n = \begin{pmatrix} 2^n & n2^{n-1} & \dfrac{n(n-1)}{2}2^{n-2} \\ 0 & 2^n & n2^{n-1} \\ 0 & 0 & 2^n \end{pmatrix}.$$

Now

$$x(n, x_0) = P J^n P^{-1} x_0$$

$$= 2^{n-4} \begin{pmatrix} n^2 - 5n + 16 & n^2 + 3n & -n^2 + 5n \\ 4n & 4n + 16 & -4n \\ n^2 - n & n^2 + 7n & -n^2 + n + 16 \end{pmatrix} \begin{pmatrix} 1 \\ 1 \\ 1 \end{pmatrix}$$

$$= 2^{n-4} \begin{pmatrix} n^2 + 3n \\ 4n + 16 \\ n^2 + 7n + 16 \end{pmatrix}$$

Exercise 3.3.

In Problems 1 through 6, use Formula 3.3.6 to find the solution of $x(n+1) = A x(n)$, where A is given in the exercise.

1. $A = \begin{pmatrix} 2 & -1 \\ 0 & 4 \end{pmatrix}, \qquad x(0) = \begin{pmatrix} 1 \\ 2 \end{pmatrix}.$

2. $A = \begin{pmatrix} 1 & 0 \\ 1 & 2 \end{pmatrix}.$

3. $A = \begin{pmatrix} 2 & 3 & 0 \\ 4 & 3 & 0 \\ 0 & 0 & 6 \end{pmatrix}, \qquad x(0) = \begin{pmatrix} 0 \\ 1 \\ 0 \end{pmatrix}.$

4. $A = \begin{pmatrix} 2 & -1 & 0 \\ 0 & 4 & 0 \\ 2 & 5 & 3 \end{pmatrix}.$

5. $A = \begin{pmatrix} 1 & 0 & 1 \\ 1 & 2 & 3 \\ 0 & 0 & 3 \end{pmatrix}.$

6. $A = \begin{pmatrix} 1 & 1 & 0 \\ -1 & 1 & 0 \\ 1 & 0 & 1 \end{pmatrix}$, $\quad x(0) = \begin{pmatrix} 1 \\ 0 \\ 1 \end{pmatrix}$.

7. Suppose that $x(n) = u(n) + iv(n)$ is a solution of Eq. (3.3.2). Prove that $u(n)$ and $v(n)$ are also solutions of Eq. (3.3.2).

8. Utilize problem 7 to find a fundamental matrix of $x(n + 1) = Ax(n)$ with

$$A = \begin{pmatrix} 1 & 1 & 0 & 1 \\ -1 & 1 & 0 & 1 \\ 0 & 0 & 2 & 1 \\ 0 & 0 & -1 & 2 \end{pmatrix}.$$

9. Apply problem 7 to find a fundamental matrix of $x(n + 1) = Ax(n)$ with

$$A = \begin{pmatrix} 1 & 1 & 0 \\ -1 & 1 & 0 \\ 1 & 0 & 1 \end{pmatrix}.$$

10. Find the eigenvalues and the corresponding eigenvectors and generalized eigenvectors for the matrix A.

(a) $A = \begin{pmatrix} 3 & 1 \\ 0 & 3 \end{pmatrix}$. (b) $A = \begin{pmatrix} 2 & 1 & 0 \\ 0 & 2 & 1 \\ 0 & 0 & 2 \end{pmatrix}$.

(c) $A = \begin{pmatrix} 4 & 2 & 3 \\ -\frac{1}{2} & 2 & 0 \\ 0 & 0 & 3 \end{pmatrix}$. (d) $A = \begin{pmatrix} 2 & 0 & 0 & 0 \\ 0 & 2 & 1 & 0 \\ 0 & 0 & 2 & 1 \\ 0 & 0 & 0 & 2 \end{pmatrix}$.

11. Find A^n for the matrices in problem 10 using the Jordan form.

12. Use the Jordan form to find a fundamental matrix for $x(n + 1) = Ax(n)$ with

$$A = \begin{pmatrix} 3 & 2 & 1 \\ -1 & 3 & 2 \\ 1 & -3 & -2 \end{pmatrix}.$$

13. Use the Jordan form to solve $x(n + 1) = Ax(n)$ with

$$A = \begin{pmatrix} 3 & 2 & 3 \\ -1/2 & 1 & 0 \\ 0 & 0 & 2 \end{pmatrix}.$$

14. Let A and B be two similar matrices with $P^{-1} A P = B$.

(i) Show that A and B have the same eigenvalues.

(ii) Show that if ξ is an eigenvector of B, the $P\xi$ is an eigenvector of A.

15. Suppose that $P^{-1}A\,P = D = \operatorname{diag}(\lambda_1, \lambda_2, \ldots, \lambda_k)$, where $P = [\xi_1, \xi_2, \ldots, \xi_k]$ is a nonsingular $k \times k$ matrix. Show that $\xi_1, \xi_2, \ldots, \xi_k$ are the eigenvectors of A that correspond to the eigenvalues $\lambda_1, \lambda_2, \ldots, \lambda_k$, respectively.

16. Let A be a 4×4 matrix with an eigenvalue $\lambda = 3$ of multiplicity 4. Write all possible Jordan forms of A.

17. Show that $(P\,J\,P^{-1})^n = P\,J^n\,P^{-1}$.

18. If λ is an eigenvalue of A, and ξ is the corresponding eigenvector of A, show that $\lambda^n \xi$ is a solution of Eq. (3.3.2).

19. Let $A = \begin{pmatrix} \lambda & 1 & \ldots & 0 \\ 0 & \lambda & \ldots & 0 \\ \vdots & \vdots & & \vdots \\ 0 & 0 & \ldots & 1 \\ 0 & 0 & \ldots & \lambda \end{pmatrix}$.

Then one may write $A = \lambda\,I + N$, where

$$N = \begin{pmatrix} 0 & 1 & \ldots & 0 \\ 0 & 0 & & \vdots \\ \vdots & \vdots & & 1 \\ 0 & 0 & & 0 \end{pmatrix}.$$

Show that for any $\alpha > 0$, A is similar to a matrix

$$B = \lambda\,I + \alpha N = \begin{pmatrix} \lambda & \alpha & \ldots & 0 \\ 0 & \lambda & & \\ \vdots & \vdots & & \alpha \\ 0 & 0 & & \lambda \end{pmatrix}.$$

[Hint: Use the similarity matrix $P = \operatorname{diag}(1, \alpha, \alpha^2, \ldots, \alpha^{k-1})$.]

20. Prove that if a $k \times k$ matrix A has k distinct eigenvalues, then

(i) A has k linearly independent eigenvectors;

(ii) A is diagonalizable. (Use mathematical induction.)

21. Prove Corollary 3.26.

22. Consider the companion matrix A of Eq. (3.2.14) with the coefficients p_i's constant. Assume that the eigenvalues of A are real and distinct. Let V denote the Vandermonde matrix

$$V = \begin{pmatrix} 1 & 1 & \cdots & 1 \\ \lambda_1 & \lambda_2 & \cdots & \lambda_k \\ \vdots & \vdots & & \vdots \\ \lambda_1^{k-1} & \lambda_2^{k-1} & \cdots & \lambda_k^{k-1} \end{pmatrix}.$$

Show that $V^{-1}AV$ is a diagonal matrix.

3.4   Linear Periodic Systems

In this section, we regard the linear periodic system

$$x(n + 1) = A(n)x(n), \tag{3.4.1}$$

where for all $n \in z$, $A(n + N) = A(n)$, for some positive integer N.

We now show that the study of the periodic system (3.4.1) simplifies down to the study of an associated autonomous system. This inference is the analogue of Floquet theory in differential equations. But before we prove that analogue, we need the following theorem.

Lemma 3.28. Let B be a $k \times k$ nonsingular matrix and m be any positive integer. Then there exists some $k \times k$ matrix C such that $C^m = B$.

Proof Let

$$P^{-1}BP = J = \begin{pmatrix} J_1 & & & \\ & J_2 & & \\ & & \ddots & \\ & & & J_r \end{pmatrix}$$

be the Jordan form of B. Let us write

$$J_i + \lambda_i \left(I_i + \frac{1}{\lambda_i} N_i \right),$$

where I_i is the $s_i \times s_i$ identity matrix and

$$N_i = \begin{pmatrix} 0 & 1 & & & & 0 \\ & 0 & 1 & & & \\ & & \ddots & \ddots & & \\ & & & & & 1 \\ 0 & & & & & 0 \end{pmatrix}.$$

Observe that

$$N_i^{s_i} = 0 \tag{3.4.2}$$

To motivate our construction, we formally write

$$
\begin{aligned}
H_i &= \text{Exp}\left[\frac{1}{m}\ln J_i\right] \\
&= \text{Exp}\left[\frac{1}{m}\left\{\ln \lambda_i I_i + \ln\left(I_i + \frac{1}{\lambda_i}N_i\right)\right\}\right] \\
&= \text{Exp}\left[\frac{1}{m}\left\{\ln \lambda_i I_i + \sum_{s=1}^{\infty} \frac{(-1)^{s+1}}{s}\right\}\right]\left(\frac{N_i}{\lambda_i}\right)^s.
\end{aligned}
$$

Applying Formula (3.4.2), we obtain

$$H_i = \text{Exp}\left[\frac{1}{m}\left\{\ln \lambda_i I_i + \sum_{s=1}^{s_i-1} \frac{(-1)^{s+1}}{s}\right\}\left(\frac{N_i}{\lambda_i}\right)^s\right]. \tag{3.4.3}$$

Hence, H_i is a well defined matrix. Furthermore, $H_i^m = J_i$.

Now, if we let $H = \begin{pmatrix} H_1 & & & 0 \\ & H_2 & & \\ & & \ddots & \\ 0 & & & H_r \end{pmatrix}$, where H_i is defined in Formula

(3.4.3), then

$$
Hm = \begin{bmatrix} H_1^m & & & 0 \\ & H_2^m & & \\ & & \ddots & \\ 0 & & & H_r^m \end{bmatrix}
$$

$$= J.$$

Define $C = PHP^{-1}$. Then $C^m = PH^m P^{-1} = PJP^{-1} = B$.

Armed with this lemma, we are now prepared to introduce the primary result for this section.

Lemma 3.29. For the System (3.4.1), the following statements hold.

(i) If $\Phi(n)$ is a fundamental matrix, then so is $\Phi(n + N)$.

(ii) $\Phi(n + N) = \Phi(n)C$, for some nonsingular matrix C.

(iii) $\Phi(n + N, N) = \Phi(n, 0)$.

Proof The proof is left as an exercise for the reader (Exercise 3.4, Problem 1). There are many consequences of this lemma, including the following theorem.

Theorem 3.30. For every fundamental matrix $\Phi(n)$ of System (3.4.1), there exists a nonsingular periodic matrix $P(n)$ of period N such that

$$\Phi(n) = P(n)B^n. \tag{3.4.4}$$

Proof By Lemma 3.28, there exists some matrix B, such that $B^N = C$. Define $P(n) = \Phi(n)B^{-n}$, where $B^{-n} = (B^n)^1$.

$$
\begin{aligned}
\text{Then } P(n + N) &= \Phi(n + N)B^{-N}B^{-n} \\
&= \Phi(n)CB^{-N}B^{-n} \text{ [using part (ii) of Lemma 3.29]} \\
&= \Phi(n)B^{-n} \\
&= P(n).
\end{aligned}
$$

We now know that $P(n)$ has period N, and is clearly nonsingular. Why? From the definition of $P(n)$ it thus follows that $\Phi(n) = P(n)B^n$.

Remark 3.31 If $z(n)$ is a solution of the system

$$z(n + 1) = Bz(n) \tag{3.4.5}$$

then

$$
\begin{aligned}
x(n) &= \Phi(n)c \\
&= P(n)B^n c
\end{aligned}
$$

or

$$x(n) = P(n)z(n). \tag{3.4.6}$$

The value of this remark lies in the fact that the qualitative study of the periodic system equation (3.4.1) reduces to the study of the autonomous system (3.4.5).

The matrix $C = B^N$, which may be found using Lemma 3.29 part (ii), is referred to as a *monodromy matrix* of Eq. (3.4.1). The eigenvalues λ of B are called the *Floquet exponents* of Eq. (3.4.1); the corresponding eigenvalues λ^N of B^N are called the *Floquet multipliers* of Eq. (3.4.1). The reason we call λ^N a multiplier is that there exists a solution $x(n)$ of Eq. (3.4.1) such that $x(n + N) = \lambda x(n)$. (See Exercise 3.4, Problem 9.) Notice that the Floquet exponents (multipliers) do not depend upon the monodromy matrix chosen, that is, they do not hinge upon the particular fundamental matrix $\Phi(n)$ used to define the monodromy matrix. The following lemma explicitly states this truth.

Lemma 3.32. If $\Phi(n)$ and $\Psi(n)$ are two fundamental matrices of Eq. (3.4.1) such that

$$\Phi(n + N) = \Phi(n)C,$$

$$\Psi(n + N) = \Psi(n)E,$$

then C and E are similar (and thus they have the same eigenvalues).

Proof The reader will prove this lemma in Exercise 3.4, Problem 2.

Lemma 3.33. A complex number λ is a Floquet exponent of Eq. (3.4.1) if and only if there is a nontrivial solution of Eq. (3.4.1) of the form $\lambda^n q(n)$, where $q(n)$ is a vector function with $q(n + N) = q(n)$ for all n.

Proof First, we assume that λ is a Floquet exponent of Eq. (3.4.1). Then, we also know that $\det(B^n - \lambda^n I) = 0$. Now choose $x_0 \in R^k$, $x_0 \neq 0$, such that $(B^n - \lambda^n I)x_0 = 0$, for all n. (Why? See Exercise 3.4, Problem 4.) Hence, we have the equation $B^n x_0 = \lambda^n x_0$.

Thus, $P(n)B^n x_0 = \lambda^n P(n)x_0$, where $P(n)$ is the periodic matrix as defined in Formula (3.4.4). By Formula (3.4.4) now,

$$
\begin{aligned}
x(n,n_0,y_0) &= \Phi(n,n_0)x_0 \\
&= P(n)B^n x_0 \\
&= \lambda^n P(n)x_0 \\
&= \lambda^n q(n)
\end{aligned}
$$

and we have the desired periodic solution of Eq. (3.4.1), where $q(n) = P(n)x_0$.

Conversely, if $\lambda^n q(n)$, $q(n+N) = q(n) \neq 0$ is, a solution of Eq. (3.4.1), Theorem 3.30 implies that

$$\lambda^n q(n) = P(n)B^n x_0, \tag{3.4.7}$$

for some nonzero vector x_0. This implies that

$$\lambda^{n+N} q(n) = P(n)B^{n+N} x_0. \tag{3.4.8}$$

But from Eq. (3.4.7)

$$\lambda^{n+N} q(n) = \lambda^N P(n)B^n x_0. \tag{3.4.9}$$

Equating the right-hand side of Formulas (3.4.8) and (3.4.9) we obtain

$$P(n)B^n[B^N - \lambda^N I]x_0 = 0,$$

and thus,

$$\det[B^N - \lambda^N I] = 0.$$

This manipulation shows that λ is a Floquet exponent of Eq. (3.4.1).

Using the preceding theorem, one may easily conclude the following results.

Corollary 3.34. The following statements hold:

(i) System (3.4.1) has a periodic solution of period N if it has a Floquet multiplier $\lambda = 1$.

(ii) If there is a Floquet multiplier equal to -1, then System (3.4.1) has a periodic solution of period $2N$.

Proof Use Lemma 3.33 as you prove Corollary 3.34 in Exercise 3.

Remark Lemma 3.29 part (ii) gives us a formula to find the monodromy matrix $C = B^N$, whose eigenvalues happen to be the Floquet multipliers of Eq. (3.4.1). From Lemma 3.29,

$$C = \Phi^{-1}(n)\Phi(n + N).$$

By letting $n = 0$, we have

$$C = \Phi^{-1}(0)\Phi(N). \tag{3.4.10}$$

If we take $\Phi(N) = A(N - 1)A(N - 2)\ldots A(0)$, then $\Phi(0) = I$. Thus, Formula (3.4.10) becomes

$$C = \Phi(N),$$

or

$$C = A(N - 1)A(N - 2)\ldots A(0). \tag{3.4.11}$$

We now give an example to illustrate the above results.

Example 3.35. Consider the planar system

$$
\begin{aligned}
x(n + 1) &= A(n)x(n), \\
A(n) &= \begin{pmatrix} 0 & (-1)^n \\ (-1)^n & 0 \end{pmatrix}.
\end{aligned}
$$

Clearly, $A(n + 2) = A(n)$ for all $n \in Z$.
   Applying Formula (3.4.10),

$$B^2 = C = A(1)A(0) = \begin{pmatrix} -1 & 0 \\ 0 & -1 \end{pmatrix}.$$

Thus the Floquet multipliers are $-1, -1$. By virtue of Corollary 3.34, the system has a 4-periodic solution. Note that since $A(n)$ has the constant eigenvalues $-1, 1$, $\rho(A(n)) = 1$.
   The above example may suggest that there is some kind of relationship between the eigenvalues of $A(n)$ and its Floquet multipliers. To dispel any such thoughts we offer the following example.

Example 3.36. Consider System (3.2.1) with

$$A(n) = \begin{pmatrix} 0 & \frac{2+(-1)^n}{2} \\ \frac{2-(-1)^n}{2} & 0 \end{pmatrix}.$$

This is a system of period 2. The eigenvalues of A are $\pm\frac{\sqrt{3}}{2}$, and hence $\rho(A) = \frac{\sqrt{3}}{2} < 1$. Now

$$B^2 = C = A(1)A(0) = \begin{pmatrix} \frac{1}{4} & 0 \\ 0 & \frac{9}{4} \end{pmatrix}.$$

Thus, the Floquet multipliers are $\frac{1}{4}$ and $\frac{9}{4}$. Hence, $\rho(B) = \frac{3}{2}$.

Exercise 3.4.

1. Prove Lemma 3.29.

2. Prove Lemma 3.32.

3. Prove Corollary 3.34.

4. Suppose that $(B - \lambda I)x_o = 0$ for some $x_o \in R^k$, $x_o \neq 0$. Prove that $(B^n - \lambda^n I)x_o = 0$ for all $n \in Z^I$.

5. Let $a_1(n)$, $a_2(n)$ be N- periodic functions and let $\Psi_1(n)$, $\Psi_2(n)$ be solutions of

$$x(n + 2) + a_1(n)x(n + 1) + a_2(n)x(n) = 0 \qquad (3.4.12)$$

such that $\Psi_1(0) = 1$, $\Psi_1(1) = 0$, $\Psi_2(0) = 0$, and $\Psi_2(1) = 1$. Show that the Floquet multipliers satisfy the equation $\lambda^2 + b\lambda + c = 0$, where

$$b = -[\Psi_1(N) + \Psi_2(N + 1), \qquad c = \prod_{i=0}^{N-1} a_2(i).$$

[Hint: First change the equation to a system and then show that the monodromy matrix is equal to $\Phi(N)$.]

6. In Problem 5, let $a_2(n) \equiv 1$. Show that the product of the Floquet multipliers is equal to 1.

7. In Problem 5, let $a_2(n) \equiv 1$. Show that if $b = 2$ there is at least one solution of period $2N$ while for $b = -2$ there is at least one solution of period N.

8. In Problem 5 it is clear that if $\lambda = 1$, then Eq. (3.4.12) has a periodic solution of period N. Show that $x(n+2)+a_1(n)x(n+1)+a_2(n)x(n) = 0$ has a periodic solution of period $2N$ if and only if $\lambda = -1$.

9. Show that there exists a solution $x(n)$ of Eq. (3.4.1) that satisfies $x(n + N) = \lambda x(n)$ if and only if λ is a Floquet multiplier.

3.5 Applications

3.5.1 Markov Chains

In 1906, the Russian mathematician A.A. Markov developed the concept of Markov chains. We can describe a Markov chain as follows: Suppose that we conduct some experiment with a set of k outcomes or states $S = \{s_1, s_2, \ldots, s_k\}$. The experiment is repeated such that the probability (p_{ij}) of the state s_i, $1 \leq i \leq k$, occurring on the $(n + 1)$th repetition depends only on the state s_j occurring on the nth repetition of the experiment. In other words, the system has no memory: the future state depends only on the present state. In probability theory language, $p_{ij} = p(s_i|s_j)$ is the probability of s_i occurring on the next repetition, given that s_j occurred on the last repetition. Given that s_j has occurred in the last repetition, one of $s_1, s_2, \ldots, s_k$ must occur in the next repetition. Thus,

$$p_{1j} + p_{2j} + p_{3j} + \cdots + p_{kj} = 1, \qquad 1 \le j \le k. \qquad (3.5.1)$$

Let $p_i(n)$ denote the probability that the state s_i will occur on the nth repetition of the experiment, $1 \le i \le k$. Since only one of the states s_i must occur on the nth repetition, it follows that

$$p_1(n) + p_2(n) + \cdots + p_k(n) = 1. \qquad (3.5.2)$$

To derive the mathematical model of this experiment, we must define $p_i(n + 1)$, $1 \le i \le k$, as the probability that the state s_i occurs on the $(n + 1)$th repetition of the experiment. There are k ways that this can happen. The first case is when repetition n gives us s_1, and repetition $(n + 1)$ produces s_i. Since the probability of getting s_1 on the nth repetition is $p_1(n)$, and the probability of having s_i after s_1 is p_{i1}, it follows (by the multiplication principle) that the probability of the first case occurring is $p_{i1}p_1(n)$. The second case is when we get s_2 on repetition n and s_i on repetition $(n + 1)$. The probability of the occurrence of the second case is $p_{i2}p_2(n)$. Repeating this for cases 3, 4, $\ldots, k$, and for $i = 1, 2, \ldots, k$, we obtain the k-dimensional system

$$
\begin{aligned}
p_1(n + 1) &= p_{11}p_1(n) + p_{12}p_2(n) + \cdots + p_{1k}p_k(n) \\
p_2(n + 1) &= p_{21}p_1(n) + p_{22}p_2(n) + \cdots + p_{2k}p_k(n)
\end{aligned}
$$

$$\vdots$$

$$p_k(n + 1) = p_{k1}p_1(n) + p_{k2}p_2(n) + \cdots + p_{kk}p_k(n)$$

or, in vector notation,

$$p(n + 1) = Sp(n), \qquad n = 1, 2, 3 \ldots, \qquad (3.5.3)$$

where $p(n) = (p_1(n), p_2(n), \ldots, p_k(n))^T$ is the *probability vector* and $S = (p_{ij})$ is a $k \times k$ *transition matrix*.

The matrix S belongs to a special class of matrices called the *Markov Matrices*. A matrix $A = (a_{ij})$ is said to be *nonnegative* (positive) if $a_{ij} \ge 0$ (> 0) for all entries a_{ij} of A. A nonnegative $k \times k$ matrix A is said to be *Markov* (or *stochastic*) if $\sum_{i=1}^{k} a_{ij} = 1$ for all $j = 1, 2, \ldots, k$. It can be shown that $|\lambda| \le 1$ for all the eigenvalues λ of a Markov matrix (Exercise 3.5, Problem 3). Furthermore, $\lambda = 1$ is an eigenvalue of a Markov matrix (Exercise 3.5, Problem 1). If we define *the spectral radius* of a $k \times k$ matrix A with eigenvalues, $\lambda_1, \lambda_2, \ldots, \lambda_k$ by $\rho(A) = \max_{1 \le i \le k} |\lambda_i|$, then $\rho(A) = 1$ if A is Markov.

3.5.2   Regular Markov Chains

A regular Markov chain is one in which S^m is positive for some positive integer m. To give a complete analysis of the eigenvalues of such matrices, we need the following theorem, indebted to M. Perron.

Perron's Theorem 3.37. Let A be a positive $k \times k$ matrix. Then $\rho(A)$ is a simple *real* eigenvalue (not repeated) of A. If λ is any other eigenvalue of A, then $|\lambda| < \rho(A)$. Moreover, an eigenvector associated with $\rho(A)$ may be assumed to be positive.

Suppose now that S is the transition matrix of a regular Markov chain with eigenvalues $\lambda_1, \lambda_2, \ldots, \lambda_k$. The $\rho(S) = 1$. If S^m is positive, then $\rho(S^m) = 1$. As a matter of fact, the eigenvalues of S^m are $\lambda_1^m, \lambda_2^m, \ldots, \lambda_k^m$. By Perron's Theorem, 1 is a simple eigenvalue of S^m. Consequently, S has exactly one simple eigenvalue, say λ_1, which equals 1; all other eigenvalues satisfy $|\lambda_i| < 1, i = 2, 3, \ldots, k$. Hence, the Jordan form of S must be of the form $J = \begin{pmatrix} 1 & 0 \\ 0 & J_* \end{pmatrix}$, where the eigenvalues of J_* are $\lambda_2, \lambda_3, \ldots, \lambda_k$.

By Corollary 3.26, $J_*^n \to 0$ as $n \to \infty$ so that $J^n \to \text{diag}(1, 0, \ldots, 0)$ as $n \to \infty$. Therefore, if $S = QJQ^{-1}$, we have

$$\lim_{n\to\infty} p(n) = \lim_{n\to\infty} S^n p(0) = \lim_{n\to\infty} QJ^nQ^{-1}p(0) = (\xi_1, 0, 0, \ldots, 0)\eta = a\xi_1,$$

$$(3.5.4)$$

where $\xi_1 = (\xi_{11}, \xi_{21}, \ldots, \xi_{k1})^T$ is the eigenvector of S that corresponds to the eigenvalue $\lambda_1 = 1$, and a is the first component of $\eta = Q^{-1}p(0)$. Since finding the matrix Q is not a simple task, we will choose instead to devise a very easy method to find the constant a. Recall that for

$$p(n) = (p_1(n), p_2(n), \ldots, p_k(n))^T$$

we have, from Formula (3.5.2), $\sum_{i=1}^{k} p_i(n) = 1$. Since $\lim_{n\to\infty} p(n) = a\xi_1$, it follows that

$$a\xi_{11} + a\xi_{21} + \cdots + a\xi_{k1} = 1.$$

Therefore,

$$a = \frac{1}{\xi_{11} + \xi_{21} + \cdots + \xi_{k1}}.$$

The following example illustrates a regular Markov chain.

Example 3.38. The simplest type of genetic inheritance of traits in animals occurs when a certain trait is determined by a specific pair of genes, each of which may be of two types, say G and g. An individual may have a GG combination, a Gg (which is genetically the same as gG) or a gg combination. An individual with GG genes is said to be *dominant*; a gg individual is referred to as *recessive*; a *hybrid* has Gg genes.

In the mating of two animals, the offspring inherits one gene of the pair from each parent: the basic assumption of genetics is that the selection of these genes is random.

Let us consider a process of continued matings. We begin with an individual of known genetic character (GG) and mate it with a hybrid. Assuming there is one offspring, we mate that offspring with a hybrid, repeating this process through a number of generations. In each generation there are three possible states, $s_1 = GG$, $s_2 = Gg$, and $s_3 = gg$. Let $p_i(n)$ represent the probability the state s_i occurs in the nth generation and p_{ij} be the probability that s_i occurs in the $(n+1)$ generation giving that s_j occurred in the nth generation.

The difference system which models this Markov chain is denoted by

$$
\begin{aligned}
p_1(n+1) &= p_{11}p_1(n) + p_{12}p_2(n) + p_{13}p_3(n), \\
p_2(n+1) &= p_{21}p_1(n) + p_{22}p_2(n) + p_{23}p_3(n), \\
p_3(n+1) &= p_{31}p_1(n) + p_{32}p_2(n) + p_{33}p_3(n).
\end{aligned}
$$

Now p_{11} is the probability of producing an offspring GG by mating GG and Gg. Clearly, the offspring receives a G gene from his parent GG with probability 1 and the other G from his parent Gg with probability $\frac{1}{2}$. By the multiplication principle, $p_{11} = 1 \times \frac{1}{2} = \frac{1}{2}$. The probability of creating an offspring GG from mating a Gg with a Gg is p_{12}. By similar analysis one may show that $p_{12} = \frac{1}{2} \times \frac{1}{2} = \frac{1}{4}$. Likewise, p_{13} is the probability of generating an offspring GG from mating a gg with a Gg. Obviously, $p_{13} = 0$. One may show, by the same process that

$$
p_{21} = \frac{1}{2}, \quad p_{22} = \frac{1}{2}, \quad p_{23} = \frac{1}{2}, \quad p_{31} = 0, \quad p_{32} = \frac{1}{4}, \quad p_{33} = \frac{1}{2}.
$$

Hence, we have

$$
p(n+1) = S\, p(n)
$$

with

$$
S = \begin{pmatrix} 0.5 & 0.25 & 0 \\ 0.5 & 0.5 & 0.5 \\ 0 & 0.25 & 0.5 \end{pmatrix}.
$$

Notice that all the entries for S^2 are positive, and thus this is a regular Markov chain. The eigenvalues of S are $\lambda_1 = 1$, $\lambda_2 = \frac{1}{2}$, and $\lambda_3 = 0$. Recall from Formula (3.5.4) that

$$
\lim_{n \to \infty} p(n) = a\xi_1.
$$

Now

$$
\xi_1 = \begin{pmatrix} 1 \\ 2 \\ 1 \end{pmatrix},
$$

and

$$
a = \frac{1}{4}
$$

implies that

$$
\lim_{n \to \infty} p(n) = \begin{pmatrix} 0.25 \\ 0.5 \\ 0.25 \end{pmatrix}.
$$

This relation dictates that as the number of repetitions approaches ∞, the probability of producing a purely dominant or a purely recessive offspring is 0.25, and the probability of creating a hybrid offspring is 0.5.

3.5.3 Absorbing Markov Chains

A state s_i in a Markov chain is said to be *absorbing* if, whenever it occurs on the nth repetition of the experiment, it then occurs on every subsequent repetition, i.e., $p_{ii} = 1$ and $p_{ij} = 0$ for $i \neq j$. A Markov chain is said to be *absorbing* if it has at least one absorbing state, and from every state it is possible to go to an absorbing state. In an absorbing Markov chain, a state which is not absorbing is called *transient*.

Example 3.39 (Drunkard's Walk). A man walks along a four block stretch. He starts at corner x. With probability $\frac{1}{2}$ he walks one block to the right, and with probability $\frac{1}{2}$ walks one block to the left. When he comes to the next corner he again randomly chooses his direction. He continues until he reaches corner 5 which is a bar or corner 1 which is his home. (See Fig. 3.2.) If he reaches either home or the bar, he stays there. This is clearly an absorbing Markov chain.

Let us rename the states so that the absorbing states at 1 and 5 are last, and referred to as s_4 and s_5, and the transient states 2, 3, and 4 will be called s_1, s_2 and s_3, respectively. Accordingly, $p_1(n)$, $p_2(n)$, $p_3(n)$, $p_4(n)$, and $p_5(n)$ will be, respectively, the probability of reaching s_1, s_2, s_3, s_4, and s_5 after n walks. The difference equation that represents this Markov chain is $p(n + 1) = S\, p(n)$, where the transition matrix is

$$S = \left(\begin{array}{ccc|cc} 0 & \frac{1}{2} & 0 & 0 & 0 \\ \frac{1}{2} & 0 & \frac{1}{2} & 0 & 0 \\ 0 & \frac{1}{2} & 0 & 0 & 0 \\ \cdots & \cdots & \cdots & \cdots & \cdots \\ \frac{1}{2} & 0 & 0 & 1 & 0 \\ 0 & 0 & \frac{1}{2} & 0 & 1 \end{array} \right) = \left(\begin{array}{cc} T & 0 \\ Q & I \end{array} \right).$$

Let $u(n) = (p_1(n), p_2(n), p_3(n))^T$ and $v(n) = (p_4(n), p_5(n))^T$. Then

$$\left(\begin{array}{c} u(n+1) \\ v(n+1) \end{array} \right) = \left(\begin{array}{cc} T & 0 \\ Q & I \end{array} \right) \left(\begin{array}{c} u(n) \\ v(n) \end{array} \right)$$

or

$$u(n+1) = T\, u(n), \tag{3.5.5}$$

$$v(n+1) = v(n) + Q\, T(n). \tag{3.5.6}$$

Therefore,

$$u(n) = T^n u(0). \tag{3.5.7}$$

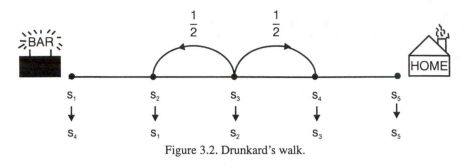

Figure 3.2. Drunkard's walk.

Substituting from Formula (3.5.7) into Formula (3.5.6) yields

$$v(n + 1) = v(n) + Q\,T^n u(0). \qquad (3.5.8)$$

By Formula (3.2.13), it follows that the solution of Eq. (3.5.8) is given by

$$v(n) = v(0) + \sum_{r=0}^{n-1} QT^r u(0). \qquad (3.5.9)$$

The eigenvalues of T are

$$0, -\sqrt{\frac{1}{2}}, \text{ and } \sqrt{\frac{1}{2}}.$$

Hence, by Corollary 3.26, the $\lim_{n\to\infty} T^n = 0$. In this case one may show that

$$\sum_{r=0}^{\infty} T^r = \lim_{n\to\infty} \sum_{r=0}^{n-1} T^r = (I - T)^{-1}. \quad \text{(Exercise 3.5, Problem 4)}$$

Using Formula (3.5.9), we generate

$$\lim_{n\to\infty} v(n) = v(0) + Q(I - T)^{-1} u(0).$$

Now

$$(I - T)^{-1} = \begin{pmatrix} \frac{3}{2} & 1 & \frac{1}{2} \\ 1 & 2 & 1 \\ \frac{1}{2} & 1 & \frac{3}{2} \end{pmatrix}.$$

Assume that the man starts midway between the home and the bar, that is, at state s_2. Then

$$u(0) = \begin{pmatrix} 0 \\ 1 \\ 0 \end{pmatrix}$$

and

$$v(0) = \begin{pmatrix} 0 \\ 0 \end{pmatrix}.$$

In this case

$$\lim_{n \to \infty} v(n) = \begin{pmatrix} \frac{1}{2} & 0 & 0 \\ 0 & 0 & \frac{1}{2} \end{pmatrix} \begin{pmatrix} \frac{3}{2} & 1 & \frac{1}{2} \\ 1 & 2 & 1 \\ \frac{1}{2} & 1 & \frac{3}{2} \end{pmatrix} \begin{pmatrix} 0 \\ 1 \\ 0 \end{pmatrix} = \begin{pmatrix} \frac{1}{2} \\ \frac{1}{2} \end{pmatrix}.$$

Thus, the probability that the man ends up in his home is 0.5. The probability that he ends up in the bar is also 0.5. Common sense could have probably told us this in the first place, but not every situation will be this facile.

3.5.4   A Trade Model [4]

Example 3.40. Consider a model of the trade between two countries, restricted by the following assumptions:

(i) National income = consumption outlays + net investment + exports − imports.

(ii) Domestic consumption outlays = Total consumption − imports.

(iii) Time is divided into periods of equal length, denoted by $n = 0, 1, 2, \ldots$.

Let for country $j = 1, 2$,

$$
\begin{aligned}
y_j(n) &= \text{national income in period } n \\
c_j(n) &= \text{total consumption in period } n \\
i_j(n) &= \text{net investment in period } n \\
x_j(n) &= \text{exports in period } n \\
m_j(n) &= \text{imports in period } n \\
d_j(n) &= \text{consumption of domestic products in period } n.
\end{aligned}
$$

For country 1 we then have

$$
\begin{aligned}
y_1(n) &= c_1(n) + i_1(n) + x_1(n) - m_1(n), \\
d_1(n) &= c_1(n) - m_1(n)
\end{aligned}
$$

which, combining those two equations, gives

$$y_1(n) = d_1(n) + x_1(n) + i_1(n). \tag{3.5.10}$$

Likewise, for country 2, we have

$$y_2(n) = d_2(n) + x_2(n) + i_2(n). \tag{3.5.11}$$

We now make the following reasonable assumption: the domestic consumption $d_j(n)$ and the imports $m_j(n)$ of each country at period $(n + 1)$ are proportional to the country's national income $y_i(n)$ one time period earlier. Thus,

$$d_1(n + 1) = a_{11} y_1(n), \qquad m_1(n + 1) = a_{21} y_1(n) \tag{3.5.12}$$

$$d_2(n + 1) = a_{22}y_2(n), \qquad m_2(n + 1) = a_{12}y_2(n). \qquad (3.5.13)$$

The constants a_{ij} are called *marginal propensities*. Furthermore, $a_{ij} > 0$, for $i, j = 1, 2$. Since we are considering a world with only two countries, the exports of one must be equal to the imports of the other, i.e.,

$$m_1(n) = x_2(n), \qquad m_2(n) = x_1(n). \qquad (3.5.14)$$

Substituting from Eqs. (3.5.12), (3.5.13), and (3.5.14), into Eqs. (3.5.10) and (3.5.11) leads to

$$\begin{pmatrix} y_1(n + 1) \\ y_2(n + 1) \end{pmatrix} = \begin{pmatrix} a_{11} & a_{12} \\ a_{21} & a_{22} \end{pmatrix} \begin{pmatrix} y_1(n) \\ y_2(n) \end{pmatrix} + \begin{pmatrix} i_1(n) \\ i_2(n) \end{pmatrix}. \qquad (3.5.15)$$

Let us further assume that the net investments $i_1(n) = i_1$ and $i_2(n) = i_2$ are constants. Then Eq. (3.5.15) becomes

$$\begin{pmatrix} y_1(n + 1) \\ y_2(n + 1) \end{pmatrix} = \begin{pmatrix} a_{11} & a_{12} \\ a_{21} & a_{22} \end{pmatrix} \begin{pmatrix} y_1(n) \\ y_2(n) \end{pmatrix} + \begin{pmatrix} i_1 \\ i_2 \end{pmatrix}. \qquad (3.5.16)$$

By the Variation of Constants Formula (3.2.13), we obtain

$$\begin{aligned} y(n) &= A^n y(0) + \sum_{r=0}^{n-1} i A^{n-r-1} \\ &= A^n y(0) + \sum_{r=0}^{n-1} A^r i. \end{aligned} \qquad (3.5.17)$$

To have a stable economy, common sense dictates that the sum of the domestic consumption $d_j(n + 1)$ and the imports $i_j(n + 1)$ in period $(n + 1)$ must be less than the national income $y_j(n)$ in period n; that is

$$d_j(n + 1) + i_j(n + 1) < y_j(n), \qquad j = 1, 2.$$

or

$$a_{11} + a_{21} < 1, \qquad a_{12} + a_{22} < 1. \qquad (3.5.18)$$

Under Conditions (3.5.18), one may show that for all the eigenvalues λ of A, $|\lambda| < 1$ (Exercise 3.5, Problem 2).

This implies from Corollary 3.26 that $A^n \to 0$ as $n \to \infty$. This fact further generates the so-called *Neumann's expansion*: (Exercise 3.5, Problem 4)

$$\lim_{n \to \infty} \sum_{r=0}^{n-1} A^r = \sum_{r=0}^{\infty} A^r = (I - A)^{-1}. \qquad (3.5.19)$$

It follows from Formula (3.5.17) that

$$\lim_{n \to \infty} y(n) = (I - A)^{-1} i.$$

This equation says that the national income of countries 1 and 2 approach equilibrium values independent of the initial values of the national incomes $y_1(0)$, $y_2(0)$.

However, as we all know, international economics involves many more factors than we can account for here. But in Exercise 3.5, Problem 11, the student will be allowed to create a model for the economic interaction between 3 countries.

Figure 3.3. Heat transfer.

3.5.5   The Heat Equation

Example 3.41. Consider the distribution of heat through a thin bar composed of a homogeneous material. Let $x_1, x_2, \ldots, x_k$ be k equidistant points on the bar. Let $T_i(n)$ be the temperature at time $t_n = (\Delta t)n$ at the point x_i, $1 \leq i \leq k$. Denote the temperatures at the left and the right ends of the bar at time t_n by $T_0(n)$, $T_{k+1}(n)$, respectively. (See Fig. 3.3.)

Assume that the sides of the bar are sufficiently well insulated that no heat energy is lost through them. The only thing, then, that affects the temperature at the point x_i is the temperature of the points next to it, which are x_{i-1}, x_{i+1}. Assume that the left end of the bar is kept at b degrees centigrade and the right end of the bar at c degrees centigrade. These conditions imply that $x_0(n) = b$ and $x_{k+1}(n) = c$, for $n \geq 0$.

We assume that the temperature at a given point x_i is determined only by the temperature at the nearby point x_{i-1} and x_{i+1}. Then according to Newton's Law of Cooling, the change in temperature $T_i(n+1) - T_i(n)$ at a point x_i from time n to $n+1$ is directly proportional to the temperature difference between the point x_i and the nearby point x_{i-1} and x_{i+1}. In other words

$$
\begin{aligned}
T_i(n+1) - T_i(n) &= \alpha([T_{i-1}(n) - T_i(n)] + [T_{i+1}(n) - t_i(n)]) \\
&= \alpha[T_{i+1}(n) - 2T_i(n) + T_{i-1}(n)]
\end{aligned} \tag{3.5.20}
$$

or

$$
T_i(n+1) = \alpha T_{i-1}(n) + (1 - 2\alpha)T_i(n) + \alpha T_{i+1}(n), \qquad i = 2, 3, \ldots, k-1.
$$

Similarly, one may also derive the following two equations:

$$
\begin{aligned}
T_1(n+1) &= (1 - 2\alpha)T_1(n) + \alpha T_2(n) + \alpha b, \\
T_k(n+1) &= \alpha T_{k-1}(n) + (1 - 2\alpha)T_k(n) + \alpha c.
\end{aligned}
$$

This correlation may be written in the compact form

$$
T(n+1) = AT(n) + g
$$

where

$$A = \begin{pmatrix} (1-2\alpha) & \alpha & 0 & \cdots & 0 \\ \alpha & (1-2\alpha) & \alpha & & \vdots \\ 0 & \alpha & (1-2\alpha) & \ddots & \\ \vdots & \vdots & \vdots & \ddots & \alpha \\ 0 & 0 & 0 & \alpha & (1-2\alpha) \end{pmatrix}, \quad g = \begin{pmatrix} \alpha b \\ 0 \\ 0 \\ \vdots \\ \alpha c \end{pmatrix}.$$

This is a tridiagonal *Toeplitz*[4] matrix. Its eigenvalues may be found by the formula [5]

$$\lambda_n = (1-2\alpha) + \alpha \cos\left(\frac{n\pi}{k+1}\right), \quad n = 1, 2, \ldots, k.$$

Hence $|\lambda| < 1$ for all eigenvalues λ of A. Corollary 3.26 then implies that

$$\lim_{n\to\infty} A^n = 0.$$

From the Variation of Constants Formula (3.2.12), it follows that

$$T(n) = A^n T(0) + \sum_{r=1}^{n-1} A^r g.$$

Thus, $\lim_{n\to\infty} T(n) = (I-A)^{-1}g$. Finally, this equation points out that the temperature at the point x_i, $1 \le i \le k$, approaches the ith component of the vector $(I-A)^{-1}g$, regardless of the initial temperature at the point x_i.

Consider the above problem, with $k = 3$, $\alpha = 0.4$, $T_0(n) = 10$ degrees centigrade, $T_4(n) = 20$ degrees centigrade.

Then

$$A = \begin{pmatrix} 0.2 & 0.4 & 0 \\ 0.4 & 0.2 & 0.4 \\ 0 & 0.4 & 0.2 \end{pmatrix}, \quad g = \begin{pmatrix} 4 \\ 0 \\ 8 \end{pmatrix},$$

$$(I-A)^{-1} = \begin{pmatrix} 0.8 & -0.4 & 0 \\ -0.4 & 0.8 & -0.4 \\ 0 & -0.4 & 0.8 \end{pmatrix} = \begin{pmatrix} \frac{15}{8} & \frac{5}{4} & \frac{5}{8} \\ \frac{5}{4} & \frac{5}{2} & \frac{5}{4} \\ \frac{13}{8} & \frac{5}{4} & \frac{15}{8} \end{pmatrix}.$$

---

[4]A is a Toeplitz matrix if it is of the form $\begin{pmatrix} a_0 & a_1 & a_2 & \cdots & a_{k-1} \\ a_{-1} & a_0 & a_1 & & \vdots \\ a_{-2} & a_{-1} & a_0 & & a_2 \\ \vdots & \vdots & & & a_1 \\ a_{-k+1} & \cdots & a_{-2} & a_{-1} & a_0 \end{pmatrix}.$

Hence

$$\lim_{n\to\infty} T(n) = \begin{pmatrix} \frac{15}{8} & \frac{5}{4} & \frac{5}{8} \\ \frac{5}{4} & \frac{5}{2} & \frac{5}{4} \\ \frac{13}{8} & \frac{5}{4} & \frac{15}{8} \end{pmatrix} \begin{pmatrix} 4 \\ 0 \\ 8 \end{pmatrix} = \begin{pmatrix} \frac{25}{2} \\ 15 \\ \frac{43}{2} \end{pmatrix}.$$

Remark Let $\Delta x = x_i - x_{i-1}$ and $\Delta t = t_{i-1} - t_i$. If we assume that the constant of proportionality α depends on both Δt and Δx, then we may write

$$\alpha = \left[\frac{\Delta t}{(\Delta x)^2} \right]^{\beta}, \tag{3.5.21}$$

where β is a constant that depends on the material of the bar. Formula (3.5.21) simply states that the smaller the value of Δt, the smaller should be the change in the temperature at a given point. Moreover, the smaller the separation of points, the larger should be their influence on the temperature changes in nearby points. Using Formula (3.5.21) in Eq. (3.5.20) yields

$$\frac{x_i(n+1) - x_i(n)}{\Delta t} = \beta \left[\frac{x_{i+1}(n) - 2x_i(n) + x_{i-1}(n)}{(\Delta x)^2} \right]. \tag{3.5.22}$$

If we let $\Delta t \to 0$, $\Delta x \to 0$ as $n \to \infty$ and $i \to \infty$, $x_i = (\Delta x)i = x$, and $t_i = (\Delta t)i = t$, then Eq. (3.5.22) gives the partial differential equation

$$\frac{\partial T(x,t)}{\partial t} = \beta \frac{\partial^2 T(x,t)}{\partial x^2}. \tag{3.5.23}$$

Equation (3.5.23) is known as the *heat equation* [6].

Exercise 3.5.

1. Show that if A is a $k \times k$ Markov matrix, then it has an eigenvalue equal to 1. [Hint: Consider $A^T \xi = \xi$ with $\xi = (1, 1, \ldots, 1)^T$.]

2. Let $A = (a_{ij})$ be a $k \times k$ positive matrix such that $\sum_{j=1}^{k} a_{ij} < 1$ for $i = 1, 2, \ldots, k$. Show that $|\lambda| < 1$ for all eigenvalues λ of A.

3. Let A be a $k \times k$ Markov matrix. Show that $\rho(A) = 1$. (Hint: Use Problem 1.)

4. Let A be a $k \times k$ matrix with $|\lambda| < 1$ for all eigenvalues λ of A. Show that

   (i) $(I - A)$ is nonsingular. (Hint: Consider $(I - A)x = 0$.)

   (ii) $\sum_{i=0}^{\infty} A^i = (I - A)^{-1}$. [Hint: Use $(I - A)(I + A + A^2 + \cdots + A^{n-1}) = I - A^n$).]

5. Modify Example 3.29 by first mating a recessive individual (genes gg) with a dominant individual (genes GG). Then, continuing to mate the offspring with a dominant individual, write down the difference equation that describes the probabilities of producing individuals with genes GG, Gg, and gg. Find $\lim_{n\to\infty} p(n)$ and then interpret your results.

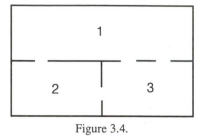

Figure 3.4.

6. In the dark ages, Harvard, Yale, and MIT admitted only male students. Assume that, at the time, 80% of the sons of Harvard men went to Harvard and the rest went to MIT, 40% of the sons of MIT men went to MIT, and the rest split evenly between Harvard and Yale; and of the sons of Yale men, 70% went to Yale, 20% to Harvard, and 10% to MIT. Find the transition matrix R of this Markov chain. Find the long term probabilities that the descendants of Harvard men will go to Yale. (Assume that we start with N men, and each man sends one son to college.)

7. A New York governor tells person A his intention either to run or not to run in the next presidential election. Then A relays the news to B, who in turn relays the message to C, and so forth, always to some new person. Assume there is a probability α that a person will change the answer from yes to no when transmitting it to the next person and a probability β that he will change it from no to yes. Write down the state transition matrix of this process, then find its limiting state. Note that the initial state is the governor's choice.

8. A psychologist conducts and experiment in which 20 rats are placed at random in a compartment that has been divided into rooms labed 1, 2, and 3 as shown in Fig. 3.4. Observe that there are four doors in the arrangement. There are three possible states for the rats: they can be in rooms 1, 2, or 3. Let us assume that the rats move from room to room. A rat in room 1 has the probabilities $p_{11} = 0$, $P_{21} = \frac{1}{3}$, and $p_{31} = \frac{2}{3}$ of moving to the various rooms based on the distribution of doors. Predict the distribution of the rats in the long run. What is the limiting probability that a given marked rat will be in room 2?

9. In Example 3.30 (Drunkard's walk), assume that the probability of a step to the right is $\frac{2}{3}$, and a step to the left is $\frac{1}{3}$. Write down the transition matrix and determine $\lim_{n\to\infty} p(n)$.

10. Three tanks fight a three-way duel. Tank A has probability $\frac{1}{2}$ of destroying the tank at which it fires. Tank B has probability $\frac{1}{3}$ of destroying the tank at which it fires, and Tank C has probability $\frac{1}{6}$ of destroying the tank at which it fires. The tanks fire simultaneously and each tank fires at the strongest opponent not yet destroyed. Form a Markov chain by taking as states the subsets of the sets of tanks. Set up a Markov chain model and interpret your

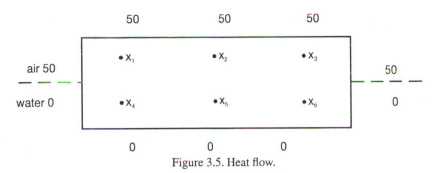

Figure 3.5. Heat flow.

results. (Hint: take as states ABC, AC, BD, A, B, C, and none, indicating the tanks that could survive starting in state ABC. You may omit the state AB because it can't be reached by ABC.)

11. In the trade model (Example 3.40) let $a_{11} = 0.4$, $a_{21} = 0.5$, $a_{12} = 0.3$, $a_{22} = 0.6$, $i_1 = 25$ billion dollars, and $i_2 = 20$ billion dollars. If $y_1(n)$ and $y_2(n)$ denote the national income of countries 1 and 2 in year n, respectively, and $y_1(0) = 500$ billion dollars and $y_2(0) = 650$ billion dollars, find $y_1(3)$ and $y_2(3)$. What are the equilibrium national incomes for nations 1 and 2?

12. Develop a mathematical model for a foreign trade model among three countries using an argument similar to that used in Example 3.31.

13. In Example 3.41, let $k = 4$, $\alpha = 0.2$, and $x_0(n) = T_5(n) = 0°C$. Compute $T_i(n)$, $1 \leq i \leq 4$, for $n = 1, 2, 3$, then find $\lim_{n \to \infty} T_i(n)$.

14. Suppose we have a grid of six points on a bar as shown in Fig. 3.5. Part of the bar is in air that is kept at a constant temperature of 50 degrees, and part of the bar is submerged in a liquid that is kept at a constant temperature of 0 degrees. Assume that the temperature at the point x_i, $1 \leq i \leq 6$, depends only on the temperature of the four nearest points, that is, the points above, below, to the left, and to the right.

   (i) Write a mathematical model that describes the flow of heat in this bar.

   (ii) Find the equilibrium temperature at the six points x_i.

References

[1] R.P. Agarwal, *Difference Equations and Inequalities*, Marcel Dekker Inc., New York, 1992.

[2] W.G. Kelley and A.C. Peterson, *Difference Equations, An Introduction with Applications*, Academic, New York, 1991.

[3] V. Lakshmikantham and D. Trigiante, *Theory of Difference Equations: Numerical Methods and Applications*, Academic, New York, 1988.

[4] S. Goldberg, *Introduction to Difference Equations*, Wiley & Sons, New York, 1958.

[5] J.M. Ortega, *Matrix Theory, A Second Course*, Plenum, New York, 1987.

[6] G.D. Smith, *Numerical Solution of Partial Differential Equations: Finite Difference Methods*, Third Edition, Claredon, Oxford, 1985.

4

Stability Theory

In Chapter 1, we studied the stability properties of first order difference equations. In this chapter, we will develop the theory for k-dimensional systems of first order difference equations. As shown in Chapter 3, this study includes difference equations of any order. Here, we are interested in the qualitative behavior of solutions without actually computing them. Realizing that most of the problems that arise in practice are nonlinear and mostly unsolvable, this investigation is of vital importance to scientists, engineers, and applied mathematicians.

In this chapter, we adapt the differential methods and techniques of Liapunov [1], Perron [2] and many others to difference equations. First, we give definitions of various notions of stability and some simple examples to illustrate them in Section 4.1. In Section 4.2, we study the stability of linear and periodic systems. Iterative methods are given as an application of stability theory. Special methods for investigating scalar equations are established in Section 4.3. In Section 4.4, we study the geometrical properties of planar linear systems by means of the phase space analysis. Section 4.5 describes the method of linear approximation in stability theory. Section 4.5 also introduces to the reader the basic theory of the direct method of Liapunov, by far the most advanced topic in this chapter. Due to the enormity of the existing literature on Liapunov Theory, we have limited our exposition to autonomous equations.

4.1   Preliminaries

We start this section by introducing the notion of norms of vectors and matrices.

(i) the L_1 norm

$$\|x\|_1 = \sum_{i=1}^{k} |x_i|$$

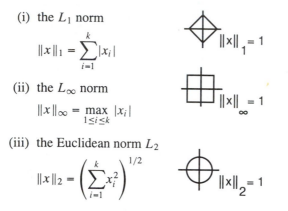

$\|x\|_1 = 1$

(ii) the L_∞ norm

$$\|x\|_\infty = \max_{1 \le i \le k} |x_i|$$

$\|x\|_\infty = 1$

(iii) the Euclidean norm L_2

$$\|x\|_2 = \left(\sum_{i=1}^{k} x_i^2\right)^{1/2}$$

$\|x\|_2 = 1$

Figure 4.1. A circle in different norms.

Definition 4.1. A real valued function on a vector space V is called a *norm* and is denoted by $\|\ \|$, if the following properties hold:

(i) $\|x\| \ge 0$ and $\|x\| = 0$ only if $x = 0$;

(ii) $\|\alpha x\| = |\alpha| \|x\|$ for all $x \in V$ and scalars α;

(iii) $\|x + y\| \le \|x\| + \|y\|$ for all $x, y \in V$.

The three most commonly used norms on R^k (k-dimensional complex space) are shown in Fig. 4.1.

We remark here that all norms on R^k are equivalent in the sense that if $\|\ \|$, $\|\ \|'$ are any two norms then there exists constants $\alpha, \beta > 0$ such that

$$\alpha\|x\| \le \|x\|' \le \beta\|x\|.$$

Thus if $\{x_n\}$ is a sequence in R^k, then $\|x_n\| \to 0$ as $n \to \infty$ if and only if $\|x_n\|' \to 0$ as $n \to \infty$.

Corresponding to each vector norm $\|\ \|$ on R^k one may define an operator norm $\|\ \|$ on a $k \times k$ matrix A as

$$\|A\| = \max_{\|x\| \ne 0} \frac{\|Ax\|}{\|x\|}. \tag{4.1.1}$$

It may be shown easily that

$$\|A\| = \max_{\|x\| \le 1} \|Ax\| = \max_{\|x\| = 1} \|Ax\|. \tag{4.1.1}'$$

Using this definition one may easily compute $\|A\|$ relative to the above three norms as shown in Table 4.1. (For a proof see [3]).

Table 4.1. Vector and matrix norms.

norm	$\|x\|$	$\|A\|$
L_1	$\sum_{i=1}^{k} \|x_i\|$	$\max_{1 \leq j \leq k} \sum_{i=1}^{k} \|a_{ij}\|$
L_∞	$\max_{1 \leq i \leq k} \|x_i\|$	$\max_{1 \leq i \leq k} \sum_{j=1}^{k} \|a_{ij}\|$
L_2	$\left(\sum_{i=1}^{k} x_i^2\right)^{\frac{1}{2}}$	$\left[\rho(A^T A)\right]^{\frac{1}{2}}$

From Eq. (4.1.1) we may deduce that for any operator norm on A, (Exercise 4.1, Problem 5)

$$\rho(A) \leq \|A\|, \tag{4.1.2}$$

where $\rho(A) = \max\{|\lambda| : \lambda \text{ is an eigenvalue of } A\}$ is the spectral radius of A.

Now if A is a Markov matrix (i.e., $\sum_{j=1}^{k} a_{ij} = 1, 1 \leq i \leq k, a_{ij} \geq 0$), then it follows from Table 4.1 that $\|A\|_\infty = 1$ which by Inequality (4.1.2) yields

$$\rho(A) \leq 1. \tag{4.1.3}$$

Now let us consider the vector difference equation

$$x(n + 1) = f(n, x(n)), \qquad x(n_0) = x_0, \tag{4.1.4}$$

where $x(n) \in R^k$, $f : Z^+ \times R^k \to R^k$. We assume that $f(n, x)$ is continuous in x. Recall that Eq. (4.1.4) is said to be *autonomous* or time-invariant if the varible n does not appear explicitly in the right-hand side of the equation ($f(n, x(n)) \equiv f(x(n))$). It is said to be *periodic* if for all $n \in Z$, $f(n + N, x) = f(n, x)$ for some positive integer N.

A point x^* in R^k is called an *equilibrium point* of Eq. (4.1.4) if $f(n, x^*) = x^*$ for all $n \geq n_0$. In most of the literature x^* is assumed to be the origin 0 and is called the zero solution. The justification for this assumption is as follows: Let $y(n) = x(n) - x^*$. Then Eq. (4.1.4) becomes

$$y(n + 1) = f(n, y(n) + x^*) - x^* = g(n, y(n)). \tag{4.1.5}$$

Notice that $y = 0$ corresponds to $x = x^*$. Since in many cases it is not convenient to make this change of coordinates, we will not assume that $x^* = 0$ unless it is more convenient to do so.

Recall that in Chapter 3 we dealt with the existence and uniqueness of solutions of linear systems, that is, when $f(n, x(n)) = A(n)x(n)$, where $A(n)$ is a $k \times k$ matrix. The existence and uniqueness of solutions of Eq. (4.1.4) may be established in a similar fashion (Exercise 4.1, Problem 19).

We are now ready to introduce the various stability notions of the equilibrium point x^* of Eq. (4.1.4).

Definition 4.2. The equilibrium point x^* of Eq. (4.1.4) is said to be

(i) *Stable* (S) if given $\varepsilon > 0$ and $n_0 \geq 0$ there exists $\delta = \delta(\varepsilon, n_0)$ such that $\|x_0 - x^*\| < \delta$ implies $\|x(n, n_0, x_0) - x^*\| < \varepsilon$ for all $n \geq n_0$, *uniformly stable* (US) if δ may be chosen independent of n_0, *unstable* if it is not stable;

(ii) *Attractive* (A) if there exists $\mu = \mu(n_0)$ such that $\|x_0 - x^*\| < \mu$ implies $\lim_{n \to \infty} x(n, n_0, x_0) = x^*$, *uniformly attractive* (UA) if the choice of μ is independent of n_0. The condition for *uniform attractivity* may be paraphrased by saying there exists $\mu > 0$ such that for every ε and n_0 there exists $N = N(\varepsilon)$ independent of n_0, such that $\|x(n, n_0, x_0) - x^*\| < \varepsilon$ for all $n \geq n_0 + N$ whenever $\|x_0 - x^*\| < \mu$.

(iii) *Asymptotically stable* (AS) if it is stable and attractive, and *uniformly asymptotically stable* (UAS) if it is uniformly stable and uniformly attractive;

(iv) *Exponentially stable* (ES) if there exist $\delta > 0$ $M > 0$, $\eta \in (0, 1)$ such that $\|x(n, n_0, x_0) - x^*\| \leq M\|x_0 - x^*\|\eta^{n-n_0}$, whenever $\|x_0 - x^*\| < \delta$.

(v) A solution $x(n, n_0, x_0)$ is bounded if for some positive constant M, $\|x (n, n_0, x_0)\| \leq M$ for all $n \geq n_0$ where M may depend on each solution.

If in part (iii) $\mu = \infty$ as in part (iv) $\delta = \infty$, the corresponding stability property is said to be *global*. In Fig. 4.2, we suppress the (time) n and show only the movement of a solution that starts inside a ball of radius δ. The figure illustrates that all future states $x(n, n_0, x_0)$, $n \geq n_0$ will stay within the ε ball. This diagram is called a *phase space* portrait, and will be used extensively in later sections. In Fig. 4.3, the time n is considered part of a three-dimensional coordinate system that provides another perspective on stability. Figure 4.4 depicts the uniform asymptotic stability of the zero solution.

Note that in the above definitions, some of the stability properties automatically imply one or more of the others. Figure 4.5 shows the hierarchy of the stability notions.

Important Remark

In general, none of the arrows in Fig. 4.5 may be reversed. However, we will show in the sequel that

(i) For *linear systems*, $UAS \leftrightarrow ES$.

(ii) For *autonomous (time-invariant) systems*, $S \leftrightarrow US$, $AS \leftrightarrow UAS$, and $A \leftrightarrow UA$. (Exercise 4.2, Problem 5). This is also true for periodic systems but the proof is omitted (see [4] for a proof).

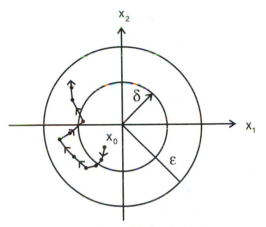

Figure 4.2. Stable equilibrium in phase space.

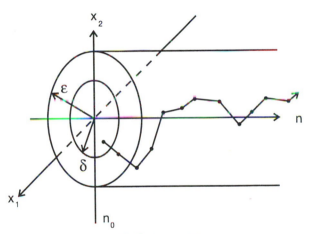

Figure 4.3. Stable equilibrium.

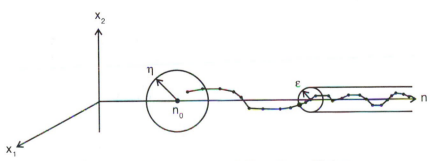

Figure 4.4. Uniformly asymptotically stable equilibrium.

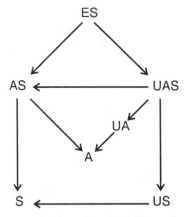

Figure 4.5. Hierarchy of stability notions.

The following examples serve to illustrate the definitions.

1. The solution of the scalar equation $x(n+1) = x(n)$ is given by $x(n, n_0, x_0) = x_0$, hence the zero solution is uniformly stable, but not asymptotically stable.

2. The solutions of the scalar equation $x(n + 1) = a(n)x(n)$ are

$$x(n, n_0, x_0) = \left[\prod_{i=n_0}^{n-1} a(i)\right] x_0. \tag{4.1.6}$$

Hence one may conclude the following:

(i) The zero solution is stable if and only if

$$\left|\prod_{i=n_0}^{n-1} a(i)\right| \le M(n_0) \equiv M, \tag{4.1.7}$$

where M is a positive constant that depends on n_0 (Exercise 4.1, Problem 9). This condition holds if $a(i) = (1 + \eta^i)$, where $0 < \eta < 1$.

To show this we write the solution as $x(n, n_0, x_0) = \Phi(n)x_0$, where $\Phi(n) = \prod_{i=n_0}^{n-1}(1 + \eta^i)$. Since $1 + \eta^i < \mathrm{Exp}(\eta^i)$, it follows that

$$\Phi(n) \le \mathrm{Exp}\left(\sum_{i=n_0}^{n-1} \eta^i\right) \le \mathrm{Exp}\left(\sum_{i=n_0}^{\infty} \eta^i\right) \le \mathrm{Exp}\left(\frac{\eta^{n_0}}{1 - \eta}\right)$$
$$= M(n_0) = M.$$

Given $\varepsilon > 0$ and $n_0 \ge 0$, if we let $\delta = \varepsilon/2M$, then $|x_0| < \delta$ implies $|x(n, n_0, x_0)| = \Phi(n)x_0 < \varepsilon$.

(ii) The zero solution is uniformly stable if and only if

$$\left|\prod_{i=n_0}^{n-1} a(i)\right| \leq M, \tag{4.1.8}$$

where M is a positive constant independent of n_0 (Exercise 4.1, Problem 10). This condition holds if $a(i) = \sin(i + 1)$.

(iii) The zero solution is asymptotically stable if and only if

$$\lim_{n \to \infty} \left|\prod_{i=n_0}^{n-1} a(i)\right| = 0 \text{ (Exercise 4.1, Problem 12)}. \tag{4.1.9}$$

This condition clearly holds if $a(i) = \frac{i+1}{i+2}$. The solution is given by $x(n, n_0, x_0) = (n_0 + 1)/(n + 1)x_0$. Thus, the zero solution is uniformly stable and asymptotically stable (globally), but not uniformly asymptotically stable (Why? See Exercise 4.1, Problem 13).

(iv) The zero solution is uniformly asymptotically stable (and thus exponentially stable) if and only if

$$\left|\prod_{i=n_0}^{n-1} a(i)\right| \leq M \eta^{n-n_0}, \tag{4.1.10}$$

for some $M > 0, 0 < \eta < 1$. This may be satisfied if $a(i) = 1/i$ (Exercise 4.1, Problem 15).

Now we give two important examples. In the first example we show that the zero solution is stable but not uniformly stable. In the second example the zero solution is attractive but not stable (personal communication by Professor B. Aulbach).

Example 4.3. The solution of the equation $x(n + 1) = \left(\frac{n+1}{2}\right)[x(n)]^2$ is given by

$$x(n, n_0, x_0) = \left(\frac{n}{2}\right)\left(\frac{n-1}{2}\right)^2\left(\frac{n-2}{2}\right)^4, \ldots, \left(\frac{n_0 + 1}{2}\right)^{2^{n-n_0-1}}(x_0)^{2^{n-n_0}},$$
$$x(n_0) = x_0.$$

If $|x_0|$ is sufficiently small, than $\lim_{n \to \infty} x(n) = 0$. Thus, the zero solution is attractive. However, it is not uniformly attractive. For if $\delta > 0$ is given and n_0 is chosen such that $(n_0 + 1)\delta^2 \geq 2$, then for $|x_0| = \delta$,

$$|x(n_0 + 1, n_0, x_0)| = \left(\frac{n_0 + 1}{2}\right)|x_0|^2 \geq 1.$$

Let us now check the stability of the zero solution. Given $\varepsilon > 0$ and $n_0 \geq 0$, let $\delta - \varepsilon/(n_0 + 1)$. If $|x_0| < \delta$, then $|x(n, n_0, x_0)| < \varepsilon$ for all $n \geq n_0$ (See Exercise 4.1, Problem 16). Since δ depends on the choice of n_0, the zero solution is stable but not uniformly stable.

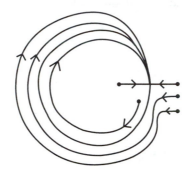

Figure 4.6. Attractive but not stable equilibrium.

Example 4.4. Consider the difference equation (in polar coordinates)

$$r(n+1) = \sqrt{r(n)}, \qquad r > 0,$$
$$\theta(n+1) = \sqrt{2\pi\,\theta(n)}, \qquad 0 \le \theta \le 2\pi.$$

We claim that the equilibrium point $(1,0)$ is attractive but not stable. To show this observe that

$$r(n) = (r_0)^{\frac{1}{2}n}, \qquad r_0 = r(0),$$
$$\theta(n) = (2\pi)^{\left(1-\left(\frac{1}{2}\right)^n\right)} \cdot \theta_0^{\frac{1}{2}^n}, \qquad \theta_0 = \theta(0).$$

Clearly $\lim_{n\to\infty} r(n) = 1$ and $\lim_{n\to\infty} \theta(n) = 2\pi$. Now if $r_0 \neq 0$, $\theta_0 = 0$, then $(r(n), \theta(n)) = (r_0^{\frac{1}{2}^n}, 0)$ which converges to the equilibrium point $(1,0)$. However, if $\theta_0 = 2\pi - \delta$, $0 < \delta < 1$, then the orbit of (r_0, θ_0) will spiral around the circle clockwise to converge to the equilibrium point $(1,0)$. Hence the equilibrium point $(1,0)$ is attractive but not stable. (See Fig. 4.6.)

Exercise 4.1.

1. Compute $\|A\|_1$, $\|A\|_\infty$, $\|A\|_2$, and $\rho(A)$ for the following matrices:

(a) $\begin{bmatrix} 2 & 1 \\ 1 & 2 \end{bmatrix}$,   (b) $\begin{bmatrix} 1 & 1 & 2 \\ 0 & 2 & -1 \\ 0 & 3 & 0 \end{bmatrix}$,   (c) $\begin{bmatrix} 2 & 1 & 0 \\ 0 & 2 & 0 \\ 0 & 3 & 4 \end{bmatrix}$.

2. Give an example of a matrix A such that $\rho(A) \neq \|A\|_\infty, \|A\|_1, \|A\|_2$.

3. Let

$$A = \begin{pmatrix} \lambda & 1 & 0 \\ 0 & \lambda & 1 \\ 0 & 0 & \lambda \end{pmatrix}.$$

Show that for each $\varepsilon > 0$ there exists a diagonal matrix D, such that $\|D^{-1}AD\| \le |\lambda| + \varepsilon$, for any operator norm $\| \; \|$. [Hint: Use $D = $ diag $(1, \varepsilon, \varepsilon^2)$.]

4. Generalize Problem 3 to any $k \times k$ matrix A in the Jordan form diag$(J_1, J_2, \ldots, J_r)$.

5. Prove that $\rho(A) \le \|A\|$ for any operator norm $\| \; \|$ on A.

6. Show that for any two norms $\| \; \|$, $\| \; \|'$ on R^k there are contents $\alpha, \beta > 0$ such that $\alpha \|x\| \le \|x\|' \le \beta \|x\|$.

7. Deduce from Problem 6 that for any sequence $\{x(n)\}$, $\|x(n)\| \to 0$ as $n \to \infty$ if and only if $\|x(n)\|' \to 0$.

8. Meditate upon the scalar equation $x(n + 1) = ax(n)$. Prove that

   (i) If $|a| < 1$, the zero solution is uniformly asymptotically stable.

   (ii) If $|a| = 1$, the zero solution is uniformly stable.

   (iii) If $|a| > 1$, the zero solution is not stable.

9. (a) Prove that the zero solution of the scalar equation $x(n+1) = a(n)x(n)$ is stable if and only if

$$\left| \prod_{i=n_0}^{n-1} a(i) \right| \le M(n_0),$$

   where M depends on n_0.

   (b) Show that the zero solution of the equation $x(n+1) = (1+\eta^n)x(n)$, $0 < \eta < 1$, is stable.

10. (a) Prove that the zero solution of the equation $x(n + 1) = a(n)x(n)$ is uniformly stable if and only if

$$\left| \prod_{i=n_0}^{n-1} a(i) \right| \le M,$$

   where M is a positive constant independent of n_0.

   (b) Show that the zero solution of the equation $x(n + 1) = \sin(n + 1)x(n)$ is uniformly stable.

11. Show that the zero solution of the equation $x(n + 1) = \frac{n+1}{n+2}x(n)$ is asymptotically stable.

12. Prove that the zero solution of the equation $x(n + 1) = a(n)x(n)$ is asymptotically stable if and only if

$$\lim_{n \to \infty} \left| \prod_{i=n_0}^{n-1} a(i) \right| = 0.$$

13. Show that the zero solution of the equation in Problem 11 is not uniformly asymptotically stable.

14. Prove that the zero solution of the equation $x(n) = a(n)x(n)$ is uniformly asymptotically stable if and only if

$$\left| \prod_{i=n_0}^{n-1} a(i) \right| \leq M \, \eta^{n-n_0},$$

for some $M > 0, 0 < \eta < 1$.

15. Show that the zero solution of the scalar equation $x(n+1) = (1/n)x(n), n \geq 1$, is uniformly asymptotically stable.

16. It was shown in Example 4.3 that the zero solution of the equation $x(n+1) = ((n + 1)/2)[x(n)]^2$ is attractive. Give a detailed argument to show that the zero solution is, however, not uniformly attractive.

17. Establish the existence and uniqueness of solutions of Eq. (4.1.4).

4.2   Stability of Linear Systems

In this section, we investigate the stability of the linear nonautonomous (time-variant) system given by

$$x(n + 1) = A(n)x(n), n \geq n_0 \geq 0. \tag{4.2.1}$$

It is always assumed that $A(n)$ is nonsingular for all $n \geq n_0$.

The results obtained for system (4.2.1) will be specialized subsequently for the autonomous (time-variant) linear system

$$x(n + 1) = Ax(n). \tag{4.2.2}$$

If $\Phi(n)$ is any fundamental matrix of System (4.2.1) or (4.2.2), then recall that $\Phi(n, m) = \Phi(n)\Phi^{-1}(m)$ is the state transition matrix. Furthermore, for System (4.2.2), $\Phi(n, m) = \Phi(n - m) = e^{A(n-m)}$. In the following result we express the conditions for stability in terms of a fundamental matrix $\Phi(n)$ of System (4.2.1).

Theorem 4.5. Consider System (4.2.1). Then its zero solution is

(i) stable if and only if there exists a positive constant M such that

$$\|\Phi(n)\| \leq M \text{ for } n \geq n_0 \geq 0; \tag{4.2.3}$$

(ii) uniformly stable if and only if there exists a positive constant M such that

$$\|\Phi(n, m)\| \leq M \text{ for } n_0 \leq m \leq n < \infty; \tag{4.2.4}$$

(iii) asymptotically stable if and only if

$$\lim_{n \to \infty} \|\Phi(n)\| = 0; \tag{4.2.5}$$

(iv) uniformly asymptotically stable if and only if there exist positive constants M and $\eta \in (0, 1)$ such that

$$\|\Phi(n, m)\| \leq M\eta^{n-m} \text{ for } n_0 \leq m \leq n < \infty. \tag{4.2.6}$$

Proof Without loss of generality we may assume that $\Phi(n_0) = I$, since Conditions 4.2.3 through 4.2.6 hold true for every fundamental matrix if they hold for one. Thus $x(n, n_0, x_0) = \Phi(n)x_0$.

(i) Suppose that Inequality (4.2.3) holds. Then $\|x(n, n_0, x_0)\| \leq M\|x_0\|$. So for $\varepsilon > 0$, let $\delta < M/\varepsilon$. Then $\|x_0\| < \delta$ implies $\|x(n, n_0, x_0)\| < \varepsilon$, and consequently, the zero solution is stable. Conversely, suppose that $\|x(n, n_0, x_0)\| = \|\Phi(n)x_0\| < \varepsilon$ whenever $\|x_0\| \leq \delta$. Observe that $\|x_0\| \leq \delta$ if and only if $\frac{1}{\delta}\|x_0\| \leq 1$. Hence

$$\|\Phi(n)\| = \sup_{\|\xi\| \leq 1} \|\Phi(n)\xi\| = \frac{1}{\delta} \sup_{\|x_0\| \leq \delta} \|\Phi(n)x_0\| \leq \frac{\varepsilon}{\delta} = M.$$

Parts (ii) and (iii) remain as Exercise 4.2, Problems 9 and 10.

(iv) Suppose finally that Inequality (4.2.6) holds. The zero solution of system (4.2.1) would then be uniformly stable by part (ii). Furthermore, for $\varepsilon > 0, 0 < \varepsilon < M$, take $\mu = 1$ and N such $\eta^N < \varepsilon/M$. Hence, if $\|x_0\| < 1$, then $\|x(n, n_0, x_0)\| = \|\Phi(n, n_0)x_0\| \leq M\eta^{n-n_0} < \varepsilon$ for $n \geq n_0 + N$. The zero solution would be uniformly asymptotically stable. Conversely, suppose that the zero solution is uniformly asymptotically stable. It is also then uniformly stable and thus by part of our Theorem 4.4 (ii), $\|\Phi(n, m)\| \leq M$ for $0 \leq n_0 \leq m \leq n < \infty$. From uniform attractivity, there exist $\eta > 0$ such that for ε with $0 < \varepsilon < 1$ there exists N such that $\|\Phi(n, n_0)x_0\| < \varepsilon$ for

$n \geq n_0 + N$ whenever $\|x_0\| < \mu$. Then for $n \epsilon [n_0 + mN, n_0 + (m+1)N]$, $m > 0$, we have

$$
\begin{aligned}
\|\Phi(n, n_0)\| &\leq \|\Phi(n, n_0 + mN)\| \, \|\Phi(n_0 + mN, n_0 + (m-1)N) \\
&\quad \| \cdots \|\Phi(n_0 + N, n_0)\| \\
&\leq M\varepsilon^m \leq \frac{M}{\varepsilon} \left(\varepsilon^{\frac{1}{N}} \right)^{(m+1)N} = \tilde{M}\eta^{(m+1)N}, \\
&\quad \text{where } \tilde{M} = \frac{M}{\varepsilon}, \eta = \varepsilon^{\frac{1}{N}}, \\
&\leq \tilde{M}\eta^{(n-n_0)}, \text{ for } mN \leq n - n_0 \leq (m+1)N.
\end{aligned}
$$

This concludes the proof of the theorem.

The following result arises as an immediate consequence of the above theorem. [See Remark 4.3 part (i).]

Corollary 4.6. For System (4.2.1) the following statements hold:

(i) The zero solution is stable if and only if all solutions are bounded.

(ii) The zero solution is exponentially stable if and only if it is uniformly asymptotically stable.

Proof Statements (i) and (ii) follow immediately from Conditions (4.2.4) and (4.2.6), respectively, (Exercise 4.2, Problem 6).

The following is another *important consequence* of Theorem 4.5:

Corollary 4.7. For System (4.2.1), every local stability property of the zero solution implies the corresponding global stability property.

Proof Use Theorem 4.5. (Exercise 4.2, Problem 6).

We not give a simple but powerful criteria for uniform stability and uniform asymptotic stability.

Theorem 4.8. [5]

(i) If $\sum_{i=1}^{k} |a_{ij}(n)| \leq 1, 1 \leq j \leq k, n \geq n_0$, then the zero solution of System (4.2.1) is uniformly stable.

(ii) If $\sum_{i=1}^{k} |a_{ij}(n)| \leq 1 - v$, for some $v > 0, 1 \leq j \leq k, n \geq n_0$, then the zero solution is uniformly asymptotically stable.

Proof

(i) From condition (i), in Theorem 4.7, $\|A(n)\|_1 \leq 1$ for all $n \geq n_0$. Thus,

$$
\|\Phi(n, m)\|_1 = \left\| \prod_{i=m}^{n-1} A(i) \right\|_1 \leq \|A(n-1)\|_1 \|A(n-2)\|_1, \ldots, \|A(m)\|_1 \leq 1.
$$

This now implies uniform stability by Theorem 4.5, part (ii).

(ii) The proof of statement (ii) is so similar to the proof of statement (i) that we will omit it here.

In the next theorem, we summarize the main stability results for linear autonomous systems (4.4.2).

Theorem 4.9. The following statements hold.

(i) The zero solution of Eq. (4.2.2) is stable if and only if $\rho(A) \leq 1$ and the eigenvalues of unit modules are semisimple.

(ii) The zero solution of Eq. (4.2.2) is asymptotically stable if and only if $\rho(A) < 1$.

Proof

(i) Let $A = PJP^{-1}$, where $J = \text{diag}(J_1, J_2, \ldots, J_r)$ is the Jordan form of A and

From Theorem 4.4 the zero solution of Eq. (4.2.2) is stable if and only if $\|A^n\| = \|P J^n P^{-1}\| \leq M$ or $\|J^n\| \leq \tilde{M}$, where $\tilde{M} = M/(\|P\| \|P^{-1}\|)$. Now $J^n = \text{diag}(J_1^n, J_2^n, \ldots, J_r^n)$, where

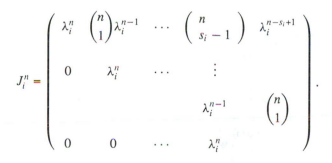

Obviously, J_i^n becomes unbounded if $|\lambda_i| > 1$ or $|\lambda_i| = 1$ and J_i is not 1×1. If $|\lambda_i| < 1$, then $J_i^n \to 0$ as $n \to \infty$. To prove this conclusion it suffices to show that $|\lambda_i|^n n^\ell \to 0$ as $n \to \infty$ for any positive integer ℓ. This conclusion follows from L'Hopital rule since $|\lambda_i|^n n^\ell = n^\ell e^{(\ln |\lambda_i|)n}$. (Exercise 4.2, Problem 8).

(ii) The proof of statement (ii) has already been established by the above argument. This completes the proof of the theorem.

We now use the above result to investigate the stability of the periodic system

$$x(n + 1) = A(n)x(n), \qquad A(n + N) = A(n) \qquad (4.2.7)$$

Recall from Chapter 3 that if $\Phi(n, n_0)$ is a fundamental matrix of Eq. (4.2.7), then there exists a constant matrix B, whose eigenvalues are called the Floquet exponents, and a period matrix $P(n, n_0)$ such that $\Phi(n, n_0) = P(n, n_0)B^{n-n_0}$ where $P(n+N, n_0) = P(n, n_0)$. Thus if B^n is bounded, then so is $\Phi(n, n_0)$ and if $B^n \to 0$ as $n \to \infty$, then it follows that $\Phi(n, n_0) \to 0$ as $n \to \infty$. This proves the following result.

Theorem 4.10. The zero solution of Eq. (4.2.7) is

(i) stable if and only if the Floquet exponents have modulus less than or equal to 1; those with modulus of 1 are semisimple;

(ii) asymptotically stable if and only if all the Floquet exponents lie inside the unit disk.

For practical purposes, the following corollary is of paramount importance.

Corollary 4.11. The zero solution of Eq. (4.2.7) is

(i) stable if each eigenvalue of the matrix $C = A(N-1)A(N-2), \ldots, A(0)$ has modulus less than or equal to 1; those solutions with modulus of value 1 are semisimple;

(ii) asymptotically stable if each eigenvalue of $C = A(N-1)A(N-2), \ldots, A(0)$ has modulus less than one.

Let us summarize our learning thus far. First, for the autonomous (time-invariant) linear system $x(n + 1) = Ax(n)$, the eigenvalues of A determine the stability properties of the system (Theorem 4.9). But, for a *periodic system* $x(n + 1) = A(n)x(n)$, the eigenvalues of $A(n)$ do not play any role in the determination of the stability properties of the system. Instead, the *Floquet multipliers* of $A(n)$ determine those properties. The following example should dispel any wrong ideas concerning the role of eigenvalues in a nonautonomous system.

Example 4.12. Let us again consider the periodic system in Example 3.36 where

$$A(n) = \begin{pmatrix} 0 & \dfrac{2 + (-1)^n}{2} \\ \dfrac{2 + (-1)^n}{2} & 0 \end{pmatrix}.$$

Here the eigenvalues of A are $\pm\sqrt{3}/2$ and thus $\rho[A(n)] < 1$. By applying Corollary 4.11, one may quickly check the stability of this system. We have

$$C = A(1)A(0) = \begin{pmatrix} 0 & 3/2 \\ 1/2 & 0 \end{pmatrix}\begin{pmatrix} 0 & 1/2 \\ 3/2 & 0 \end{pmatrix} = \begin{pmatrix} 1/4 & 0 \\ 0 & 9/4 \end{pmatrix}.$$

Hence, by Corollary 4.11, all solutions are unbounded.

For the eager reader, perpetually searching for a challenge, we might determine the stability by explicitly transcribing the fundamental matrix as follows:

$$\Phi(n) = \begin{pmatrix} \dfrac{2^{1-n} - (-2)^{1-n}}{2} & \dfrac{\left(\frac{3}{2}\right)^n - \left(-\frac{3}{2}\right)^n}{2} \\[4mm] \dfrac{2^{-n} - (-2)^{-n}}{2} & \dfrac{\left(\frac{3}{2}\right)^n - \left(-\frac{3}{2}\right)^n}{2} \end{pmatrix}.$$

Hence, all the solutions are unbounded. Consequently, the zero solution is unstable. This example demonstrates without any doubt that eigenvalues do not provide any information about the stability of nonautonomous difference systems.

Example 4.13. (Iterative Methods). Consider the system of linear algebraic equations

$$Ax = b \tag{4.2.8}$$

where $A = (a_{ij})$ is a $k \times k$ matrix.

Iterative methods are used widely to solve Eq. (4.2.8) numerically. We generate a sequence $x(n)$ using the difference equation

$$x(n + 1) = Bx(n) + d \tag{4.2.9}$$

where the choice of B and d depend on the iterative method used. The iterative method [Eq. (4.2.9)] is *consistent* with Eq. (4.2.8) if a solution x^* of Eq. (4.2.8) is an equilibrium point of Eq. (4.2.9), i.e., if

$$Bx^* + d = x^*. \tag{4.2.10}$$

We now describe one such consistent method, *Jacobi Iterative Method*. Assuming the diagonal elements a_{ii} of A are nonzero, then $D = \operatorname{diag}(a_{11}, a_{22}, \ldots, a_{kk})$ is nonsingular. In Eq. (4.2.9), define

$$B = I - D^{-1}A, \qquad d = D^{-1}b. \tag{4.2.11}$$

This method is consistent (Exercise 4.2, Problem 11). If A is nonsingular, then x^* is unique. The associated error equation may be derived by letting $e(n) = x(n) - x^*$. Equations (4.2.9) and (4.2.10) then yield the equation

$$e(n + 1) = Be(n). \tag{4.2.12}$$

The quantity $e(n)$ represents the error in approximating the solution x^* by the nth iterate $x(n)$ of Eq. (4.2.9). Clearly, $x(n) \to x^*$ as $n \to \infty$ if and only if $e(n) \to 0$ as $n \to \infty$. In other words, the iterative method (Eq. (4.2.9)) converges to the solution x^* of Eq. (4.2.8) if and only if the zero solution of Eq. (4.2.12) is asymptotically stable. Using Theorem 4.9, we conclude that the iterative method converges if and only if $\rho(B) < 1$.

Exercise 4.2.

1. Determine whether the zero solution of the system $x(n + 1) = Ax(n)$ is stable, asymptotically stable, or unstable if the matrix A is

   (a) $\begin{pmatrix} 1 & 0 \\ -2 & 1 \end{pmatrix}$,   (b) $\begin{pmatrix} \frac{5}{12} & 0 & \frac{1}{2} \\ -1 & -\frac{1}{2} & \frac{5}{4} \\ \frac{1}{3} & 0 & 0 \end{pmatrix}$

   (c) $\begin{pmatrix} -1 & 5 \\ -0.5 & 2 \end{pmatrix}$,   (d) $\begin{pmatrix} 1.5 & 1 & -1 \\ -1.5 & -0.5 & 1.5 \\ 0.5 & 1 & 0 \end{pmatrix}$.

2. Give another example (see Example 4.11) of a matrix $A(n)$ such that $\rho[A(n)] < 1$ and the zero solution of $x(n + 1) = A(n)$ is unstable.

3. Give an example of a stable matrix A (i.e., $\rho(A) < 1$) with $\|A\| > 1$, for some matrix norm $\| \; \|$.

4. Consider the autonomous (time-invariant) system (4.2.2). Prove the following statements.

   (i) The zero solution is stable if and only if it is uniformly stable.

   (ii) The zero solution is asymptotically stable if and only if it is uniformly asymptotically stable.

5. Use Theorem 4.8 to determine whether or not the zero solution of $x(n+1) = A(n)x(n)$ is uniformly stable or uniformly asymptotically stable, where $A(n)$ is the matrix

   (a) $\begin{pmatrix} -1 & \frac{1}{4}\cos(n) \\ 0 & \frac{1}{2}\sin(n) \end{pmatrix}$ (b) $\begin{pmatrix} \frac{n}{n+1} & 0 \\ -1 & 1 \end{pmatrix}$

   (c) $\begin{pmatrix} \frac{1}{n+1} & 0 & \frac{1}{2}\sin(n) \\ \frac{1}{4} & \frac{1}{2}\sin(n) & \frac{1}{4}\cos(n) \\ \frac{1}{5} & 0 & 0 \end{pmatrix}$ (d) $\begin{pmatrix} \frac{n+2}{n+1} & 0 & 0 \\ 0 & 1 & 0 \\ \frac{1}{n+1} & 0 & 1 \end{pmatrix}$.

6. Prove Corollary 4.7.

7. Prove Corollary 4.6.

8. Show that if $|\lambda| < 1$, then $\lim_{n \to \infty} |\lambda|^n n^s = 0$ for any given positive integer s.

9. Prove that the zero solution of System (4.2.1) is uniformly stable if and only if there exists $M > 0$ such that $\|\Phi(n, m)\| \le M$, for $n_0 \le m \le n < \infty$.

10. Prove that the zero solution of System (4.2.1) is asymptotically stable if and only if $\lim_{n \to \infty} \|\Phi(n)\| = 0$.

11. Show that the Jacobi Iterative Method is consistent.

12. Consider Eqs. (4.2.8) and (4.2.9) with the assumption that the diagonal elements of A are nonzero. Let L be the lower triangular part of A and U be the strictly upper triangular part of A (i.e., the main diagonal of U is zero). Then $A = L + U$. The Gauss Siedel iterative method defines $B = -L^{-1}U$, and $d = L^{-1}b$.

In problems 13 and 14 we are considering the k-dimensional system $x(n + 1) = A(n)x(n)$.

13. Define $H(n) = A^T(n)A(n)$. Prove the *Lagrange identity*.

$$\|x(n + 1)\|_2^2 = x^T(n + 1)x(n + 1) = x^T(n)H(n)x(n). \qquad (4.2.12)$$

14.* (a) Show that all eigenvalues of $H(n)$ as defined in Problem 14 are real and nonnegative.

(b) Let the eigenvalues of $H(n)$ be ordered as $\lambda_1(n) \le \lambda_2 \le \cdots \le \lambda_k(n)$. Show that for all $x \in R^n$,

$$\lambda_1(n)x^T x \le x^T H(n)x \le \lambda_k(n)x^T. \qquad (4.2.13)$$

(c) Use Lagrange Identity (4.2.12) in Formula (4.2.13) to show that

$$\prod_{i=n_0}^{n-1} \lambda_1(i)x^T(n_0)x(n_0) \le x^T(n)x(n) \le \prod_{i=n_0}^{n-1} \lambda_k(i)x^T(n_0)x(n_0).$$

(d) Show that $\displaystyle\prod_{i=n_0}^{n-1} \lambda_1(i) \le \|\Phi(n, n_0)\|^2 \le \prod_{i=n_0}^{n-1} \lambda_k(i). \qquad (4.2.14)$

4.3   Scalar Equations

Consider the kth order scalar equation

$$x(n+k) + p_1 x(n+k-1) + p_2 x(n+k-2) + \cdots + p_k x(n) = 0 \qquad (4.3.1)$$

where the p_i's are real numbers.

It follows from Lemma 2.24 and Corollary 2.25 that the zero solution of Eq. (4.3.1) is asymptotically stable if and only if $|\lambda| < 1$ for every characteristic root λ of Eq. (4.3.1), that is, for every zero of the characteristic polynomial

$$p(\lambda) = \lambda^k + p_1 \lambda^{k-1} + \cdots + p_k. \qquad (4.3.2)$$

Furthermore, the zero solution of Eq. (4.3.1) is stable if and only if $|\lambda| \leq 1$ and those characteristic roots λ with $|\lambda| = 1$ are simple. On the other hand, if there is a multiple characteristic root λ with $|\lambda| = 1$, then the zero solution of Eq. (4.3.1) is unstable according to Lemma 2.24.

We present the Schur–Cohn criterion which defines the conditions for the characteristic roots of Eq. (4.3.1) to fall inside the unit circle. Hence, this criterion is also a criterion for asymptotic stability. But before starting the Schur–Cohn criterion we need to introduce few preliminaries.

First let us define the *inners* of a matrix $B = (b_{ij})$. The inners of a matrix are the matrix itself and all the matrices obtained by omitting successively the first and last rows and the first and last columns. For example, the inners for the following matrices are highlighted

A 3 × 3 matrix   **A 4 × 4 matrix**

$$\begin{pmatrix} b_{11} & b_{12} & b_{13} \\ b_{21} & \boxed{b_{22}} & b_{23} \\ b_{31} & b_{32} & b_{33} \end{pmatrix} \qquad \begin{pmatrix} b_{11} & b_{12} & b_{13} & b_{14} \\ b_{21} & \boxed{\begin{matrix} b_{22} & b_{23} \\ b_{32} & b_{33} \end{matrix}} & & b_{24} \\ b_{31} & & & b_{34} \\ b_{41} & b_{42} & b_{43} & b_{44} \end{pmatrix}$$

A 5 × 5 matrix

$$\begin{pmatrix} b_{11} & b_{12} & b_{13} & b_{14} & b_{15} \\ b_{21} & b_{22} & b_{23} & b_{24} & b_{25} \\ b_{31} & b_{32} & \boxed{b_{33}} & b_{34} & b_{35} \\ b_{41} & b_{42} & b_{43} & b_{44} & b_{45} \\ b_{51} & b_{52} & b_{53} & b_{54} & b_{55} \end{pmatrix}$$

A matrix B is said to be *positive innerwise* if the determinants of all of its inners are positive.

Theorem 4.12. (Jury [6]). The zeros of the characteristic polynomial (4.3.2) lie inside the unit circle if and only if the following hold:

(i) $p(1) > 0$,

(ii) $(-1)^k p(-1) > 0$,

(iii) the $(k-1)x(k-1)$ matrices

$$B_{k-1}^{\pm} = \begin{pmatrix} 1 & 0 & \cdots & & 0 \\ p_1 & 1 & \cdots & & 0 \\ \vdots & & & & \vdots \\ p_{k-3} & & & & \\ p_{k-2} & p_{k-3} & \cdots & p_1 & 1 \end{pmatrix} \pm \begin{pmatrix} 0 & 0 & \cdots & 0 & p_k \\ 0 & 0 & \cdots & p_k & p_{k-1} \\ \vdots & \vdots & & & \vdots \\ 0 & p_k & & & p_3 \\ p_k & p_{k-1} & \cdots & p_3 & p_2 \end{pmatrix}$$

are positive innerwise.

Example 4.13. Consider the equation

$$x(n+2) + p_1 x(n+1) + p_2 x(n) = 0. \tag{4.3.3}$$

Its characteristic polynomial is given by $p(\lambda) = \lambda^2 + p_1\lambda + p_2$.

Let us attempt to discover the conditions that make the zero solution of Eq. (4.3.3) asymptotically stable. Using the Schur–Cohn criterion, we require

$$p(1) = 1 + p_1 + p_2 > 0 \text{ and } (-1)^2 p(-1) = 1 - p_1 + p_2 > 0.$$

This implies that $1 + p_2 > 0$. From condition (iii) we stipulate that $1 - p_2 > 0$ or $p_2 < 1$.

Thus a necessary and sufficient condition for the zero solution of Eq. (4.3.3) to be asymptotically stable is

(i) $1 + p_1 + p_2 > 0$,

(ii) $1 - p_1 + p_2 > 0$,

(iii) $p_2 < 1$.

We will now consider a special class of kth order scalar equations.

Example 4.14. Consider the kth order scalar equation

$$x(n+k) + px(n+k-1) + qx(n) = 0. \tag{4.3.4}$$

Find conditions under which the zero solution of Eq. (4.3.4) is asymptotically stable or unstable.

Solution The characteristic polynomial of Eq. (4.3.4) is $p(\lambda) = \lambda^k + p\lambda^{k-1} + q$.

We must show that all the zeros z of $p(\lambda)$ lie inside the unit circle. The most effective method would be to apply Rouche's Theorem from Complex Analysis [7].

Theorem 4.15 [Rouche's Theorem]: If the complex functions $f(z)$ and $g(z)$ are analytic inside and on a simple closed curve γ and if $|g(z)| < |f(z)|$ on γ, then $f(z)$ and $f(z) + g(z)$ have the same number of zeros inside γ.

We take γ to be the unit circle $|z| = 1$. The function $g(z) = pz^{k-1}$, while $f(z) = z^k$.

Then $|g(z)| \leq |p| + |q|$, and $|f(z)| = 1$. Now if

$$|p| + |q| < 1 \tag{4.3.5}$$

then $|g(z)| < f(z)$ on γ. By Rouche's Theorem, it follows that $p(z) = f(z) + g(z)$, and $f(z) = z^k$ have the same number of zeros inside the unit circle. But $f(z) = z^k = 0$ has k repeated zeros $z = 0$ inside the unit circle. Hence all the k zeros of $p(\lambda)$ lie inside the unit circle, and consequently, the zero solution of Eq. (4.3.4) is asymptotically stable.

The reader will extend this result to the general kth order scalar equation (Exercise 4.3, Problem 5).

To find conditions under which the zero solution of Eq. (4.3.4) is unstable, we let $f(z) = pz^{q-1}$ and $g(z) = z^k + q$. Then on the unit circle $|z| = 1$, $|f(z)| = |p|$, $|g(z)| \leq 1 + |q|$. Suppose that

$$|p| - |q| > 1. \tag{4.3.6}$$

Then on the unit circle, $|g(z)| < |f(z)|$. By Rouche's Theorem, $f(z)$ and $p(z) = f(z) + g(z)$ have the same number of zeros inside the unit circle, namely, none. Thus, under Condition (4.3.6), the zero solution of Eq. (4.3.4) is unstable.

Levin and May [8] considered the equation

$$x(n + k) - x(n + k - 1) + qx(n) = 0. \tag{4.3.7}$$

They concluded that the zero solution of Eq. (4.3.7) is asymptotically stable if and only if

$$0 < q < 2\cos\frac{(k - 1)\pi}{2k - 1} \tag{4.3.8}$$

where $k = 1, 2, 3, \ldots$.

Example 4.15. Consider the second order difference equation

$$x(n + 2) - x(n + 1) + \frac{\alpha - 1}{\alpha}x(n) = 0. \tag{4.3.9}$$

Applying Condition (4.3.8), one concludes that the zero solution of Eq. (4.3.9) is asymptotically stable if and only if

$$0 < \frac{\alpha - 1}{\alpha} < 2\cos\left(\frac{\pi}{3}\right) = 1$$

that is, if and only if $\alpha > 1$.

Exercise 4.3

1. Show that the zero solution of $x(n + 3) + p_1 x(n + 2) + p_2 x(n + 1) p_3 x(n) = 0$ is asymptotically stable if and only if $|p_3 + p_1| < 1 + p_2$ and $|p_2 - p_3 p_1| < 1 - p_3^2$.

2. Show that the zero solution of $x(n + 4) + p_1 x(n + 3) + p_2 x(n + 2) + p_3 x(n + 1) + p_4 x(n) = 0$ is asymptotically stable if and only if $|p_4| < 1$, $|p_3 + p_1| < 1, +p_2 + p_4$, and $|p_2(1 - p_4) + p_4(1 - p_4^2) + p_1(p_4 p_1 - p_3)| < p_4 p_2(1 - p_4) + (1 - p_4^2) + p_3(p_4 p_1 - p_3)$.

3. Extend the result in Problem 2 to the fifth order equation

   $$x(n + 5) + p_1 x(n + 4) + p_2 x(n + 3) + p_3 x(n + 2) + p_4 x(n + 1) + p_5 x(n) = 0.$$

4.* (Project) Extend the result in Problem 3 to the general kth order Eq. (4.3.1).

5. For what values α is the zero solution of $x(n + 3) - x(n + 2) + \frac{\alpha - 1}{\alpha} x(n) = 0$ asymptotically stable?

6. Consider the equation $x(n + k) - x(n + k - 1) + \frac{\alpha - 1}{\alpha} x(n) = 0$.

   (i) Show that the zero solution is asymptotically stable if and only if

   $$(L)1 < \alpha < 1 + \left[2\cos\left[\frac{(k - 1)\pi}{2k - 1}\right] \bigg/ \left(1 - 2\cos\left[\frac{(k - 1)\pi}{2k - 1}\right]\right)\right].$$

   (ii) Show that in (L) as k increases to ∞, α decreases monotonically to 1.

7.* (Project) Consider the equation $x(n + k) + p_1 x(n + k - 1) + p_2 x(n + k - 2) +, \ldots, p_k x(n) = 0$.

   (a) Use Roche's Theorem to determine sufficient conditions for the asymptotic stability of the zero solution.

   (b) Use Rouche's Theorem to find sufficient conditions for the instability of the zero solution. (Hint: See Example 4.14.)

Jury Test for Stability [6]

Consider the polynomial

$$p(\lambda) = \lambda^k + p_1 \lambda^{k-1} + \cdots + p_k. \qquad (*)$$

Define the following parameters

$$c_n = \begin{vmatrix} p_k & p_{k-1-n} \\ 1 & p_{n+1} \end{vmatrix}, \qquad n = 0, 1, \ldots, k-1, \text{ with } p_0 = 1.$$

$$d_n = \begin{vmatrix} c_{k-1} & c_{k-2-n} \\ c_0 & c_{n+1} \end{vmatrix}, \qquad n = 0, 1, \ldots, n-2.$$

The pattern continues until finally we get the last three terms in the (last) row $2k - 3$.

$$q_n = \begin{vmatrix} b_3 & b_{2-n} \\ b_0 & b_{n+1} \end{vmatrix}, \qquad n = 0, 1, 2.$$

The Jury table is constructed as follows: The first row consists of the coefficients of $p(\lambda)$ arranged in ascending order of powers of λ. The second row is a reverse order of the first row. All even-numbered rows are simply the reverse of the preceding odd-numbered row.

Row	λ^0	λ^1	λ^2	$\cdots$	λ^n	$\cdots$	λ^{k-2}	λ^{k-1}	λ^k
1	p_k	p_{k-1}	p_{k-2}		p_{k-n}	$\cdots$	p_2	p_1	1
2	1	p_1	p_2		p_n	$\cdots$	p_{k-2}	p_{k-1}	p_k
3	c_{k-1}	c_{k-2}	c_{k-3}		c_{k-n-1}	$\cdots$	c_1	c_0	
4	c_0	c_1	c_2		c_n	$\cdots$	c_{k-2}	c_{k-1}	
5	d_{k-2}	d_{k-3}	d_{k-4}		d_{k-n-2}	$\cdots$	d_0		
6	d_0	d_1	d_2		d_k	$\cdots$	d_{k-2}		
$\vdots$									
$2k-5$	b_3	b_2	b_1	b_0					
$2k-4$	b_0	b_1	b_2	b_3					
$2k-3$	q_2	q_1	q_0						

Theorem [6]. The roots of the polynomial (*) all lie inside the unit disk if and only if the following conditions hold:

(i) $p(1) > 0, (-1)^k p(-1) > 0$.

(ii) $|p_k| < 1$
$|c_{k-1}| > |c_0|$
$|d_{k-2}| > |d_0|$

$\vdots$

$|q_2| > |q_0|$.

In problems 8 through 10 apply the *Jury Stability Test* to check the stability of the equation.

8. The equation in problem 1.

9. The equation in problem 2.

10. The equation in problem 3.

11. Apply either Theorem 4.12 or the Jury Stability Test to show that the zero solution of the difference equation $\Delta x(n) = -qx(n-1), q > 0$, is asymptotically stable if and only if $q < 1$.

12.* (Hard). Prove that the zero solution of the difference equation $\Delta x(n) = -qx(n-k), q > 0, k > 1$ is asymptotically stable if $qk < 1$ (compare to [9]).

4.4   Phase Space Analysis

In this section, we will study the stability properties of the second order linear autonomous (time- invariant) systems,

$$x_1(n+1) = a_{11}x_1(n) + a_{12}x_2(n),$$

$$x_2(n+1) = a_{21}x_1(n) + a_{22}x_2(n),$$

or

$$x(n+1) = Ax(n) \tag{4.4.1}$$

where

$$A = \begin{pmatrix} a_{11} & a_{12} \\ a_{21} & a_{22} \end{pmatrix}.$$

Recall that x^* is an equilibrium point of System (4.4.1) if $Ax^* = x^*$ or $(A - I)x^* = 0$. So if $(A - I)$ is nonsingular, then $x^* = 0$ is the only equilibrium point of System (4.4.1). On the other hand, if $(A - I)$ is singular, then there is a family of equilibrium points, as illustrated in Fig. (4.7e). In the later case we let $y(n) = x(n) - x^*$ in Eq. (4.4.1) to obtain the system $y(n+1) = Ay(n)$ which is identical to System (4.3.1). Thus the stability properties of any equilibrium point $x^* \neq 0$ are the same as those of the equilibrium point $x^* = 0$. Henceforth, we will assume that $x^* = 0$ is the only equilibrium point of System (4.4.1).

Let $J = p^{-1}AP$ be the Jordan form of A. Then J may have the following canonical forms

$$\begin{pmatrix} \lambda_1 & 0 \\ 0 & \lambda_2 \end{pmatrix} \qquad \begin{pmatrix} \lambda & 1 \\ 0 & \lambda \end{pmatrix} \qquad \begin{pmatrix} \alpha & \beta \\ -\beta & \alpha \end{pmatrix}$$

(a)	(b)	(c)
distinct real eigenvalues λ_1, λ_2	repeated real eigenvalue λ	complex conjugate eigenvalues $\lambda = \alpha \pm i\beta$

If we let

$$y(n) = P^{-1}x(n),$$

or

$$x(n) = Py(n), \qquad (4.4.2)$$

then System (4.3.1) becomes

$$y(n + 1) = Jy(n). \qquad (4.4.3)$$

If $x(0) = x_0$ is an initial condition for System (4.4.1), then $y(0) = y_0 = P^{-1}x_0$ would be the corresponding initial condition for System (4.4.3). Notice that the qualitative properties of the equilibrium points of System (4.4.1) and System (4.4.3) are identical.

Our program is to sketch the phase space of System (4.4.3) in cases (a), (b), and (c). Starting with an initial value

$$y_0 = \begin{pmatrix} y_{10} \\ y_{20} \end{pmatrix}$$

in the $y_1 - y_2$ plane, we trace the movement of the points $y(1), y(2), y(3), \dots$. Essentially, we draw the orbit $\{y(n, 0, y_0) | n \geq 0\}$. The arrows signify increasing time n.

Case (a) In this case the system becomes

$$y_1(n + 1) = \lambda_1 y_1(n),$$
$$y_2(n + 1) = \lambda_2 y_2(n).$$

Hence

$$\begin{pmatrix} y_1(n) \\ y_2(n) \end{pmatrix} = \begin{pmatrix} \lambda_1^n y_{10} \\ \lambda_2^n y_{20)} \end{pmatrix},$$

and thus

$$\frac{y_2(n)}{y_1(n)} = \left(\frac{\lambda_2}{\lambda_1}\right)^n \left(\frac{y_{20}}{y_{10}}\right).$$

If $\lambda_1 > \lambda_2$, then $\lim_{n\to\infty} y_2(n)/y_1(n) = 0$, and if $\lambda_1 < \lambda_2$, then $\lim_{n\to\infty} y_2(n)/y_1(n) = \infty$ (see Fig. 4.7a, 4.7b, 4.7c, 4.7d, 4.7e).

Case (b) In this case,

$$\begin{pmatrix} y_1(n) \\ y_2(n) \end{pmatrix} = J^n \begin{pmatrix} y_{10} \\ y_{20} \end{pmatrix} = \begin{pmatrix} \lambda^n & n\lambda^{n-1} \\ 0 & \lambda^n \end{pmatrix} \begin{pmatrix} y_{10} \\ y_{20} \end{pmatrix}$$

or

$$y_1(n) = \lambda^n y_{10} + n\lambda^{n-1} y_{20},$$
$$y_2(n) = \lambda^n y_{20}.$$

Thus

$$\lim_{n\to\infty} \frac{y_2(n)}{y_1(n)} = 0.$$

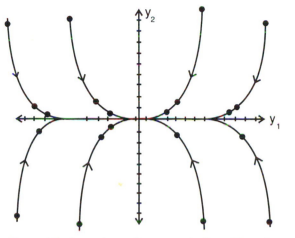

Figure 4.7a. $\lambda_1 < \lambda_2 < 1$, asymototically stable node.

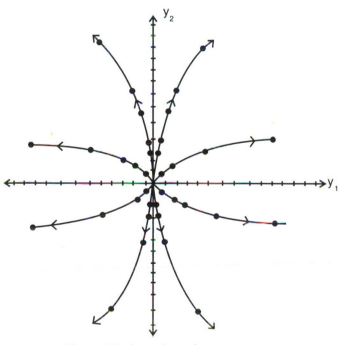

Figure 4.7b. $\lambda_1 > \lambda_2 > 1$, unstable node.

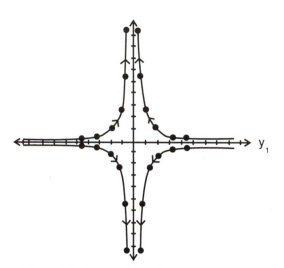

Figure 4.7c. $\lambda_1 < 1, \lambda_2 > 1$, saddle (unstable).

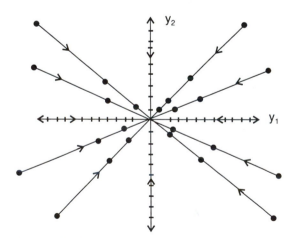

Figure 4.7d. $\lambda_1 = \lambda_2 < 1$, asymptotically stable node.

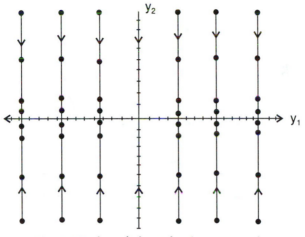
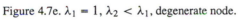

Figure 4.7e. $\lambda_1 = 1$, $\lambda_2 < \lambda_1$, degenerate node.

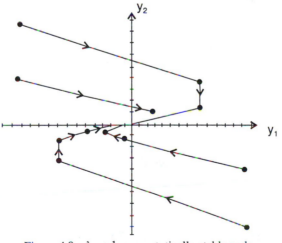

Figure 4.8a. $\lambda < 1$, asymptotically stable node.

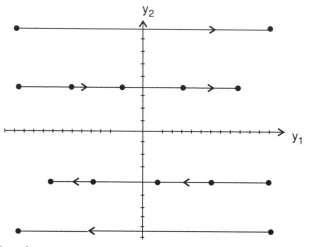

Figure 4.8b. $\lambda = 1$, degenerate case (unstable). All points in the y_1 axis are equilibrium points.

(See Fig. 4.8a, 4.8b)

Case (c) In this case, the matrix A has two complex conjugate eigenvalues

$$\lambda_1 = \alpha + i\beta,$$

and

$$\lambda_2 = \alpha = i\beta, \qquad \beta \neq 0.$$

The eigenvector corresponding to $\lambda_1 = \alpha + i\beta$ is given by $\xi_1 = \binom{1}{i}$, and the solution may be given by

$$\binom{1}{i}(\alpha + i\beta)^n = \binom{1}{i}$$

$$|\lambda_1|n(\cos n\omega + i \sin n\omega),$$

$$= |\lambda_1|^n \binom{\cos n\omega}{-\sin n\omega} + i|\lambda_1|^n \binom{\sin n\omega}{\cos n\omega}$$

where $\omega = \tan^{-1}(\beta/\alpha)$.

A general solution may then be given by

$$\binom{y_1(n)}{y_2(n)} = |\lambda_1|^n \binom{c_1 \cos n\omega + c_2 \sin \omega}{-c_1 \sin n\omega + c_2 \cos n\omega}.$$

Given the initial values $y_1(0) = y_{10}$ and $y_2(0) = y_{20}$, one may find $c_1 = y_{10}$ and $c_2 = y_{20}$. The solution is denoted by

$$y_1(n) = |\lambda_1|^n (y_{10} \cos n\omega + y_{20} \sin n\omega),$$
$$y_2(n) = |\lambda_1|^n (-y_{10} \sin n\omega + y_{20} \cos n\omega).$$

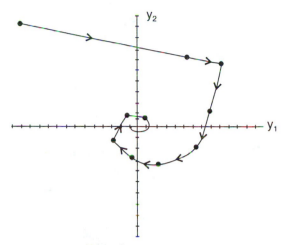

Figure 4.8c. $|\lambda| < 1$, asymptotically stable focus.

If we let $\cos \gamma = y_{10}/r_0$ and $\sin \gamma = y_{20}/r_0$, where $r_0 = \left(y_{10}^2 + y_{20}^2\right)^{1/2}$, we have $y_1(n) = |\lambda_1|^n r_0 \cos(n\omega - \gamma)$ and $y_2(n) = |\lambda_1|^n r_0 \sin(n\omega - \gamma)$. Using polar coordinates we may now write the solution as

$$r(n) = r_0|\lambda_1|^n, \qquad \theta(n) = -(n\omega - \gamma).$$

If $|\lambda_1| < 1$, we have an asymptotically stable *focus*, as illustrated by in Fig. 4.8c. If $|\lambda_1| > 1$, we find an unstable *focus*, as shown in Fig. 4.8d. When $|\lambda_1| = 1$, we obtain a center where orbits are circles with radii

$$r_0 = \sqrt{y_{10}^2 + y_{20}^2}.$$

(Fig. 4.8e).

Using Eq. (4.4.2), one may sketch the corresponding phase space portraits in the $x_1 - x_2$ plane for system equation (4.4.1). The following example illustrates this method.

Example 4.16. Sketch the phase space portrait of the system

$$x(n + 1) = Ax(n), \quad \text{where } A = \begin{pmatrix} 1 & 1 \\ 0.25 & 1 \end{pmatrix}.$$

Solution The eigenvalues of A are $\lambda_1 = 1.5$ and $\lambda_2 = 1/2$; the corresponding eigenvectors are $\xi_1 = \binom{2}{1}$ and $\xi_2 = \binom{2}{-1}$, respectively. Thus

$$P^{-1}AP = J = \begin{pmatrix} 1.5 & 0 \\ 0 & 0.5 \end{pmatrix}, \quad \text{where } P = \begin{pmatrix} 2 & 2 \\ 1 & -1 \end{pmatrix}.$$

Figure 4.9a shows the phase space portrait for $y(n + 1) = Jy(n)$. To find the corresponding phase space portrait of our problem, we let $x(n) = Py(n)$. We define the relationship between the $y_1 - y_2$ system and the $x_1 - x_2$ system by

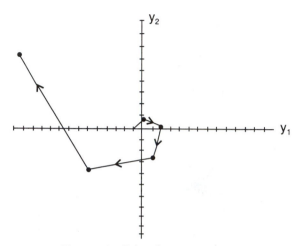

Figure 4.8d. $|\lambda| > 1$, unstable focus.

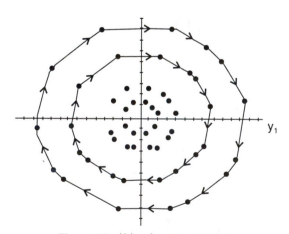

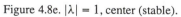

Figure 4.8e. $|\lambda| = 1$, center (stable).

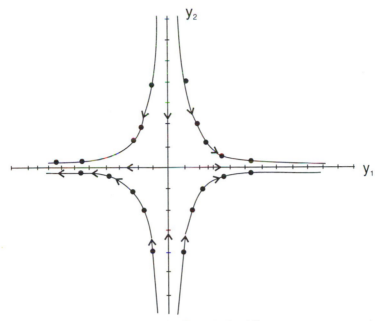

Figure 4.9a. Canonical saddle.

noticing that $\binom{1}{0}$ in the $y_1 - y_2$ system corresponds to $p\binom{1}{0} = \binom{2}{1}$ in the $x_1 - x_2$ system, and $\binom{0}{1}$ in the $y_1 - y_2$ system corresponds to the point $P\binom{0}{1} = \binom{2}{-1}$ in the $x_1 - x_2$ system. The y_1 axis is rotated by $\theta_1 = \tan^{-1}(0.5)$ to the x_1 axis, and the y_2 axis is rotated by $\theta_2 = \tan^{-1}(-0.5)$ to the x_2 axis. Furthermore, the initial point

$$\binom{y_{10}}{y_{20}} = \binom{1}{0}$$

for the canonical system corresponds to the initial point

$$\binom{x_{10}}{x_{20}} = P\binom{1}{0} = \binom{2}{1}.$$

The phase space portrait of our system is shown in Fig. 4.9b. Basically, the axis x_1 is $c\xi_1 = \binom{2c}{c}$, $c \in R$, and the axis x_2 is $c\xi_2 = \binom{2c}{-c}$, $c \in R$.

Example 4.17. Sketch the phase space portrait of the system $x(n + 1) = Ax(n)$ with

$$A = \begin{pmatrix} 1 & 3 \\ -1 & 1 \end{pmatrix}.$$

Solution The eigenvalues of A are $\lambda_1 = 1+\sqrt{3}i$ and $\lambda_2 = 1-\sqrt{3}i$. The eigenvector corresponding to λ_1 is

$$\xi_1 = \begin{pmatrix} \sqrt{3} \\ i \end{pmatrix} = \begin{pmatrix} \sqrt{3} \\ 0 \end{pmatrix} + i\begin{pmatrix} 0 \\ 1 \end{pmatrix}.$$

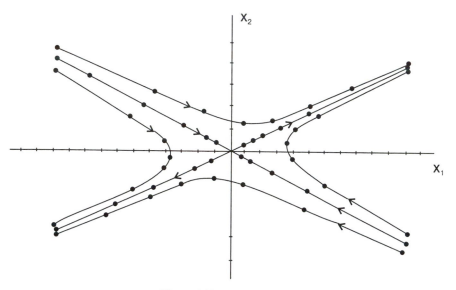

Figure 4.9b. Actual saddle.

If we let

$$p = \begin{pmatrix} \sqrt{3} & 0 \\ 0 & 1 \end{pmatrix},$$

then

$$p^{-1}AP = J = \begin{pmatrix} 1 & \sqrt{3} \\ -\sqrt{3} & 1 \end{pmatrix},$$

which is in the canonical form (c). (See Exercise 4.1.) Hence, the solution of $y(n+1) = Jy(n)$ is

$$r(n) = r_0 |\lambda_1|^n = \sqrt{y_{10}^2 + y_{10}^2 (2)^n}$$

and

$$\theta(n) = \alpha - n\omega,$$

where

$$\alpha = \tan^{-1} \left(\frac{y_{20}}{y_{10}} \right), \qquad \omega = \tan^{-1} \left(\sqrt{3} \right) = \frac{\pi}{3}.$$

Figure 4.10a depicts the orbit of $(\frac{-1}{16}, 0)$. The solution is given by $r(n) = \frac{1}{16}(2)^n = 2^{n-4}$, $\theta(n) = \pi - (n\pi)/3$. The corresponding orbit in the original system has an initial point

$$x_0 = \begin{pmatrix} \sqrt{3} & 0 \\ 0 & 1 \end{pmatrix} \begin{pmatrix} \frac{-1}{16} \\ 0 \end{pmatrix} = \begin{pmatrix} \frac{\sqrt{3}}{16} \\ 0 \end{pmatrix}$$

and is depicted in Fig. 4.10b. Notice that no axis rotation has occurred here.

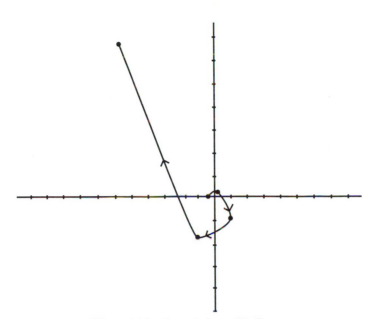

Figure 4.10a. Canonical unstable focus.

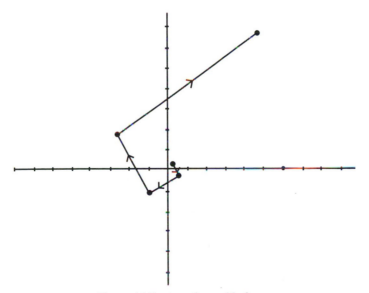

Figure 4.10b. Actual unstable focus.

Exercise 4.4.

1. Sketch the phase space and determine the stability of the equation $x(n+1) = Ax(n)$, where A is given by

   (a) $\begin{pmatrix} 0.5 & 0 \\ 0 & 0.5 \end{pmatrix}$,   (b) $\begin{pmatrix} 0.5 & 0 \\ 0 & 2 \end{pmatrix}$,

   (c) $\begin{pmatrix} 2 & 1 \\ 0 & 2 \end{pmatrix}$,   (d) $\begin{pmatrix} -0.5 & 1 \\ 0 & -0.5 \end{pmatrix}$.

2. Sketch the phase space and determine the stability of the system $x(n+1) = Ax(n)$, where A is given by

   (a) $\begin{pmatrix} 0 & 2 \\ -2 & 0 \end{pmatrix}$,   (b) $\begin{pmatrix} 0.6 & -0.5 \\ 0.5 & 0.6 \end{pmatrix}$,

   (c) $\begin{pmatrix} 1 & 0.5 \\ -0.5 & 1 \end{pmatrix}$,   (d) $\begin{pmatrix} 0.6 & 0.8 \\ -0.8 & 0.6 \end{pmatrix}$.

In Exercises 3 through 6, sketch the phase space and determine the stability of the system $x(n+1) = Ax(n)$ if A is given by

3. $A = \begin{pmatrix} 1 & 1 \\ -1 & 3 \end{pmatrix}$.

4. $A = \begin{pmatrix} -2 & 1 \\ -1 & 3 \end{pmatrix}$.

5. $A = \begin{pmatrix} -2 & 1 \\ -7 & 3 \end{pmatrix}$.

6. $A = \begin{pmatrix} 1 & 2 \\ -1 & -1 \end{pmatrix}$.

7. If the eigenvalues of a real 2×2 matrix A are $\alpha + i\beta$, $\alpha - i\beta$, show that the Jordan Canonical form of A is

$$\begin{pmatrix} \alpha & \beta \\ -\beta & \alpha \end{pmatrix}.$$

4.5   Stability by Linear Approximation

The linearization method is the oldest method in stability theory. Scientists and engineers frequently use this method in the design and analysis of control systems and feedback devices. The mathematicians Liapunov and Perron originated the linearization method, each with their own unique approaches, in their work with the stability theory of differential equations. In this section, we adapt Perron's approach to our study of the nonlinear systems of difference equations.

$$y(n + 1) = A(n)y(n) + g(n, y(n)) \qquad (4.5.1)$$

using its linear component

$$x(n + 1) = A(n)x(n) \qquad (4.5.2)$$

where $A(n)$ is a $k \times k$ nonsingular matrix for all $n \in Z^+$ and $g : Z^+ \times G \to R^k, G \subset R^k$, is a continuous function. One may perceive, System (4.5.1) as a perturbation of System (4.5.2). The function $g(n, y(n))$ represents the perturbation due to noise, inaccuracy in measurements, or other outside disturbances. System (4.5.1) may arise from the linearization of nonlinear equations of the form

$$y(n + 1) = f(n, y(n)) \qquad (4.5.3)$$

where $f : Z^+ \times G \to R^k, G \subset R^k$, is continuously differentiable at an equilibrium point y^* (i.e., $\partial f/\partial y_i | y^*$ exists and continuous on an open neighborhood of y^* for $1 \leq i \leq k$.) For simplicity, we shall assume in this section that $y^* = 0$, unless otherwise stated. Therefore, we also assume throughout this section that $f(n, 0) = 0$ for all $n \in Z^+$. We now describe the linearization method applied to System (4.5.3). Let us write $f = (f_1, f_2, \ldots, f_k)^T$. The Jacobian matrix of f is defined as

$$\frac{\partial f(n, y)}{\partial y}\bigg|_{y=0} = \frac{\partial f(n, 0)}{\partial y} = \begin{pmatrix} \dfrac{\partial f_1(n, 0)}{\partial y_1} & \dfrac{\partial f_1(n, 0)}{\partial y_2} & \cdots & \dfrac{\partial f_1(n, 0)}{\partial y_k} \\[2mm] \dfrac{\partial f_2(n, 0)}{\partial y_1} & \dfrac{\partial f_2(n, 0)}{\partial y_2} & \cdots & \dfrac{\partial f_2(n, 0)}{\partial y_k} \\[2mm] \vdots & \vdots & & \vdots \\[2mm] \dfrac{\partial f_k(n, 0)}{\partial y_1} & \dfrac{\partial f_n(n, 0)}{\partial y_2} & \cdots & \dfrac{\partial f_k(n, 0)}{\partial y_k} \end{pmatrix}.$$

For simplicity $\frac{\partial f(n,0)}{\partial y}$ is denoted by $f'(n, 0)$.

Let $\partial f(n, 0)/\partial y = A(n)$ and $g(n, y) = f(n, y) - A(n)y(n)$. Then System (4.5.3) may be written in the form of System (4.5.1).

We say that $g(n, y) = o(y)$ ("little oh of y") as $\|y\| \to 0$ if, given $\varepsilon > 0$, there is $\delta > 0$ such that $\|g(n, y)\| \leq \varepsilon \|y\|$ whenever $\|y\| < \delta$ and $n \varepsilon Z^+$.

An important special case of System (4.5.3) is the autonomous system

$$y(n + 1) = f(y(n)) \qquad (4.5.4)$$

which may be written as

$$y(n + 1) = Ay(n) + g(y(n)), \qquad (4.5.5)$$

where $A = f'(0)$ is the Jacobian matrix of f at 0 and $g(y) = f(y) - Ay$. Since f is differentiable at 0, it follows from Real Analysis that $g(y) = o(y)$ as $\|y\| \to 0$. Equivalently,

$$\lim_{\|y\| \to 0} \frac{\|g(y)\|}{\|y\|} = 0.$$

Before commencing our stability analysis we must consider a simple but important lemma. This lemma is the discrete analogue of the so-called Gronwall inequality which is used, along with its variations, extensively in differential equations.

Lemma 4.20 (Discrete Gronwall Inequality). Let $z(n)$ and $h(n)$ be two sequences of real numbers, $n \geq n_0 \geq 0$ and $h(n) \geq 0$. If

$$z(n) \leq M \left[z(n_0) + \sum_{j=n_0}^{n-1} h(j) z(j) \right], \qquad \text{for some } M > 0,$$

then

$$z(n) \leq z(n_0) \prod_{j=n_0}^{n-1} [(1 + Mh(j)], \qquad n \geq 0, \qquad (4.5.6)$$

$$z(n) \leq z(n_0) \exp \left[\sum_{j=n_0}^{n-1} Mh(j) \right], \qquad n \geq n_0. \qquad (4.5.7)$$

Proof Let

$$u(n) = M \left[u(n_0) + \sum_{j=n_0}^{n-1} h(j) u(j) \right], \qquad u(n_0) = z(n_0). \qquad (4.5.8)$$

Since $h(j) \geq 0$ for all $j \geq n_0$, it follows that $z(n) \leq u(n)$ for all $n \geq n_0$. From Eq. (4.5.8) we have $u(n + 1) - u(n) = Mh(n)u(n)$ or $u(n + 1) = [1 + Mh(n)]u(n)$. By Formula (2.1.4) we obtain

$$u(n) = \prod_{j=n_0}^{n-1} [1 + Mh(j)] u(n_0).$$

This proves Formula (4.5.6). The conclusion of Formula (4.5.7) follows by noting that $1 + Mh(j) \leq \exp(Mh(j))$.

Theorem 4.21. Assume that $g(n, y) = o(\|y\|)$ as $\|y\| \to 0$. If the zero solution of the homogeneous System (4.5.2) is uniformly asymptotically stable, then the zero solution of the nonlinear System (4.5.1) is exponentially stable.

Proof From Eq. (4.2.6) it follows that $\|\Phi(n, m)\| \leq M\eta^{(n-m)}, n \geq m \geq n_0$, for some $M \geq 1$ and $\eta \in (0, 1)$. By the Variation of Constants Formula (3.2.12), the solution of Eq. (4.5.5) is given by

$$y(n, n_0, y_0) = \Phi(n, n_0) y_0 + \sum_{j=n_0}^{n-1} \Phi(n, j + 1) g(n, y(j)).$$

Thus

$$\|y(n)\| \leq M\eta^{(n-n_0)}\|y_0\| + M\eta^{-1}\sum_{j=n_0}^{n-1}\eta^{(n-j)}\|g(n, y(j))\|. \tag{4.5.9}$$

For a given $\varepsilon > 0$ there is $\delta > 0$ such that $\|g(n, y)\| < \varepsilon$ whenever $\|y\| < \delta$. So as long as $\|y(j)\| < \delta$, Eq. (4.5.9) becomes

$$\eta^{-n}\|y(n)\| \leq M\left[\eta^{-n_0}\|y_0\| + \sum_{j=n_0}^{n-1}\varepsilon\eta^{-j-1}\|y(j)\|\right]. \tag{4.5.10}$$

Letting $z(n) = \eta^{-n}\|y(n)\|$, and then applying the Gronwall inequality (4.5.6), one obtains

$$\eta^{-n}\|y(n)\| \leq \eta^{-n_0}\|y_0\|\sum_{j=n_0}^{n-1}[1 + \varepsilon\eta^{-1}M].$$

Thus,

$$\|y(n)\| \leq \|y_0\|(\eta + \varepsilon M)^{(n-n_0)}. \tag{4.5.11}$$

Choose $\varepsilon < (1 - \eta)/M$. Then $\eta + \varepsilon M < 1$. Thus $\|y(n)\| \leq \|y_0\| < \delta$ for all $n \geq n_0 \geq 0$. Thereby, Formula (4.5.10) holds and consequently, by virtue of Formula (4.5.11), we obtain exponential stability.

Corollary 4.22. If $\rho(A) < 1$, then the zero solution of Eq. (4.5.5) is exponentially stable.

Proof Using Theorem 4.8, the corollary follows immediately from Theorem 4.20.

Corollary 4.23. If $\|f'(0)\| < 1$, then the zero solution of Eq. (4.5.5) is exponentially stable.

Proof Since $\rho(f'(0))\| \leq \|f'(0)\|$, the proof follows from Corollary 4.2.1.

A Remark about Corollaries 4.22 and 4.23

It is possible that $\|A\| \geq 1$ but $\rho(A) < 1$. For example for

$$A = \begin{pmatrix} 0.5 & 1 \\ 0 & 0.5 \end{pmatrix}, \qquad \|A\|_2 = \sqrt{\rho(A^T A)} = \frac{3}{4} + \frac{\sqrt{2}}{2} > 1,$$

$$\|A\|_\infty = \frac{3}{2}, \qquad \|A\|_1 = \frac{3}{2}.$$

However, $\rho(A) = \frac{1}{2}$. With the above matrix A, the zero solution of the system $x(n + 1) = Ax(n) + g(x(n))$ is exponentially stable provided that $g(x) = o(x)$ as $\|x\| \to 0$. Obviously, Corollary 4.23 fails to help us in determining the stability of the system. However, even with all its shortcomings, Corollary 4.23 is surprisingly popular among scientists and engineers.

It is also worthwhile to mention that if $\rho(A) < 1$, there exists a nonsingular matrix Q such that $\|QAQ^{-1}\| < 1$ [9]. One may define a new norm on A, $\|A\| = \|QAQ^{-1}\|$, and then apply Corollary 4.20 in a more useful way.

Let us return to our example where

$$A = \begin{pmatrix} 0.5 & 1 \\ 0 & 0.5 \end{pmatrix}.$$

Let

$$Q = \begin{pmatrix} 1 & 0 \\ 0 & \alpha \end{pmatrix}.$$

Then

$$Q^{-1} = \begin{pmatrix} 1 & 0 \\ 0 & 1/\alpha \end{pmatrix}$$

and

$$Q^{-1}AQ = \begin{pmatrix} 0.5 & \alpha \\ 0 & 0.5 \end{pmatrix}.$$

$\|QAQ^{-1}\|_1 = \alpha + 0.5$. If we choose $\alpha < 0.5$, then $\|QAQ^{-1}\|_1 < 1$.

The above procedure may be generalized to any Jordan Block

$$A = \begin{pmatrix} \lambda & 1 & \cdots & 0 \\ \vdots & \lambda & & \vdots \\ \vdots & \vdots & & 1 \\ 0 & 0 & & \lambda \end{pmatrix}.$$

In this case, we let $Q = \mathrm{diag}(1, \alpha, \alpha^2, \ldots, \alpha^{k-1})$, where k is the order of A.

Hence,

$$QAQ^{-1} = \begin{pmatrix} \lambda & \alpha & \cdots & 0 \\ \vdots & \lambda & & \vdots \\ \vdots & \vdots & & \alpha \\ 0 & 0 & & \lambda \end{pmatrix},$$

and $\|QAQ^{-1}\|_1 = |\lambda| + |\alpha|$ (see Exercise 4.1, Problems 3 and 4). Consequently, if $|\lambda| < 1$, one may choose an α such that $|\lambda| + |\alpha| < 1$, so that under the norm $\|A\| = \|QAQ^{-1}\|_1$, $\|A\| < 1$. We now give two examples to illustrate the preceding conclusions.

Example 4.24. Investigate the stability of the zero solution of the planar system

$$y_1(n+1) = ay_2(n)/[1 + y_1^2(n)],$$

$$y_2(n+1) = by_1(n)/[1 + y_2^2(n)]. \tag{4.5.12}$$

Solution Let $f = (f_1, f_2)^T$, where $f_1 = ay_2(n)/[1 + y_1^2(n)]$, and $f_2 = by_1(n)/[1 + y_2^2(n)]$. Then the Jacobian matrix

$$\frac{\partial f}{\partial y}\bigg|(0, 0) = \begin{pmatrix} \frac{\partial f_1(0,0)}{\partial y_1} & \frac{\partial f_1(0,0)}{\partial y_2} \\ \frac{\partial f_2(0,0)}{\partial y_1} & \frac{\partial f_2(0,0)}{\partial y_2} \end{pmatrix} = \begin{pmatrix} 0 & a \\ b & 0 \end{pmatrix}.$$

Hence system (∗) may be written as

$$\begin{pmatrix} y_1(n+1) \\ y_2(n+1) \end{pmatrix} = \begin{pmatrix} 0 & a \\ b & 0 \end{pmatrix} \begin{pmatrix} y_1(n) \\ y_2(n) \end{pmatrix} + \begin{pmatrix} -ay_2(n)y_1^2(n)/[1 + y_1^2(n)] \\ -by_2^2(n)y_1(n)/[1 + y_2^2(n)] \end{pmatrix}$$

or as

$$y(n+1) = Ay(n) = g(y(n)).$$

The eigenvalues of A are $\lambda_1 = \sqrt{ab}$, $\lambda_2 = -\sqrt{ab}$. Hence, if $|ab| < 1$, the zero solution of the linear part $x(n + 1) = Ax(n)$ is asymptotically stable. Since $g(y)$ is continuously differentiable at $(0, 0)$, $g(y)$ is $o(y)$. Corollary 4.21 then implies that the zero solution of Eq. (4.5.12) is exponentially stable.

Example 4.25 (Pielou logistic delay equation [10]). In Example 2.37 we investigated the Pielou logistic equation

$$x(n+1) = \frac{\alpha x(n)}{1 + \beta x(n)}.$$

If we now assume that there is a delay of time 1 in the response of the growth rate per individual to density change, then we obtain the difference delay equation

$$y(n+1) = \frac{\alpha y(n)}{1 + \beta y(n-1)}, \qquad \alpha > 1, \beta > 0 \qquad (4.5.13)$$

An example of a population that can be modeled by Equation (4.5.13) is the blow fly (Lucilia cuprina) (See [11]). Find conditions on α, β for which the positive equilibrium point $y^* = \frac{\alpha-1}{\beta}$ is asymptotically stable.

Solution

Method (1): Let $\bar{y}(n) = y(n) - (\alpha - 1)/\beta$. Then Eq. (4.5.13) becomes

$$\bar{y}(n+1) = \frac{\alpha\bar{y}(n) - (\alpha - 1)\bar{y}(n-1)}{\alpha + \beta\bar{y}(n-1)}. \qquad (4.5.14)$$

The equilibrium point $\bar{y}^*(n) = 0$ of Eq. (4.5.14) corresponds to $y^* = (\alpha - 1)/\beta$. To change Eq. (4.5.14) to a planar system, we let

$$x_1(n) = \bar{y}(n-1) \text{ and } x_2(n) = \bar{y}(n).$$

Then

$$\begin{pmatrix} x_1(n+1) \\ x_2(n+1) \end{pmatrix} = \begin{pmatrix} x_2(n) \\ \dfrac{\alpha x_2(n) - (\alpha - 1)x_1(n)}{\alpha + \beta x_1(n)} \end{pmatrix}. \qquad (4.5.15)$$

By linearizing Eq. (4.5.15) around (0,0) we give it the new form

$$x(n+1) = Ax(n) + g(x(n)),$$

where

$$A = \begin{pmatrix} 0 & 1 \\ \dfrac{1-\alpha}{\alpha} & 1 \end{pmatrix}$$

and

$$g(x) = \begin{pmatrix} 0 \\ \dfrac{\beta(\alpha-1)x_1^2 - \alpha\beta x_1 x_2}{\alpha(\alpha + \beta x_1)} \end{pmatrix}.$$

The eigenvalues of A are

$$\lambda_{1,2} = \frac{1}{2} \pm \sqrt{\frac{4-3\alpha}{\alpha}}.$$

We have the following two cases to consider:

Case 1: $4 - 3\alpha \geq 0$. In this case we have $1 < \alpha \leq 4/3$. Now

$$\rho(A) = |\lambda| = \frac{1}{2} + \frac{1}{2}\sqrt{\frac{4-3\alpha}{\alpha}}.$$

Let $h(\alpha) = (4 - 3\alpha)/\alpha$. Since $h(1) = 1$ and h is a decreasing function ($h'(\alpha) < 0$), it follows that $0 \leq h(\alpha) < 1$, for $1 < \alpha \leq 4/3$. This implies that $\rho(A) < \frac{1}{2} + \frac{1}{2} = 1$.

Case 2: $4 - 3\alpha < 0$, or $\alpha > 4/3$. In this case we have

$$\rho(A) = |\lambda_1| = \left| \frac{1}{2} + \frac{1}{2}i\sqrt{\frac{3\alpha - 4}{\alpha}} \right| = \sqrt{1 - \frac{1}{\alpha}} < 1.$$

Therefore $\rho(A) < 1$ for all $\alpha > 1$. Since $g(x)$ is continuously differentiable at $(0, 0)$, the zero solution of Eq. (4.5.15) is uniformly asymptotically stable. Consequently, the equilibrium point $x^* = (\alpha - 1)/\beta$ of Eq. (4.5.13) is asymptotically stable.

Method (2): Letting $y(n) = (\alpha - 1)/\beta \, \mathrm{Exp}(x(n))$ in Eq. (4.5.13) we obtain the new equation

$$\mathrm{Exp}\,(x(n+1)) = \mathrm{Exp}\,(x(n))/[\{1 + (\alpha - 1)\,\mathrm{Exp}\,(x(n-1))\}/\alpha].$$

Taking the logarithm of both sides we get

$$x(n + 1) - x(n) + \frac{\alpha - 1}{\alpha} f[x(n - 1)] = 0$$

or

$$x(n + 2) - x(n + 1) + \frac{\alpha - 1}{\alpha} f[x(n)] = 0, \qquad (4.5.16)$$

where

$$f(x) = \frac{\alpha}{\alpha - 1} \ln \left[\frac{(\alpha - 1)e^x + 1}{\alpha} \right].$$

The Taylor's Expansion of f around 0 is given by $f(x) = x + g(x)$, where $g(x)$ is a polynomial in x which contains terms of degree higher than or equal 2. Thus $g(x) = o(x)$. The linearized equation of Eq. (4.5.16) is denoted by

$$x(n + 2) - x(n + 1) + \frac{\alpha - 1}{\alpha} x(n) = 0. \qquad (4.5.17)$$

In Example 4.13 we showed that the zero solution Eq. (4.5.17) is asymptotically stable. Corollary 4.22 then implies that the zero solution of Eq. (4.5.16) is asymptotically stable. Since the equilibrium point $y^* = (\alpha - 1)/\beta$ corresponds to the zero solution of Eq. (4.5.16), it then follows that $y^* = (\alpha - 1)/\beta$ is an symptotically stable equilibrium point of Eq. (4.5.14).

Our final result deals with the cases $\rho(A) = 1$ and $\rho(A) > 1$.

Theorem 4.2. The following statements hold:

(i) If $\rho(A) = 1$, then the zero solution of Eq. (4.5.5) may be stable or unstable.

(ii) If $\rho(A) > 1$, and $g(x)$ is $o(x)$ as $\|x\| \to 0$, then the zero solution of Eq. (4.5.5) is unstable.

Proof

(i) See Example 4.36

(ii) See [12].

Exercise 4.5.

1. Determine the stability of the zero solution of the equation

$$x(n + 2) - \frac{1}{2} x(n + 1) + 2x(n + 1)x(n) + \frac{13}{16} x(n) = 0.$$

2. Judge the stability of the zero solution of the equation

$$x(n + 3) - x(n + 1) + 2x_1^2(n) + 3x_1(n) = 0.$$

3. (a) Hunt down the equilibrium points of the system
$$x_1(n + 1) = x_1(n) - x_2(n)(1 - x_2(n)),$$
$$x_2(n + 1) = x_1(n),$$
$$x_3(n + 1) = \frac{1}{2}x_3(n).$$

   (b) Determine the stability of all the equilibrium points in part (a).

4. Investigate the stability of the zero solution of the system
$$x_1(n + 1) = \frac{1}{2}x_1(n) - x_2^2(n) + x_3(n),$$
$$x_2(n + 1) = x_1(n) - x_2(n) + x_3(n),$$
$$x_3(n + 1) = x_1(n) - x_2(n) + \frac{1}{2}x_3(n).$$

5. Linearize the equation
$$x_1(n + 1) = \sin(x_2) - 0.5x_1(n),$$
$$x_2(n + 1) = x_2/(0.6 + x_1(n)),$$

   around the origin and then determine if the zero solution is stable.

6. (a) Find the equilibrium points of the system
$$x_1(n + 1) = \cos x_1(n) - x_2(n),$$
$$x_2(n + 1) = x_1(n).$$

   (b) Is the point $(\pi/2, -\pi/2)$ asymptotically stable?

7. Determine conditions for the asymptotic stability of the zero solution of the system
$$x_1(n + 1) = ax_1(n)/[1 + x_2(n)],$$
$$x_2(n + 1) = [bx_2(n) - x_1(n)][1 + x_1(n)].$$

8. Show that if the zero solution of Eq. (4.5.2) is uniformly stable, (uniformly asymptotically stable), then the zero solution of Eq. (4.5.1) is also uniformly stable (uniformly asymptotically stable) provided that
$$\|g(n, y(n))\| \le a_n \|y(n)\|, \qquad \text{where } a_n > 0 \text{ and } \sum_{n=0}^{\infty} a_n < \infty.$$

9. Suppose that the zero solution of $x(n + 1) = Ax(n)$ is asymptotically stable. Prove that the zero solution of $y(n + 1) = [A + B(n)]y(n)$ is asymptotically stable if $\sum_{n=0}^{\infty} \|B(n)\| < \infty$.

4.6   Liapunov's Direct or Second Method

In his famous memoir, published in 1892, the Russian mathematician A.M. Liapunov introduced a new method for investigating the stability of nonlinear differential equations. This method, known as Liapunov's Direct Method, allows one
to investigate the qualitative nature of solutions without actually determining the
solutions themselves. Therefore, we regard it as one of the major tools in stability
theory. The method hinges upon finding certain real-valued functions which are
named after Liapunov. The major drawback, in the Direct Method, however, lies
in determining the appropriate Liapunov function for a given equation.

In this section we adapt the Liapunov's direct method to difference equations.
We begin our study with the autonomous difference equation

$$x(n + 1) = f(x(n))   \tag{4.6.1}$$

where $f : G \rightarrow R, G \subset R^k$, is continuous. We assume that $x*$ is an equilibrium
point of Eq. (4.6.1), that is, $f(x*) = x*$.

Let $V : R^k \rightarrow R$ be defined as a real valued function. The variation of V relative
to Eq. (4.6.1) would then be defined as

$$\Delta V(x) = V(f(x)) - V(x)$$

and

$$\Delta V(x(n)) = V\big(f(x(n))\big) - V(x(n)) = V(x(n + 1)) - V(x(n)).$$

Notice that if $\Delta V(x) \leq 0$, then V is nonincreasing along solutions of Eq. (4.6.1).
The function V is said to be a *Liapunov function* on a subset H of R^k if

(i) V is continuous on H and

(ii) $\Delta V(x) \leq 0$ whenever x and $f(x) \in H$.

Let $B(x, \gamma)$ denote the open ball in R^k of radius γ and center x defined by $B(x, \gamma) =$
$\{y \in R^k | \|y - x\| < \gamma\}$. For the sake of brevity, $B(0, \gamma)$ will henceforth be denoted
by $B(\gamma)$. We say that the real-valued function V is *positive definite* at $x*$ if

(i) $V(x*) = 0$ and

(ii) $V(x) > 0$ for all $x \in B(x*, \gamma)$, for some $\gamma > 0$.

We now present to the reader an informal geometric discussion of the first
Liapunov stability theorem. For simplicity, we will assume that our system is
planar with $x* = 0$ as the equilibrium point. Suppose that Eq. (4.6.1) has a positive
definite Liapunov function V defined on $B(\eta)$. Figure 4.11a then illustrates the
graph of V in a 3-dimensional coordinate system, while Fig. 4.11b gives the level
curves $V(x_1, x_2) = c$ of V in the plane. If we now let $\varepsilon > 0$, $B(\varepsilon)$ then contains
one of the level curves of V, say $V(x) = \tilde{c}_2$. The level curve $V(x) = \tilde{c}_2$ contains the
ball $B(\delta)$, for some δ with $0 < \delta \leq \varepsilon$. If a solution $x(n, 0, x_0)$ starts at $x_0 \in B(\delta)$,

then $V(x_0) \leq \tilde{c}_2$. Since $\Delta V \leq 0$, V is a monotonic nonincreasing function along solutions of Eq. (4.6.1). Hence, $V(x(n)) \leq V(x_0) \leq \tilde{c}_2$ for all $n \geq 0$. Thus, the solution $x(n, 0, x_0)$ will stay forever in the ball $B(\varepsilon)$. Consequently, the zero solution is stable. The above argument contains the essence of the proof of the first Liapunov stability theorem.

Theorem 4.27 (Liapunov Stability Theorem). If V is a Liapunov function for Eq. (4.6.1) on a neighborhood H of the equilibrium point x^*, and V is positive definite with respect to x^*, then x^* is stable. If, in addition, $\Delta V(x) < 0$, whenever x, $f(x) \in H$ and $x \neq x^*$, then x^* is asymptotically stable. Moreover, if $G = H = R^k$ and

$$V(x) \to \infty \text{ as } \|x\| \to \infty \tag{4.6.2}$$

then x^* is globally asymptotically stable.

Proof Choose $\alpha_1 > 0$ such that $B(x^*, \alpha_1) \subset G \cap H$. Since f is continuous, there is $\alpha_2 > 0$ such that if $x \in B(x^*, \alpha_2)$ then $f(x) \in B(x^*, \alpha_1)$. Let $0 < \varepsilon \leq \alpha_2$ be given. Define $\psi(\varepsilon) = \min \{V(x) | \varepsilon \leq \|x - x^*\| \leq \alpha_1\}$. By the Intermediate Value Theorem, there exists $0 < \delta < \varepsilon$ such that $V(x) < \psi(\varepsilon)$ whenever $\|x - x^*\| < \delta$.

Realize now that if $x_0 \in B(x^*, \delta)$, then $x(n) \in B(x^*, \varepsilon)$ for all $n \geq 0$. This claim is true because, if not, there exists $x_0 \in B(x^*, \delta)$ and a positive integer m such that $x(r) \in B(x^*, \varepsilon)$ for $1 \leq r \leq m$ and $x(m + 1) \notin B(x^*, \varepsilon)$. Since $x(m) \in B(x^*, \varepsilon) \subset B(x^*, \alpha_2)$, it follows that $x(m+1) \in B(x^*, \alpha_1)$. Consequently, $V(x(m + 1)) \geq \psi(\varepsilon)$. However, $V(x(m + 1)) \leq, \ldots, \leq V(x_0) < \psi(\varepsilon)$ and we thus have a contradiction. This establishes stability.

To prove asymptotic stability, assume that $x_0 \in B(x^*, \delta)$. Then $x(n) \in B(x^*, \varepsilon)$ holds true for all $n \geq 0$. If $\{x(n)\}$ does not converge to x^*, then it has a subsequence $\{x(n_i)\}$ that converges to $y \in R^k$. Let $E \subset B(x^*, \alpha_1)$ be an open neighborhood of y with $x^* \notin E$. Having already defined on E the function $h(x) = V(f(x))/V(x)$, we may consider h as well-defined, continuous and $h(x) < 1$ for all $x \in E$. Now if $\eta \in (h(y), 1)$ then there exists $\delta > 0$ such that $x \in B(y, \delta)$ implies $h(x) \leq \eta$. Thus for sufficiently large n_i,

$$V(f(x(n_i))) \leq \eta V(x(n_i - 1)) \leq \eta^2 V(x(n_i - 2)), \ldots, \leq \eta^{n_i} V(x_0).$$

Hence,

$$\lim_{n_i \to \infty} V(x(n_i)) = 0.$$

But, since $\lim_{n_i \to \infty} V(x(n_i)) = V(y)$, this statement implies that $V(y) = 0$, and consequently $y = x^*$.

To prove the global asymptotic stability, it suffices to show that all solutions are bounded, and then repeat the above argument. Begin by assuming there exists an unbounded solution $x(n)$, and then some subsequence $\{x(n_i)\} \to \infty$ as $n_i \to \infty$. By Condition (4.6.2), this assumption implies that $V(x(n_i)) \to \infty$ as $n_i \to \infty$, which is a contradiction since $V(x_0) > V(x(n_i))$ for all i. This concludes the proof.

The result on boundedness has its own independent importance, so we give it its due respect by stating it here as a separate theorem.

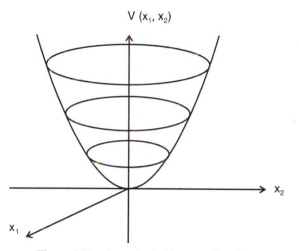

Figure 4.11a. A quadratic Liapunov function.

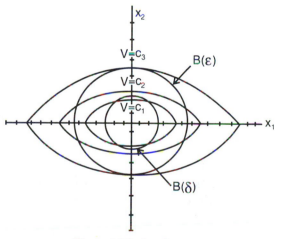

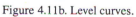

Figure 4.11b. Level curves.

Theorem 4.28. If V is a Liapunov function on the set $\{x \in R^k | \|x\| > \alpha$ for some $\alpha > 0\}$, and if Condition (4.6.2) holds, then all solutions of Eq. (4.6.1) are bounded.

Proof (**Exercise 4.6, Problem 1**).

Example 4.29. Consider the following second order difference equation

$$x(n+1) = \frac{\alpha x(n-1)}{1 + \beta x^2(n)}, \beta > 0.$$

This equation is often called an *equation with delay*. There are three equilibrium points, namely, $x^* = 0$ and

$$x^* = \pm\sqrt{\frac{(\alpha - 1)}{\beta}} \text{ if } \alpha > 1.$$

Let us first change the equation into a system by letting $y_1(n) = x(n-1)$ and $y_2(n) = x(n)$. Then we obtain the system

$$y_1(n+1) = y_2(n),$$
$$y_2(n+1) = \frac{\alpha y_1(n)}{1 + \beta y_2^2(n)}.$$

Consider the stability of the equilibrium point $(0, 0)$. Our first choice of a Liapunov function will be $V(y_1, y_2) = y_1^2 + y_2^2$. This is clearly continuous and positive definite on R^2.

$$\Delta V(y_1(n), y_2(n)) = y_1^2(n+1) + y_2^2(n+1) - y_1^2(n) - y_2^2(n).$$

Thus,

$$\Delta V(y_1(n), y_2(n)) = \left(\frac{\alpha^2}{[1 + \beta y_2^2(n)]^2} - 1\right) y_1^2(n) \leq (\alpha^2 - 1)y_1^2(n). \quad (4.6.3)$$

If $\alpha^2 \leq 1$, then $\Delta V \leq 0$. In this case $x^* = 0$ would be the only equilibrium point, and by Theorem 4.26, the origin is stable (Fig. 4.7). Since $\lim_{\|x\| \to \infty} V(x) = -\infty$, Theorem 4.28 implies that all solutions are bounded. Since $\Delta V = 0$ for all points on the y axis, Theorem 4.27 fails to determine asymptotic stability for this equation.

This situation is typical in most of the problems encountered in applications in science and engineering. Therefore, a finer and more precise analysis is required. This need leads us to the so-called *LaSalle's Invariance Principle*, which will be presented shortly.

To prepare for the introduction of our major theorem, we ought to familiarize ourselves with some vital terminology:

(i) For a subset $G \subset R^k$, x is a *limit point* of G if there exists a sequence $\{x_i\}$ in G with $x_i \to x$ as $i \to \infty$.

(ii) The *closure* $\overline{G}$ of G is defined to be the union of G and all of its limit points.

(iii) After considering Eq. (4.6.1), the positive orbit $O^+(x_0)$ is defined as $O^+(x) = \{x(n, 0, x_0)|n \in Z^+\}$. We will denote $O^+(x_0)$ by $O(x_0)$.

(iv) *The limit set $\Omega(x_0)$*, also referred to as the *positive limit set*, of x_0 is the set of all positive limit points of x_0. Thus, $\Omega(x_0) = \{y \in R^k|x(n_i) \to y$ as $n_i \to \infty$ for some subsequence $\{n_i\}$ of $Z^+\}$.

(v) A set A is positively *invariant* if $O(x_0) \subset A$ for every $x_0 \in A$. One may easily show that both $O(x_0)$ and $\Omega(x_0)$ are (positively) invariant, as is requested in Exercise 4.6, Problem 3. The nagging question still persists as to whether or not $\Omega(x_0)$ is nonempty for a given $x_0 \in R^k$. This next Lemma satisfies that question.

Lemma 4.30. If $0(x_o)$ is bounded, then $\Omega(x_o)$ is a nonempty closed and bounded (compact) set.

Proof (**Exercise 4.6, Problem 4**).

Let V be a positive Liapunov function on a subset G of R^k. Define

$$E = \{x \in \overline{G}|\Delta V(x) = 0\}.$$

Let M be the maximal invariant subset of E, that is, define M as the union of all invariant subsets of E.

Theorem 4.31 (La Salle's Invariance Principle [12]). If there exists a positive definite Liapunov function V for Eq. (4.6.1) on $G \subset R^k$, then each solution of Eq. (4.6.1) which remains in G for all $n \geq 0$ is either unbounded or approaches the set M.

Proof Assume primarily that G is bounded. Let $x(n)$ be a solution that remains in G for all $n \geq 0$. These conditions imply that $x(n)$ is bounded; we may assume that $\lim_{n_i \to \infty} x(n_i) = y \in \overline{G}$, for some subsequence of integers $\{n_i\}$. Since $V(x(n))$ is nonincreasing and bounded below, it follows that $\lim_{n \to \infty} V(x(n)) = c$, for some number c. This implies that

$$\lim_{n \to \infty} \Delta V(x(n)) = \lim_{n \to \infty} [V(x(n+1)) - V(x(n))] = 0.$$

By the continuity of ΔV, we get

$$\lim_{n_i \to \infty} \Delta V(x(n_i)) = \Delta V(y).$$

Thus $\Delta V(y) = 0$, and, consequently, $y \in E$. This fact leads us to realize

$$\Omega(x(n)) \subset E, \text{ and therefore } \Omega(x(n)) \subset M.$$

On the other hand if G is unbounded, then it is possible to have an unbounded solution $x(n)$ with $\lim_{n \to \infty} x(n) = \infty$. This occurs if $\lim_{x \to \infty} V(x) \neq \infty$, that is if V is bounded above.

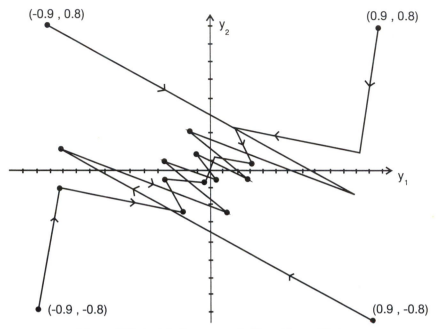

(-0.9 , 0.8)   (0.9 , 0.8)

y_2

y_1

(-0.9 , -0.8)   (0.9 , -0.8)

Figure 4.12. A globally asymptotically stable equilibrium.

Example 4.29 revisited. Let us re-examine Example 4.27 in light of LaSalle's invariance principle. We will consider two cases:

Case 1. $\alpha^2 = 1$. The set E consists of all the points on the x and y axes. We have two subcases to consider. Subcase (i): $\alpha = 1$. If $y_1(0) = a$ and $y_2(0) = 0$, then $y_1(1) = 0$ and $y_2(1) = a$, and $y_1(2) = a$, $y_2(2) = 0$. Therefore, any solution starting on either axis is of period 2, and $M = E$. Subcase (ii): $\alpha = -1$. Then $0^+(a, 0) = \{(a, 0), (0, -a), (-a, 0), (0, a)\}$. Thus any solution starting on either axis is of period 4, and $M = E$ again. Hence all solutions converge to either $(a, 0)$, $(-a, 0)$, $(0, a)$, or $(0, -a)$. Clearly, the zero solution is not asymptotically stable.

Case 2. Because $\alpha^2 < 1$, $E =$ the y axis and $M = \{(0, 0)\}$. Thus, all solutions converge to the origin. Hence the origin is globally asymptotically stable. Figure 4.12 depicts the phase portrait for $\alpha = 0.5$. Notice the difference in the way solutions in quadrants I and III begin, compared to the way the solutions in quadrants II and IV commence.

Case 3. $\alpha^2 > 1$. In this case, La Salle's invariance principle does not aid us in determining the stability of the solution. In other words, the stability is indeterminable.

Sometimes, we may simplify the difference equation by applying a simple basic transformation to the system. For instance, one might translate the system into polar coordinates (r, θ) where $x_1 = r \cos \theta$, $x_2 = r \sin \theta$. The following example demonstrates the effectiveness of this method.

Example 4.32. Consider the difference system.

$$x_1(n + 1) = x_1^2(n) - x_2^2(n),$$
$$x_2(n + 1) = 2x_1(n)x_2(n).$$

Let $x_1(n) = r(n)\cos\theta(n)$, and $x_2(n) = r(n)\sin\theta(n)$.

Then

$$r(n + 1)\cos\theta(n + 1) = r^2(n)\cos^2\theta(n) - r^2(n)\sin^2\theta(n)$$
$$= r^2(n)\cos 2\theta(n), \qquad (4.6.4)$$

and

$$r(n + 1)\sin\theta(n + 1) = 2r^2\sin\theta(n)\cos\theta(n)$$
$$= r^2(n)\sin 2\theta(n). \qquad (4.6.5)$$

Dividing Eq. (4.6.4) by Eq. (4.6.5) we get

$$\theta(n + 1) = 2\theta(n).$$

Substituting this into Eq. (4.6.4) we obtain

$$r(n + 1) = r^2(n).$$

We may write this solution as $r(n) = [r(0)]^{2^n}$ and $\theta(n) = 2^n\theta(0)$.

The equilibrium points are $(0, 0)$ and $(1, 0)$.

For $r(0) < 1$, $\lim_{n\to\infty} r(n) = 0$. Thus solutions starting inside the unit disk spiral toward the origin. Consequently, the origin is asymptotically stable (not globally), as shown in Fig. 4.13.

For $r(0) > 1$, $\lim_{n\to\infty} r(n) = \infty$, and hence solutions that start outside the unit disk spiral away from the unit circle to ∞. This occurrence makes the equilibrium point $(1, 0)$ unstable.

For $r(0) = 1$, $r(n) = 1$, for all $n \geq 0$. Therefore, the circle is an invariant set, with very complicated dynamics. For instance, the solution starting at $\left(1, \frac{\pi}{4}\right)$ will reach the equilibrium point $(1, 0)$ in three iterations: $\left(1, \frac{\pi}{4}\right)$, $\left(1, \frac{\pi}{2}\right)$ $(1, \pi)$, $(1, 0)$. However, the solution which starts at $\left(1, \frac{2\pi}{3}\right)$ is a 2 cycle. In general, $(1, \theta)$ is periodic, with period m, if and only if $2^m\theta = \theta + 2k\pi$ for some integer k, i.e., if and only if $\theta = (2k\pi)/2^m - 1$, $k = 0, 1, 2, \ldots, 2^m$. For $m = 3$, $\theta = \frac{2\pi}{7}, \frac{4\pi}{7}, \frac{6\pi}{7}, \frac{8\pi}{7}, \frac{10\pi}{7}, \frac{12\pi}{7}$. For $m = 4$, $\theta = \frac{2\pi}{15}, \frac{4\pi}{15}, \frac{2\pi}{5}, \frac{8\pi}{15}, \frac{2\pi}{3}, \frac{4\pi}{5}, \frac{14\pi}{15}, \frac{16\pi}{15}, \ldots$.

Notice here that θ is essentially the $(2^m - 1)$th root of 1. Hence, the set of periodic points $(1, \theta)$ densely fills the unit circle (Exercise 4.6, Problem 5). Furthermore, for every $m = 1, 2, \ldots$, there is a periodic point on the unit circle of that period m.

Now if $\theta = \alpha\pi$, or α irrational, then obviously, $\theta \neq (2k\pi)/2^m - 1$ for any m, and thus any solution starting at $(1, \alpha\pi)$ cannot be periodic. However, its orbit is dense within the unit circle, that is $\overline{O(x)}$ = the unit circle (Exercise 4.6, Problem 5).

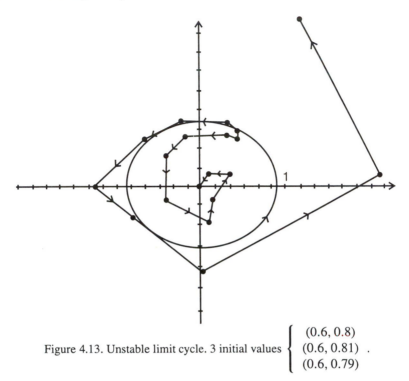

Figure 4.13. Unstable limit cycle. 3 initial values $\begin{cases} (0.6, 0.8) \\ (0.6, 0.81) \\ (0.6, 0.79) \end{cases}$.

Sometimes, using some simple intuitive observations makes it much easier to show that an equilibrium point is not asymptotically stable. The following example illustrates this remark.

Example 4.33. Consider the planar system

$$x_1(n+1) = 2x_2(n) - 2x_2(n)x_1^2(n),$$

$$x_2(n+1) = \frac{1}{2}x_1(n) + x_1(n)x_2^2(n).$$

We find three equilibrium points:

$$(0, 0), \left(\frac{1}{\sqrt{2}}, \frac{1}{\sqrt{2}}\right), \left(-\frac{1}{\sqrt{2}}, -\frac{1}{\sqrt{2}}\right).$$

Let us consider the stability of $(0, 0)$. If $V(x_1, x_2) = x_1^2 + 4x_2^2$, then

$$\begin{aligned} \Delta V\left(x_1(n), x_2(n)\right) &= 4x_2^2(n) - 8x_2^2(n)x_1^2(n) + 4x_2^2(n)x_1^4(n) + x_1^2(n) \\ &\quad + 4x_1^2(n)x_2^2(n) + 4x_1^2(n)x_2^4(n) - x_1^2(n) - 4x_2^2(n) \\ &= 4x_1^2(n)x_2^2(n)[x_1^2(n) + x_2^2(n) - 1]. \end{aligned}$$

If $x_1^2 + x_2^2 \leq 1$, then $\Delta V(x_1, x_2) \leq 0$.

For any real number a, the solution with an initial value of $x_0 = \binom{a}{0}$ is periodic with period 2 and with orbit $\left\{\binom{a}{0}, \binom{0}{a/2}\right\}$. And a solution with an initial value

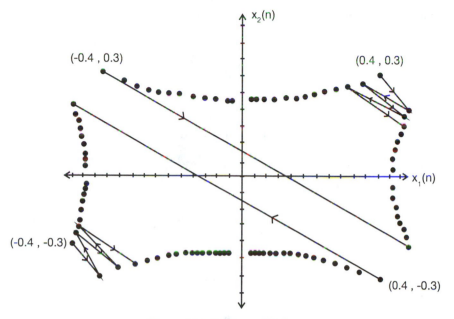

Figure 4.14. Stable equilibrium.

of $x_0 = \binom{0}{a}$ is also periodic with period 2. Hence, the zero solution cannot be asymptotically stable. However, it is stable according to Theorem 4.27. (Figure 4.14 depicts the phase space portrait near the origin.)

We now turn our attention to the dilemma of instability. We are interested in finding conditions on Liapunov functions under which the zero solution is unstable. Here is one of the widely used theorems in this area:

Theorem 4.34. If ΔV is positive definite on a neighborhood of the origin, and there exists a sequence $a_i \to 0$ with $V(a_i) > 0$, then the zero solution of Eq. (4.6.1) is unstable.

Proof Let $\Delta V(x) > 0$ for $x \in B(\eta)$, $x \neq 0$, $V(0) = 0$. We will prove Theorem 4.34 by contradiction, first assuming that the zero solution is stable. Then for $\varepsilon < \eta$, there would exist $\delta < \varepsilon$ such that $\|x_0\| < \delta$ implies $\|x(n, 0, x_0)\| < \varepsilon$, $n \in Z^+$.

Since $a_i \to 0$, pick $x_0 = a_j$ for some j with $V(x_0) > 0$, and $\|x_0\| < \delta$. Hence $\overline{0(x_0)} \subset \overline{B(\varepsilon)} \subset B(\eta)$ is closed and bounded (compact). Since its domain is compact, $V(x(n))$ is also compact, and therefore bounded above. Since $V(x(n))$ is also increasing, it follows that $V(x(n)) \to c$. Following the proof of LaSalle's Invariance Principal, it is easy to see that $\lim_{n\to\infty} x(n) = 0$. Therefore, we would be led to believe that $0 < V(x_0) < \lim_{n\to\infty} V(x(n)) = 0$. This statement is unfeasible—so the zero solution cannot be stable, as we first assumed. The zero solution of Eq. (4.6.1) is thus unstable.

Example 4.35. Consider the system

$$x_1(n + 1) = 4x_2(n) - 2x_2(n)x_1^2(n),$$
$$x_2(n + 1) = \frac{1}{2}x_1(n) + x_1(n)x_2^2(n).$$

Let $V(x_1, x_2) = x_1^2 + 16x_2^2$. Then

$$\Delta V\big(x_1(n), x_2(n)\big) = 3x_1^2(n) + 16x_1^2(n)x_2^4(n) > 0.$$

Hence, by Theorem 4.34 the zero solution is unstable.

Example 4.36. First, contemplate the system

$$x_1(n + 1) = x_1(n) + x_1^3(n)x_2^2(n),$$
$$x_2(n + 1) = x_2(n). \qquad\qquad (4.6.6)$$

Notice that $(0, 0)$ is an equilibrium of the system. Its linear component is denoted by $x(n + 1) = Ax(n)$, where

$$A = \begin{pmatrix} 1 & 0 \\ 0 & 1 \end{pmatrix},$$

and thus $\rho(A) = 1$. Let $V(x) = x_1^2 + x_2^2$ be a Liapunov function. Then

$$\Delta V[x(n)] = 2x_1^4(n)x_2^2(n) + x_1^6(n)x_2^4(n) > 0, \qquad \text{for } x_1 \neq 0, \qquad x_2 \neq 0.$$

Theorem 4.34 implies that the zero solution of System (4.6.6) is unstable.

Let us now ponder System (4.6.7), with the same linear component as system equation (4.6.6)

$$x_1(n + 1) = x_1(n) - x_1^3(n)x_2^2(n),$$
$$x_2(n + 1) = x_2(n). \qquad\qquad (4.6.7)$$

We again let $V(x) = x_1^2 + x_2^2$ be a Liapunov function for System (4.6.7). Then

$$\Delta V[x(n)] = x_1^4(n)x_2^2(n)\left[-2 + x_1^2(n)x_2^2(n)\right].$$

Hence, $\Delta V(x) \leq 0$ if $x_1^2 x_2^2 \leq 2$. It follows from Theorem 4.27 that the zero solution of System (4.6.7) is stable.

We conclude from this discussion that if $\rho(A) = 1$, then the zero solution of the nonlinear equation may be either stable or unstable, thus proving part (i) of Theorem 4.26.

We conclude this section with a brief discussion of Liapunov functions for linear autonomous systems. In Section 4.2, we noticed that the condition for asymptotic stability of the difference equation (4.2.2) is that $\rho(A) < 1$. This condition requires the computation of the eigenvalues of A. Using the second method of Liapunov, such computation deems unnecessary. Before introducing Liapunov's method, however, we need to recall the definition of a positive definite matrix. Consider the quadratic form $V(x)$ for a $k \times k$ real symmetric matrix $B = (b_{ij})$

$$V(x) = x^T B x = \sum_{i=1}^{k} \sum_{j=1}^{k} b_{ij} x_i x_j.$$

A matrix B is said to be positive definite if $V(x)$ is positive definite. Sylvester's criterion is the simplest test for positive definiteness of a matrix. It merely notes that a real symmetric matrix B is positive definite if and only if the determinants of its leading principal minors are positive, i.e., if and only if

$$b_{11} > 0, \quad \begin{vmatrix} b_{11} & b_{12} \\ b_{12} & b_{22} \end{vmatrix} > 0, \quad \begin{vmatrix} b_{11} & b_{12} & b_{13} \\ b_{21} & b_{22} & b_{23} \\ b_{31} & b_{32} & b_{33} \end{vmatrix} > 0, \ldots, \det B > 0.$$

The leading principal minors of matrix B are B itself and the minors obtained by removing successively the last row and the last column. For instance, the leading principal minors of

$$B = \begin{pmatrix} 3 & 2 & 0 \\ 2 & 5 & -1 \\ 0 & -1 & 1 \end{pmatrix}$$

are

$$B, \quad \begin{pmatrix} 3 & 2 \\ 2 & 5 \end{pmatrix}, \quad 3$$

which all have positive determinants. Hence, B is positive definite. Notice that for $x = (x_1, x_2, x_3)^T$,

$$V(x) = x^T B x = 3x_1^2 + 5x_2^2 + x_3^2 + 4x_1 x_2 - 2x_2 x_3 > 0$$

for all $x \neq 0$, and $V(0) = 0$.

On the other hand, given

$$V(x) = ax_1^2 + bx_2^2 + cx_3^2 + dx_1 x_2 + ex_1 x_3 + f x_2 x_3,$$

one may write

$$V(x) = x^T B x,$$

where

$$B = \begin{pmatrix} a & d/2 & e/2 \\ d/2 & b & f/2 \\ e/2 & f/2 & c \end{pmatrix}.$$

Hence V is positive definite if and only if B is. We now make a useful observation. Note that if B is a positive definite symmetric matrix, then all eigenvalues of B are positive (Exercise 4.6, Problem 16). Furthermore, if $\lambda_1, \lambda_2, \ldots, \lambda_k$ are the eigenvalues of B with

$$\lambda_{\min} = \min\{|\lambda_i| \mid 1 \le i \le k\},$$
$$\lambda_{\max} = \rho(A) = \max\{|\lambda_i| \mid 1 \le i \le k\}),$$

then

$$\lambda_{\min}\|x\|^2 \le V(x) \le \lambda_{\max}\|x\|^2, \quad \text{for all } x \in R^k \tag{4.6.6}$$

where $V(x) = x^T Bx$, and $\|.\|$ is the Euclidean norm (Exercise 4.6, Problem 17).

If B is a positive definite matrix, we let $V(x) = x^T Bx$ be a Liapunov function of Eq. (4.2.2). Then, relative to Eq. (4.2.2),

$$\begin{aligned} \Delta V(x(n)) &= x^T(n)A^T BAx(n) = x^T(n)B(n) \\ &= x^T(A^T BA - B)x. \end{aligned} \tag{4.6.7}$$

Thus $\Delta V < 0$ if and only if

$$A^T BA - B = -C \tag{4.6.8}$$

for some positive definite matrix C. Equation (4.6.8) is labelled the Liapunov equation of system equation (4.2.2). The above argument establishes a sufficient condition for the asymptotic stability of the zero solution of Eq. (4.2.2). It is also a necessary and vital condition as it may be seen by the following result.

Theorem 4.37. The zero solution of Eq. (4.2.2) is asymptotically stable if and only if for every positive definite symmetric matrix C, Eq. (4.6.8) has a unique solution B which is also symmetric and positive definite.

Proof Assume that the zero solution of Eq. (4.2.2) is asymptotically stable. Let C be a positive definite symmetric matrix. We will show that the Liapunov equation (4.6.8) has a unique solution B. Multiply Eq. (4.6.8) from the left by $(A^T)^r$ and from the right by A^r to obtain

$$(A^T)^{r+1} BA^{r+1-}(A^T)^r BA^r = -(A^T)^r CA^r.$$

Hence

$$\lim_{n \to \infty} \sum_{r=0}^{n} \left[(A^T)^{r+1} BA^{r+1} - (A^T)^r BA^r \right] = -\lim_{n \to \infty} \sum_{r=0}^{n} (A^T)^r CA^r$$

and

$$\lim_{n \to \infty} \left[B - (A^T)^{n+1} BA^{n+1} \right] = \sum_{r=0}^{\infty} (A^T)^r CA^r. \tag{4.6.9}$$

Using Theorem 4.8 part (ii), we conclude that $\rho(A) < 1$ and, consequently, $\rho(A^T) < 1$. This implies that $\lim_{n\to\infty}(A^T)^{n+1}BA^{n+1} = 0$. Thus Formula (4.6.9) yields

$$B = \sum_{r=0}^{\infty}(A^T)^r C A^r. \tag{4.6.10}$$

It is a mere straightforward tribe to prove that Formula (4.6.10) gives a solution of Eq. (4.6.8) (Exercise 4.6, Problem 18). But, since $\|A^T\| < 1$, and $\|A\| < 1$, it may be shown that the series in Formula (4.6.10) converges (Exercise 4.6, Problem 18). It is easy to verify that B is symmetric and positive definite (Exercise 4.6, Problem 18).

Remark. Note that from the proof preceding the statement of Theorem 4.35, the zero solution of Eq. (4.2.2) is asymptotically stable if Eq. (4.6.8) has a unique, symmetric, and positive definite matrix B for some (not all) positive definite matrix C. Indeed, one may allow C to be the identity matrix I. In this case a solution of Eq. (4.6.8) is given by

$$B = \sum_{r=0}^{\infty}(A^T)^r A^r. \tag{4.6.11}$$

Exercise 4.6.

1. Prove Theorem 4.28.

2. Show that $\Omega(x_0) = \bigcap_{i=1}^{\infty}\bigcup_{n=i}^{\infty}\{f^n(x_0)\}$.

3. Prove that for $x_0 \in R^k$ the sets $0(x_0)$ and $\Omega(x_0)$ are positively invariant.

4. Prove Lemma 4.30.

5. Consider the planar system

$$x_1(n + 1) = x_2(n)/[1 + x_1^2(n)], \qquad x_2(n + 1) = x_1(n)/[1 + x_2^2(n)].$$

   Find the equililbrium points and determine their stability. (Hint: Let $V(x) = x_1^2 + x_2^2$.)

6. In Example (4.32)

   (a) Show that the orbit starting at the point $(1, \alpha\pi)$, where α is any irrational number, is dense in the unit circle.

   (a) Show that the set of periodic points $(1, \theta)$ is dense in the unit circle.

7. Consider the planar system

$$x_1(n + 1) = g_1(x_1(n), x_2(n)),$$
$$x_2(n + 1) = g_2(x_1(n), x_2(n)),$$

with $g_1(0, 0) = g_2(0, 0)$, and

  (i) $g_1(x)g_2(x) > 0$ for $x_1x_2 > 0$,

  (ii) $g_1(x)g_2(x) > x_1x_2$,

where $x = (x_1, x_2)$ lies in a neighborhood of the origin. Show that the origin is unstable.

8. Consider Example 4.33. Determine the stability and asymptotic stability for the equilibrium points $\left(\dfrac{1}{\sqrt{2}}, \dfrac{1}{\sqrt{2}}\right)$; $\left(-\dfrac{1}{\sqrt{2}}, -\dfrac{1}{\sqrt{2}}\right)$.

9.* Consider the system

$$x_1(n + 1) = ax_2(n)/[1 + x_1^2(n)], \qquad x_2(n + 1) = bx_1(n)/[1 + x_2^2(n)].$$

  (i) Find conditions on a and b under which

    (a) the zero solution is stable, and

    (b) the zero solution is asymptotically stable.

  (ii) Find the attractor where $a^2 = b^2 = 1$.

10. Prove that the zero solution of

$$\begin{aligned} x_1(n + 1) &= x_2(n) - x_2(n)[x_1^2(n) + x_2^2(n)] \text{ and} \\ x_2(n + 1) &= x_1(n) - x_1(n)[x_1^2(n) + x_2^2(n)] \end{aligned}$$

is asymptotically stable.

11.* Suppose that

  (i) V is a Liapunov function of system equation (4.6.1) on R^k,

  (ii) $G_\lambda = \{x | V(x) < \lambda\}$ is bounded for each λ, and

  (ii) M is closed and bounded (where M is the maximal invariant set in E).

    (a) Prove that M is a global attractor, i.e., $\Omega(x_0) \subset M$ for all $x_0 \in R^k$.

    (a) Suppose that $M = \{0\}$, and in addition to the above assumptions, V is constant on M. Verify that the origin is globally asymptotically stable.

12. Show that the sets G_λ, defined in the preceding problem, are bounded if $V(x) \to \infty$ as $\|x\| \to \infty$.

13.* (Project) Suppose that $V : R^k \to R$ is a continuous function with $\Delta^2 V(x(n)) > 0$ for $x(n) \neq 0$, where $x(n)$ is a solution of Eq. (4.6.1). Prove that for any $x_0 \in R^k$, either $x(n, x_0)$ is unbounded or $\lim_{n \to \infty} x(n, x_0) = 0$.

14.* (Project) Wade through Problem 13 again, after replacing the condition $\Delta^2 V(x(n)) > 0$ by $\Delta^2 V(x(n)) < 0$.

15. Contemplate the planar system

$x(n + 1) = y(n),$

$y(n + 1) = x(n) + f(x(n)).$

If $\Delta[x(n)f(x(n))] > 0$ for all $n \in Z^+$, prove that the solutions are either unbounded or tend to the origin. (Hint: Let $V = xy$ and then use problem 13.)

16. Prove that B is a positive definite symmetric matrix, then all its eigenvalues are positive.

17. Let B be a positive definite symmetric matrix with eigenvalues $\lambda_1 \leq \lambda_2 \leq , \ldots, \leq \lambda_k$. For $V(x) = x^T Bx$, show that $|\lambda_1| \|x\|_2^2 \leq V(x) \leq |\lambda_2| \|x\|_2^2$, for all $x \in R^k$.

18. (a) Show that the matrix $B = \sum_{r=0}^{\infty} (A^T)^r C A^r$ is symmetric and positive definite if $\|A\| < 1$ and C is a positive definite symmetric matrix.

    (b) Show that the matrix B in Formula (4.6.10) is a solution of Eq. (4.6.8).

References

[1] A. Liapunov, Problème Général de la Stabilité du Movement, *Ann. of Math Study #17*, Princeton, 1947.

[2] O. Perron, Uber Stabilität und Asymptotisches Verhalten der Integrale von Differential-Gleichungssystemen, *Math. Zeit* **29** (1929), 129–160.

[3] J.M. Ortega, *Matrix Theory, A Second Course*, Plenum, New York, 1987.

[4] R.K. Miller and A.N. Michel, *Ordinary Differential Equations*, Academic, New York, 1982.

[5] D.C. Carlson, *The Stability of Finite Difference Equations*, Master's Thesis, University of Colorado, Colorado Springs, 1989.

[6] E. Jury, *Theory and Applications of the Z-transform*, Wiley, New York, 1964.

[7] R.V. Churchill and J.W. Brown, *Complex variables and Applications*, McGraw-Hill, New York, 1990.

[8] A. Levin and R.M. May, A Note on Difference Delay Equations, *Theoretical Population Biology* **9** (1976), 178–187.

[9] K.L. Cooke and I. Gyori, Numerical Approximation of the Solutions of Delay Differential Equations on an Infinite Interval Using Piecewise Constant Arguments, *Comp. Math. Appl.* **28** (1994), 81–92.

[10] E.C. Pielou, *An Introduction to Mathematical Ecology*, Wiley, New York, 1969.

[11] A.J. Nicholson, An Outline of the Dynamics of Animal Populations, *Aust. J. Zoo.* **2**, 9–65.

[12] J.P. LaSalle, The Stability and Control of Discrete Processes. *App. Math. Sci.*, **82**, (1986).

5

The Z-Transform Method

In the last four chapters, we used the so-called *time domain analysis*. In this approach we investigate the difference equations as it is, that is without transforming it into another doman. We either find solutions of the difference equations or provide information about their qualitative behavior.

An alternate approach will be developed in this chapter. The new approach is commonly known as the *transform method*. By using a suitable transform, one may reduce the study of a linear difference or differential equation to examining an associated complex function. For example, Laplace transform method is widely used in solving and analyzing linear differential equations and continuous control systems, while the Z-transform method is most suitable for linear difference equations and discrete systems. It is widely used in the analysis and design of digital control, communication, and signal processing.

The Z-transform technique is not new and may be traced back to De Moivre around the year 1730. In fact, De Moivre introduced the more general concept of "generating functions" to probability theory.

5.1 Definitions and Examples

The Z transform of a sequence $x(n)$, which is identically zero for negative integers n (i.e., $x(n) = 0$ for $n = -1, -2, \ldots$) is defined by

$$\tilde{x}(z) = Z(x(n)) = \sum_{j=0}^{\infty} x(j)z^{-j} \qquad (5.1.1)$$

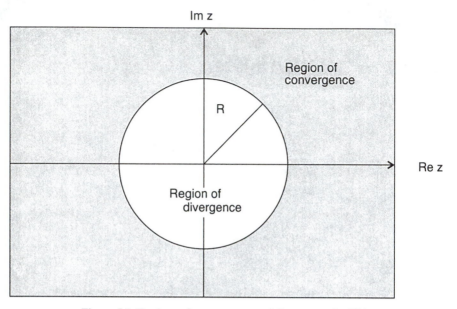

Figure 5.1. Regions of convergence and divergence for $\tilde{x}(z)$.

where z is a complex number.

The set of numbers z in the complex plane for which Series (5.1.1) converges is called the region of convergence of $x(z)$. The most commonly used method to find the region of convergence of the series (5.1.1) is the ratio test. Suppose that

$$\lim_{j \to \infty} \left| \frac{x(j+1)}{x(j)} \right| = R.$$

Then by the ratio test, the infinite series (5.1.1) converges if

$$\lim_{j \to \infty} \left| \frac{x(j+1)z^{-j-1}}{x(j)z^{-j}} \right| < 1,$$

and diverges if

$$\lim_{j \to \infty} \left| \frac{x(j+1)z^{-j}}{x(j)z^{-j}} \right| > 1,$$

Hence the series (5.1.1) converges in the region $|z| > R$ and diverges for $|z| < R$. This is depicted in Fig. 5.1, where **Re z** denotes the real axis and **Im z** represents the imaginary axis.

The number R is called the radius of convergence of series (5.1.1). If $R = 0$, the Z transform $\tilde{x}(z)$ converges everywhere with the possible exception of the origin. On the other hand if $R = \infty$, the Z transform diverges everywhere.

We now compute the Z transform of some elementary functions.

Example 5.1 Find the Z transform of the sequence $\{a^n\}$, for a fixed real number a, and its region of convergence.

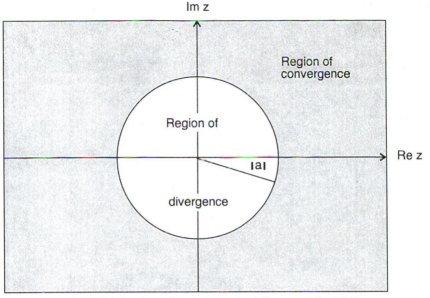

Figure 5.2. Regions of convergence and divergence for $Z(a^n)$.

Solution $Z(a^n) = \displaystyle\sum_{j=0}^{\infty} a^j z^{-j}$. The radius of convergence R of $Z(a^n)$ is given by

$$R = \lim_{j \to \infty} \left| \frac{a^{j+1}}{a^j} \right| = |a|.$$

Hence (Fig. 5.2)

$$Z(a^n) = \sum_{j=0}^{\infty} \left(\frac{a}{z} \right)^j = \frac{1}{1 - (a/z)} = \frac{z}{z - a} \quad \text{for } |z| > |a| \qquad (5.1.2)$$

A special case of the above result is when $a = 1$. In this case we have

$$Z(1) = \frac{z}{z - 1} \quad \text{for } |z| > 1.$$

Example 5.2 Find the Z transform of the sequences $\{na^n\}$ and $\{n^2 a^n\}$.

Solution Recall that an infinite series $Z(a^n)$ may be differentiated term by term any number of times in its region of convergence [2]. Now

$$\sum_{j=0}^{\infty} a^j z^{-j} = \frac{z}{(z - a)} \quad \text{for } |z| > |a|.$$

Taking the derivative of both sides yields

$$\sum_{j=0}^{\infty} -j a^j z^{-j-1} = \frac{-a}{(z - a)^2} \quad \text{for } |z| > |a|.$$

Hence

$$Z(na^n) = \sum_{j=0}^{\infty} ja^j z^{-j} = -z \sum_{j=0}^{\infty} -ja^j z^{-j-1}.$$

Therefore

$$Z(na^n) = \frac{az}{(z-a)^2} \quad |z| > 1. \tag{5.1.3}$$

Again taking the derivative of both sides of the identity

$$\sum_{j=0}^{\infty} ja^j z^{-j} = \frac{az}{(z-a)^2}, \, |z| > |a|$$

yields

$$Z(n^2 a^n) = \frac{az(z+a)}{(z-a)^3}, \quad \text{for } |z| > |a|. \tag{5.1.4}$$

Example 5.3 *The unit impulse sequence* or the *Kronecker delta sequence* is defined by

$$\delta_k(n) = \begin{cases} 1 \text{ if } n = k \\ 0 \text{ if } n \neq k \end{cases}.$$

The Z transform of this function is

$$Z(\delta_k(n)) = \sum_{j=0}^{\infty} \delta_k(j) z^{-j} = z^{-k}.$$

And if $k = 0$, we have the important special case

$$Z(\delta_0(n)) = 1. \tag{5.1.5}$$

Notice that the radius of convergence of $Z(\delta_k(n))$ is $R = 0$.

Example 5.4 Find the Z transform of the sequence $\{\sin(\omega n)\}$.

Solution Recall that the Euler Identity gives $e^{i\theta} = \cos\theta + i\sin\theta$ for any real number θ. Hence $e^{-i\theta} = \cos\theta - i\sin\theta$. Both identities yield

$$\cos\theta = \frac{e^{i\theta} + e^{-i\theta}}{2} \quad \text{and} \quad \sin\theta = \frac{e^{i\theta} + e^{-i\theta}}{2i}.$$

Thus

$$Z(\sin\omega n) = \frac{1}{2i}[Z(e^{i\omega n}) - Z(e^{-i\omega n})]$$

using Formula (5.1.2) we obtain

$$\begin{aligned} Z(\sin\omega n) &= \frac{1}{2i}\left[\frac{z}{z - e^{i\omega}} - \frac{z}{z - e^{-i\omega}}\right], \quad \text{for } |z| > 1 \\ &= \frac{z\sin\omega}{(z - e^{i\omega})(z - e^{-i\omega})} \\ &= \frac{z\sin\omega}{z^2 - (e^{i\omega} + e^{-i\omega})z + 1} \end{aligned}$$

or

$$Z(\sin \omega n) = \frac{z \sin \omega}{z^2 - 2z \cos \omega + 1}, \quad |z| > 1. \quad (5.1.6)$$

5.1.1 Properties of the Z Transform

We now establish some useful properties of the Z transform which will be needed in the sequel.

(i) *Linearity* Let $\tilde{x}(z)$ be the Z transform of $x(n)$ with radius of convergence R_1 and $\tilde{y}(z)$ be the Z transform of $y(n)$ with radius of convergence R_2. Then for any complex numbers α, β we have

$$Z[\alpha x(n) + \beta y(n)] = \alpha \tilde{x}(z) + \beta \tilde{y}(z) \text{ for } |z| > \max (R_1, R_2). \quad (5.1.7)$$

The proof of Property (5.1.7) is left to the reader as Exercise 5.1, Problem 18.

(ii) *Shifting* Let R be the radius of convergence of $\tilde{x}(z)$.

(a) Right-shifting If $x(-i) = 0$ for $i = 1, 2, \ldots, k$, then

$$Z[x(n - k)] = z^{-k}\tilde{x}(z), \quad \text{for } |z| > R. \quad (5.1.8)$$

(b) Left-shifting

$$Z[x(n + k)] = z^k \tilde{x}(z) - \sum_{r=0}^{k-1} x(r)z^{k-r} \quad \text{for } |z| > R. \quad (5.1.9)$$

The proofs are left as Exercise 5.1, Problem 16. The most commonly used cases of Formula (5.1.9) are

$$Z[x(n + 1)] = z\tilde{x}(z) - zx(0), \quad \text{for } |z| > R.$$

$$Z[x(n + 2)] = z^2\tilde{x}(z) - z^2 x(0) - zx(1), \quad \text{for} |z| > R.$$

(iii) *Initial and final-value*

(a) *Initial value theorem*

$$\lim_{|z| \to \infty} \tilde{x}(z) = x(0). \quad (5.1.10)$$

(b) *Final value theorem*

$$x(\infty) = \lim_{n \to \infty} x(n) = \lim_{z \to 1}(z - 1)\tilde{x}(z). \quad (5.1.11)$$

The proof of Formula (5.1.10) follows immediately from the definition of $\tilde{x}(z)$. To prove Formula (5.1.11) we first observe that

$$Z[x(n+1) - x(n)] = \sum_{j=0}^{\infty} [x(j+1) - x(j)]x^{-j}.$$

Using Formula (5.1.9) on the left-hand side of the above identity leads to

$$(z-1)\tilde{x}(z) = zx(0) + \sum_{j=0}^{\infty} [x(j+1) - x(j)]z^{-j}.$$

Thus

$$\begin{aligned}
\lim_{z \to 1}(z-1)\tilde{x}(z) &= x(0) + \sum_{j=0}^{\infty}[x(j+1) - x(j)] \\
&= \lim_{n \to \infty} x(n).
\end{aligned}$$

(iv) *Convolution*

A convolution * of two sequences $x(n), y(n)$ is defined by

$$x(n) * y(n) = \sum_{j=0}^{n} x(n-j)y(j) = \sum_{j=0}^{n} x(n)y(n-j).$$

Now

$$Z[x(n) * y(n)] = \sum_{m=0}^{\infty} \left[\sum_{j=0}^{m} x(m-j)y(j) \right] z^{-m}.$$

Interchanging the summation signs yields

$$Z[x(n) * y(n)] = \sum_{j=0}^{\infty} y(j) \sum_{m=j}^{\infty} x(m-j)z^{-m}.$$

And if we put $m - i = s$ we obtain

$$Z[x(n) * y(n)] = \left(\sum_{j=0}^{\infty} y(j)z^{-j} \right) \left(\sum_{s=0}^{\infty} x(s)z^{-s} \right)$$

$$Z[x(n) * y(n)] = \tilde{x}(z)\tilde{y}(z) \tag{5.1.12}$$

It is interesting to know that one may obtain Formula (5.1.12) if the convolution is defined as

$$x(n) * y(n) = \sum_{j=0}^{\infty} x(n-j)y(j).$$

(v) *Multiplication by a^n property* Suppose that $\tilde{x}(z)$ is the Z transform of $x(n)$ with radius of convergence R. Then

$$Z[a^n x(n)] = \tilde{x}\left(\frac{z}{a}\right), \qquad \text{for } |z| > |a|R. \qquad (5.1.13)$$

The proof of (5.1.13) follows easily from the definition and will be left to the reader as Exercise 5.1, Problem 19.

Example 5.5 Determine the Z transform of

$$g(n) = a^n \sin \omega n, \qquad n = 0, 1, 2, \ldots, .$$

Using Example (5.4) and Formula (5.1.13) we have

$$\tilde{g}(z) = Z(a^n \sin \omega n) = \frac{(z/a) \sin \omega}{(z/a)^2 - 2(z/a) \cos \omega + 1}$$

$$= \frac{az \sin \omega}{z^2 - 2az \cos \omega + a^2}, \qquad \text{for } |z| > |a|. \quad (5.1.14)$$

(vi) *Multiplication by n^k.* In Example 5.2 it was shown that $Z(na^n) = \frac{az}{(z-a)^2}$ which may be written in the form

$$Z(na^n) = -z \frac{d}{dz} Z(a^n).$$

Similarly Formula (5.1.4) may be written in the form

$$Z(n^2 a^n) = -z \frac{d}{dz} \left[-z \frac{d}{dz} Z(a^n) \right].$$

This may be written in the compact form

$$Z(n^2 a^n) = \left(-z \frac{d}{dz} \right)^2 Z(a^n).$$

Generally speaking we write

$$\left(-z \frac{d}{dz} \right)^k \tilde{x}(z) = \left(-z \frac{d}{dz} \left(-z \frac{d}{dz} \left(\cdots \left(-z \frac{d}{dz} \tilde{x}(z) \right) \cdots \right) \right) \right).$$

It may be shown (Exercise 5.1, Problem 7) that

$$Z[n^k x(n)] = \left(z \frac{d}{dz} \right)^k Z(x(n)). \qquad (5.1.15)$$

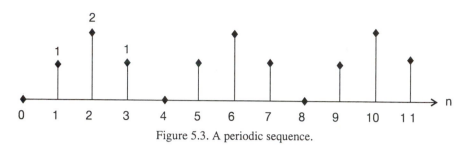

Figure 5.3. A periodic sequence.

Exercise 5.1

1. Find the Z transform and its region of convergence of the given sequence $\{x(n)\}$.

    a. $\cos \omega n.$ b. $n \sin 2n.$ c. $n.$

2. Find the Z transform and its region of convergence of the sequence

$$f(n) = \begin{cases} 1 \text{ for } n = 1, 3, 5, \\ 0 \text{ for all other values of } n \end{cases}.$$

3. Find the Z transform and its region of convergence of the sequence

$$f(n) = \begin{cases} 0 \text{ for } n = 0, -1, -2, \ldots, \\ -1 \text{ for } n = 1, \\ a^n \text{ for } n = 2, 3, 4, \ldots \end{cases}.$$

4. Let $x(n)$ be a periodic sequence of period N, i.e., $x(n + N) = x(n)$ for all $n \in Z^+$. Prove that $\tilde{x}(z) = [z^n/(z^n - 1)]\tilde{x}_1(z)$ for $|z| > 1$ where $\tilde{x}_1(z) = \sum_{j=0}^{N-1} x(j)z^{-j} (\tilde{x}_1)(z)$ is called the Z transform of the first period.

5. Determine the Z transform of the periodic sequence shown in Figure 5.3

6. Use Exercise 4 to find the Z transform and its radius of convergence for the periodic sequence of period 4

$$f(n) = \begin{cases} 1 \text{ for } n = 0, 1 \\ -1 \text{ for } n = 2, 3 \end{cases}.$$

7. Let R be the radius of convergence of $\tilde{x}(z)$. Show that

$$Z[n^k x(n)] = \left(-z\frac{d}{dz}\right)^k \tilde{x}(z) \quad \text{for } |z| > R.$$

(Hint: Use Mathematical induction on k.)

8. Prove that the Z transform of the sequence

$$x(n) = \begin{cases} (n-1)a^{n-2}, & n = 0, 1, 2, \ldots \\ 0, & n = -1, -2, \ldots \end{cases}$$

is $\tilde{x}(z) = \frac{1}{z-a^2}$ for $|z| > |a|.$

9. Find the Z transform and its region of convergence of the sequence defined by

$$x(n) = \begin{cases} \frac{(n-1)(n-2)}{2} a^{n-3}, & n = 0, 1, 2, \ldots \\ 0, & n = -1, -2, \ldots \end{cases}$$

The first backward difference for a sequence $x(n)$ is defined by $\nabla x(n) = x(n) - x(n-1)$.

10. Find $Z[\nabla x(n)]$, $Z[\Delta^2 x(n)]$.

11. Generalize the results of problem 10 and show that $Z[\nabla^k x(n)] = \left(\frac{z-1}{z}\right)^k \tilde{x}(z)$.

12. Find $Z[\Delta x(n)]$, $Z[\Delta^2 x(n)]$.

13. Show that $Z[\Delta^k x(n)] = (z-1)^k \tilde{x}(z) - z \sum_{j=0}^{k-1}(z-1)^{m-j-1}\Delta^j x(0)$.

14. Let $y(n) = \sum_{i=1}^{n} x(i)$, $n \in Z^+$. Show that $\tilde{y}(z) = \frac{z}{z-1}\tilde{x}(z)$ for $|z| > \max\{1, R\}$, where R is the radius of convergence of $\tilde{x}(z)$.

15. Let $y(n) = \sum_{i=0}^{n} i x(i)$. Prove that $\tilde{y}(z) = \frac{-z^2}{z-1}\frac{d}{dz}\tilde{x}(z)$. [Hint: $y(n) - y(n-1) = nx(n)$.]

16. Prove Formula (5.1.7) and (5.1.8).

17. Find the Z transform of

    (a) $x(n) = \sum_{r=0}^{n} a^{n-r} \sin(\omega r)$,
    (b) $\sum_{r=0}^{n} \cos \omega(n-r)$.

18. Prove Expression (5.1.7).

19. Show that $Z[a^n x(n)] = \tilde{x}\left(\frac{z}{a}\right)$ for $|z| > |a|R$, where R is the radius of convergence of $\tilde{x}(z)$.

20. Find the Z transform and its radius of convergence of the sequence $g(n) = a^n \cos(\omega n)$.

21. Use the initial value theorem to determine $x(0)$ for the sequence $\{x(n)\}$ whose Z transform is given by

    (a) $\frac{2}{z-a}$, for $|z| > a$,
    (b) $\frac{3z}{z-6}$, for $|z| > 3$.

22. Extend the initial value theorem to finding $x(1)$, $x(2)$ by proving

    (i) $x(1) = \lim_{|z|\to\infty} [z(\tilde{x}(z) - x(0))]$,

    (ii) $x(2) = \lim_{|z|\to\infty} [z(\tilde{x}(z) - zx(0) - x(1))]$.

5.2   The Inverse Z Transform and Solutions of Difference Equations

As we have mentioned in the introduction, the Z transform transforms a difference equation of an unknown sequence $x(n)$ into an algebraic equation in its Z transform $\tilde{x}(z)$. The sequence $x(n)$ is then obtained from $\tilde{x}(z)$ by a process called the inverse Z transform. This process is symbolically denoted as

$$Z^{-1}[\tilde{x}(z)] = x(n). \tag{5.2.1}$$

The question of the uniqueness of the inverse Z transform may be established as follows: Suppose that there are two sequences $x(n)$, $y(n)$ with the same Z transform, that is,

$$\sum_{i=0}^{\infty} x(i)z^{-i} = \sum_{i=0}^{\infty} y(i)z^{-i}, \quad \text{for } |z| > R.$$

Then

$$\sum_{i=0}^{\infty} [x(i) - y(i)]z^{-i} = 0, \quad \text{for } |z| > R.$$

It follows from Laurents Theorem [1] that $x(n) \equiv y(n)$. The most commonly used methods for obtaining the inverse Z transform are

1. Power Series method

2. Partial fraction method

3. Inversion integral method.

It is imperative to remind the reader when finding the inverse Z transform, it is always assumed that for any sequence $x(n)$, $x(k) = 0$ for $k = -1, -2, \ldots$.

5.2.1   The Power Series Method

In this method we obtain the inverse Z transform by simply expanding $\tilde{x}(z)$ into an infinite power series in z^{-1} in its region of convergence: $\tilde{x}(z) = \sum_{i=0}^{\infty} a_i z^{-i}$ for $|z| > R$. Then by comparing this with $Z[x(n)] = \sum_{i=0}^{\infty} x(i)z^{-i}$ for $|z| > R$, one concludes that $x(n) = a_n$, $n + 0, 1, 2, \ldots$.

If $\tilde{x}(z)$ is given in the form of a rational function $\tilde{x}(z) = g(z)/h(z)$, where $g(z)$ and $h(z)$ are polynomials in z, then we simply divide $g(z)$ by $h(z)$ to obtain a power series expansion $\tilde{x}(z)$ in z^{-1}. The only possible drawback of this method is that it does not provide us with a closed form expression of $x(n)$.

Example 5.6 Obtain the inverse Z transform of

$$\tilde{x}(z) = \frac{z(z + 1)}{(z - 1)^2}.$$

Solution We first write $x(z)$ as a ratio of two polynomials in z^{-1}:

$$\tilde{x}(z) = \frac{1 + z^{-1}}{1 - 2z^{-1} + z^{-2}}.$$

Dividing the numerator by the denominator, we have

$$\tilde{x}(z) = 1 - 3z^{-1} + 4z^{-2} - 4z^{-3} + 4z^{-4} - 4z^{-5} + \ldots.$$

Thus

$$x(0) = 1, \quad x(2) = -3, \quad x(3) = 4, \quad x(4) = -4, \ldots,$$

and in general $x(n) = 4$ if $n \geq 3$ and n is odd, $x(n) = -4$ if $n \geq 3$ and n is even.

5.2.2   The Partial Fraction Method

This method is used when the Z transform $\tilde{x}(z)$ is a rational function in z, analytic at ∞, such as

$$\tilde{x}(z) = \frac{b_0 z^m + b_1 z^{m-1} + \ldots + b_{m-1} z + b_m}{z^n + a_1 z^{n-1} + \ldots + a_{n-1} z + b_n}, \quad m \leq n. \quad (5.2.2)$$

If $\tilde{x}(z)$ in Expression (5.2.2) is expressed by a partial fraction expression,

$$\tilde{x}(z) = \tilde{x}_1(z) + \tilde{x}_2(z) + \tilde{x}_3(z) + \ldots$$

then by the linearity of the inverse Z transform one obtains

$$x(n) = Z^{-1}[\tilde{x}_1(z)] + Z^{-1}[\tilde{x}_2(z)] + Z^{-1}[\tilde{x}_3(z)] + \ldots.$$

Then a Z-transform table (Table 5.1, see end of Chapter 5) is used to find $Z^{-1}[\tilde{x}_i(z)]$, $i = 1, 2, 3, \ldots$.

Before giving some examples to illustrate this method we remind the reader that the zeros of the numerator of Expression (5.2.2) are called *zeros* of $\tilde{x}(z)$ and zeros of the denominator of Expression (5.2.2) are called *poles* of $\tilde{x}(z)$.

Remark Since $\tilde{x}(z)$ is often an improper fraction, it is more convenient to expand $\tilde{x}(z)/z$ rather than $\tilde{x}(z)$ into sums of partial fractions.

Example 5.7 (Simple poles) Solve the difference equation

$$x(n + 2) + 3x(n + 1) + 2x(n) = 0, \quad x(0) = 1, \quad x(1) = -4.$$

Solution Taking the Z transform of both sides of the equation, we get

$$\tilde{x}(z) = z(z - 1)/(z + 1)(z + 2).$$

We expand $\tilde{x}(z)/z$ into partial fractions as follows:

$$\tilde{x}(z)/z = \frac{(z - 1)}{(z + 1)(z + 2)} = \frac{a_1}{z + 1} + \frac{a_2}{z + 2}.$$

Clearing fractions, we obtain

$$z - 1 = a_1(z + 2) + a_2(z + 1).$$

This reduces to

$$z - 1 = (a_1 + a_2)z + (2a_1 + a_2).$$

Comparing coefficients of like powers of z, we get

$$a_1 + a_2 = 1,$$

$$2a_1 + a_2 = -1.$$

Hence $a_1 = -2$, $a_2 = 3$. Consequently

$$\tilde{x}(z) = \frac{-2z}{z + 1} + \frac{3z}{z + 2}.$$

Thus

$$x(n) = -2(-1)^n + 3(-2)^n.$$

Remark If $\tilde{x}(z)$ has a large number of poles, a computer may be needed to determine the constants $a_1, a_2, \ldots$.

Example 5.8 (Repeated poles) Solve the difference equation

$$x(n + 4) + 9x(n + 3) + 30x(n + 2) + 20x(n + 1) + 24x(n) = 0,$$

$$x(0) = 0, \quad x(1) = 0, \quad x(2) = 1, \quad x(3) = 10.$$

Solution Taking the Z transform, we get

$$\tilde{x}(z) = \frac{z(z - 1)}{(z + 2)^3(z + 3)}.$$

It is convenient here to expand $\tilde{x}(z)/z$ into partial fractions as follows:

$$\tilde{x}(z)/z = \frac{z - 1}{(z + 2)^3(z + 3)} = \frac{b}{z + 3} + \frac{a_1}{(z + 2)^3} + \frac{a_2}{(z + 2)^2} + \frac{a_3}{z + 2}. \tag{5.2.3}$$

This time we use a smarter method to find a_1, a_2, a_3, and a_4. To find b we multiply Eq. (5.2.3) by $(z + 3)$ and then evaluate at $z = -3$. This gives

$$b = \left.\frac{(z - 1)}{(z + 2)^3}\right|_{z=-3} = 4.$$

To find a_1 we multiply Eq. (5.2.3) by $(z + 2)^3$ to get

$$\frac{z - 1}{z + 3} = a_3(z + 2)^2 + a_2(z + 2) + a_1 + 4\frac{(z + 2)^3}{(z + 3)} \tag{5.2.4}$$

and evaluate at $z = -2$. This gives

$$a_1 = \frac{z-1}{z-3}\bigg|_{z=-2} = -3.$$

To find a_2 we differentiate Eq. (5.2.4) with respect to z to get

$$\frac{4}{(z+3)^2} = 2a_3(z+2) + a_2 + \frac{r(2z+7)(z+2)^2}{(z+3)^2} \qquad (5.2.5)$$

and again evaluate at $z = -2$. This gives

$$a_2 = \frac{d}{dz}\left(\frac{z-1}{z+3}\right)\bigg|_{z=-2} = 4.$$

Finally, to find a_3 we differentiate Eq. (5.2.5) to obtain

$$\frac{-8}{(z+3)^3} = 2a_3 + 4\frac{d^2}{dz^2}\frac{(z+2)^3}{(z+3)},$$

and if we let $z = -2$, then we have

$$a_3 = \frac{1}{2}\frac{d^2}{dz^2}\left(\frac{z-1}{z+3}\right)\bigg|_{z=-2} = -4.$$

Hence

$$\tilde{x}(z) = \frac{-4z}{z+2} + \frac{4z}{(z+2)^2} - \frac{3z}{(z+2)^3} + \frac{4z}{z+3}.$$

The corresponding sequence is (Table 5.1, at end of Chapter 5)

$$\begin{aligned}
x(n) &= -4(-2)^n - 2n(-2)^n + \frac{3}{4}n(n-1)(-2)^n + 4(-3)^n \\
&= \left(\frac{3}{4}n^2 - \frac{11}{4}n - 4\right)(-2)^n + 4(-3)^n.
\end{aligned}$$

Remark The procedure used to obtain $a_1, a_2,$ and a_3 in the preceding example can be generalized. If $\tilde{x}(z)/z$ has a pole of multiplicity m at $z = z_0$ then the corresponding terms in the partial fraction expansion can be written

$$\ldots + \frac{a_1}{(z-z_0)^m} + \ldots + \frac{a_m}{z-z_0} + \ldots$$

and $a_1, a_2, \ldots, a_m$ can be found using the formula

$$a_r = \frac{1}{(r-1)!}\frac{d^{r-1}}{dz^{r-1}}\left[(z-z_0)^m\frac{\tilde{x}(z)}{z}\right]\bigg|_{z=z_0}.$$

Example 5.9 (Complex poles): Solve the difference equation

$$x(n+3) - x(n+2) + 2x(n) = 0, \qquad x(0) = 1, \qquad x(1) = 1.$$

Solution Taking the Z transform of the equation, we get

$$\tilde{x}(z) = \frac{z^3}{(z^2 - 2z + 2)(z + 1)}.$$

Next we expand $\tilde{x}(z)/z$ as a sum of partial fraction in the form

$$\tilde{x}(z)/z = \frac{z^2}{(z^2 - 2z + 2)(z + 1)} = \frac{a_1}{[z - (1 + i)]} + \frac{a_2}{[z - (1 - i)]} + \frac{a_3}{(z + 1)}.$$

Using the method of the preceding example we obtain

$$a_3 = \left. \frac{z^2}{z^2 - 2z + 2} \right|_{z=-1} = \frac{1}{5},$$

$$a_1 = \left. \frac{z^2}{[z - (1 - i)](z + 1)} \right|_{z=1+i} = \frac{1}{2+i} = \frac{2}{5} - \frac{1}{5}i,$$

$$a_2 = \bar{a}_1 = \frac{2}{5} + \frac{1}{5}i.$$

Hence

$$\tilde{x}(z) = \frac{\frac{1}{5}z}{z + 1} + \frac{a_1 z}{z - \lambda} + \frac{\bar{a}_1 z}{z - \bar{\lambda}},$$

where $\lambda = 1 + i$. Thus

$$x(n) = \frac{1}{5}(-1)^n + a_1 \lambda^n + \bar{a}_1 \bar{\lambda}^n.$$

But

$$a_1 \lambda^n + \bar{a}_1 \bar{\lambda}^n = 2Re(a_1 \lambda^n)$$
$$= 2|\bar{a}_1|(\sqrt{2})^n \cos\left(\frac{n\pi}{4} + \arg a_1\right)$$

where $|a_1| = \frac{1}{5}\sqrt{5}$ and $\arg a_1 = \tan^{-1}(1/2) = 0.46$ radians. Thus

$$x(n) = \frac{1}{5}(-1)^n + \frac{2}{5}\sqrt{5}(\sqrt{2})^n \cos\left(\frac{n\pi}{4} + 0.46\right).$$

5.2.3   The Inversion Integral Method[1]

From the definition of the Z transform, we have

$$\tilde{x}(z) = \sum_{i=0}^{\infty} x(i)z^{-i}.$$

---

[1]Requires some knowledge of Residues in Complex Analysis. [1]

Multiplying both sides of the above equation by z^{n-1}, we get

$$
\begin{aligned}
\tilde{x}(z)z^{n-1} &= \sum_{i=0}^{\infty} x(i)z^{n-i-1} \\
&= x(0)z^{n-1} + x(1)z^{n+2} + \ldots + x(n)z^{-1} + x(n)z^{-2} + \ldots .
\end{aligned}
$$

$$(5.2.6)$$

Equation (5.2.6) gives the laurent series expansion of $\tilde{x}(z)z^{n-1}$ around $z = 0$.

Consider a circle C, centered at the origin of the z plane, that encloses all poles of $\tilde{x}(z)z^{n-1}$. Since $x(n)$ is the coefficient of z^{-1}, it follows by the Cauchy's integral formula [1].

$$
x(n) = \frac{1}{2\pi i} \oint_C \tilde{x}(z)z^{n-1}dz. \qquad (5.2.7)
$$

And by the Residue Theorem [1] we obtain

$$
x(n) = \text{sum of residues of } \tilde{x}(z)z^{n-1}. \qquad (5.2.8)
$$

Suppose that

$$
\tilde{x}(z)z^{n-1} = \frac{h(z)}{g(z)}.
$$

In evaluating the residues of $\tilde{x}(z)z^{n-1}$, there are two cases to consider.

(i) $g(z)$ has simple zeros (i.e., $\tilde{x}(z)z^{n-1}$ has simple poles) (see Fig. 5.4). In this case the residue K_i at a pole z_i is given by

$$
K_i = \lim_{z \to z_i} \left[(z - z_i) \frac{h(z)}{g(z)} \right]
$$

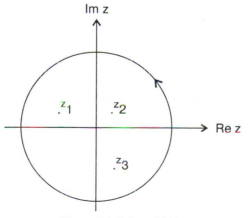

Figure 5.4. Poles of $\tilde{x}(z)$.

(ii) $g(z)$ has multiple zeros (i.e., $\tilde{x}(z)z^{n-1}$ has multiple poles). If $g(z)$ has a multiple zero z_i of order r, then the residue K_i at z_i is given by

$$K_i = \frac{1}{(r-1)!} \lim_{z \to z_i} \frac{d^{r-1}}{dz^{r-1}} \left[(z - z_i)^r \frac{h(z)}{g(z)} \right].$$

Example 5.10 Obtain the inverse Z transform of

$$\tilde{x}(z) = \frac{z(z-1)}{(z-2)^2(z+3)}.$$

Solution Notice that

$$\tilde{x}(z)z^{n-1} = \frac{(z-1)z^n}{(z-2)^2(z+3)}.$$

Thus $\tilde{x}(z)z^{n-1}$ has a simple pole at $z_1 = -3$ and a double pole at $z_2 = 2$. Thus from Formula (5.2.8), we get $x(n) = K_1 + K_2$, where K_1, K_2 are the residues of $x(z)z^{n-1}$ at z_1, z_2, respectively. Now

$$K_1 = \lim_{z \to -3} \left[\frac{(z+3)(z-1)z^n}{(z-2)^2(z+3)} \right] = \frac{-4}{25}(-3)^n,$$

$$\begin{aligned} K_2 &= \frac{1}{(2-1)!} \lim_{z \to 2} \frac{d}{dz} \left[\frac{(z-2)^2(z-1)z^n}{(z-2)^2(z+3)} \right] \\ &= \lim_{z \to 2} \frac{z^{n-1}[(z+3)(z+nz-n) - z(z-1)]}{(z+3)^2} \\ &= \frac{(8+5n)}{25}(2)^{n-1}. \end{aligned}$$

Thus

$$x(n) = \frac{-4}{25}(-3)^n + \frac{(8+5n)}{25}(2)^{n-1}, \quad n = 0, 1, 2, \ldots.$$

Example 5.11 (*Electric circuits or a ladder network*). Consider the electric network shown in Fig. 5.5. Here $i(n)$ is the current in the nth loop, R is the resistance which is assumed to be constant in every loop, and V is the voltage. By Ohm's law, the voltage (or electric potential) between the ends of a resistor R may be expressed as $V = iR$. Now Kirchhoff's[2] second law states that "in a closed circuit the impressed voltage is equal to the sum of the voltage drops in the rest of the circuit." By applying the Kirchhoff's law on the loop corresponding to $i(n+1)$ we obtain

$$R[i(n+1) - i(n+2)] + R[i(n+1) - i(n)] + Ri(n+2) = 0$$

---

[2] Gustav Kirchhoff, a German Physicist, (1824–1887) is famous for his contribution in electricity and spectroscopy.

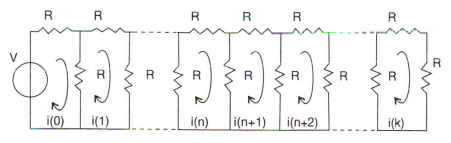

Figure 5.5. A ladder network.

or

$$i(n+2) - 3i(n+1) + i(n) = 0. \qquad (5.2.9)$$

And for the first loop on the left we have

$$V = Ri(0) + R(i(0) - i(1))$$

or

$$i(1) = 2i(0) - \frac{V}{R}. \qquad (5.2.10)$$

Taking the Z transform of Eq. (5.2.9) under the Data (5.2.10) yields the equation

$$
\begin{aligned}
\tilde{i}(z) &= \frac{z[zi(0) - 3i(0) + i(1)]}{z^2 - 3z + 1} \\
&= \left[\frac{z^2 - \left(1 + \frac{V}{Ri(0)}\right)z}{z^2 - 3z + 1} \right] i(0). \qquad (5.2.11)
\end{aligned}
$$

Let $\omega > 0$ be such that $\cosh \omega = \frac{3}{2}$. Then $\sinh \omega = \frac{\sqrt{5}}{2}$. Then Expression (5.2.11) becomes

$$
\begin{aligned}
\tilde{i}(z) = i(0)&\left[\frac{z^2 - z \cosh \omega}{z^2 - 2z \cosh \omega + 1} \right] + \left(\frac{i(0)}{2} + \frac{V}{R}\right)\left(\frac{2}{\sqrt{5}}\right) \\
&\left[\frac{z \sinh \omega}{z^2 - 2z \cosh \omega + 1} \right].
\end{aligned}
$$

Taking the inverse Z transform (Table 5.1 at end of Chapter 5) we obtain

$$i(n) = i(0)\cosh(\omega n) + \left(\frac{i(0)}{2} + \frac{V}{R}\right)\left(\frac{2}{\sqrt{5}}\right) \sinh(\omega n).$$

Exercise 5.2

1. Use partial fractions method to find the inverse Z transform of

(a) $\dfrac{z}{\left(z - \frac{1}{2}\right)(z+1)}$,   (b) $\dfrac{z(z+1)}{(z+2)^2(z-1)}$.

2. Use the Power Series Method to find the inverse Z transform of

$$\text{(a)} \quad \frac{z-2}{(z-1)(z+3)}, \qquad \text{(b)} \quad \frac{e^{-a}z}{(z-e^{-a})^2}.$$

3. Use the inversion integral method to find the inverse Z transform of

$$\text{(a)} \quad \frac{z(z-1)}{(z+2)^3}, \qquad \text{(b)} \quad \frac{z(z+2)}{\left(z-\frac{1}{2}\right)(z+i)(z-i)}.$$

4. Use the partial fraction method and the inversion integral method to find the inverse Z transform of

$$\text{(a)} \quad \frac{z(z+1)}{(z-2)^2}, \qquad \text{(b)} \quad \frac{z^2+z+1}{(z-1)(z^2-z+1)}.$$

In problems 5 through 7, use the Z-transform method to solve the given difference equation.

5. (The Fibonacci sequence) $x(n+2) = x(n+1)+x(n)$, $\quad x(0) = 0$, $\quad x(1) = 1$.

6. $x(n+2) - 3x(n+1) + 2x(n) = \delta_0(n)$, $\quad x(0) = x(1) = 0$.

7. $(n+1)x(n+1) - nx(n) = n+1$, $\quad x(0) = 0$.

8. Continued fractions: Consider the equation $x = 1 + \frac{1}{1+\frac{1}{1+\frac{1}{\cdots}}}$. Then one may write this equation as the difference equation $x(n+1) = 1 + \frac{1}{x(n)}$, $x(0) = 0$.

   (a) Solve the difference equation. (Hint: Let $x(n) = \frac{y(n+1)}{y(n)}$.)

   (b) Find $x(n)$.

9. Prove that the convolution product is commutative and associative (i.e., $x * y = y * x; x * (y * f) = (x * y) * f$).

10. Solve, using convolution, the equation $x(n+1) = 2 + 4\sum_{r=0}^{n}(n-r)x(r)$.

11. Solve the equation $x(n) = 1 - \sum_{r=0}^{n-1} e^{n-r-1}x(r)$. (Hint: Replace n by $n+1$.)

5.3   Volterra Difference Equations of Convolution Type: The Scalar Case[3]

Volterra difference equations of convolution type is of the form

$$x(n+1) = Ax(n) + \sum_{j=0}^{n} B(n-j)x(j) \qquad (5.3.1)$$

where $a \in R$ and $B : Z^+ \to R$ is a discrete function. This equation may be considered as the discrete analogue of the famous Volterra integrodifferential equation

$$x'(t) = Ax(t) + \int_0^t B(t-s)x(s)ds. \qquad (5.3.2)$$

Equation (5.3.2) has been widely used as a mathematical model in population dynamics. Both Eqs. (5.3.1) and (5.3.2) represent a system in which the future state $x(n+1)$ does not depend only on the present state $x(n)$ but also on all past states $x(n-1), x(n-2), \ldots, x(0)$. These systems are sometimes called *hereditary*. Given the initial condition $x(0) = x_0$, one can easily generate the solution $x(n, x_0)$ of Eq. (5.3.1). If $y(n)$ is any other solution of Eq. (5.3.1) with $y(0) = x_0$, then it is easy to show that $y(n) = x(n)$ for all $n \in Z^+$ (Exercise 5.3, Problem 8).

One of the most effective methods of dealing with Eq. (5.3.1) is the Z-transform method. Let us rewrite Eq. (5.3.1) in the convolution form

$$x(n+1) = Ax(n) + B * x. \qquad (5.3.1)'$$

Taking formally the Z transform of both sides of Equation (5.3.1)', we get

$$z\tilde{x}(z) - zx(0) = A\tilde{x}(z) + \tilde{B}(z)\tilde{x}(z)$$

which gives

$$[z - A - \tilde{B}(z)]\tilde{x}(z) = zx(0)$$

or

$$\tilde{x}(z) = zx(0)/[z - A - \tilde{B}(z)]. \qquad (5.3.3)$$

Let

$$g(z) = z - A - \tilde{B}(z). \qquad (5.3.4)$$

The complex function $g(z)$ will play an important role in the stability analysis of Eq. (5.3.1). Before embarking on our investigation of $g(z)$ we need to present few definitions and preliminary results.

Definition 5.12 Let E be the space of all infinite sequences of complex numbers (or real numbers): $x = (x(0), x(1), x(2), \ldots)$. There are three commonly used norms that may be defined on subsets of E. These are

---

[3]Requires some elements of complex analysis [1].

(i) the l_1 norm: $\| x \|_1 = \sum_{i=0}^{\infty} |x(i)|$,

(ii) the l_2 or Euclidean norm: $\| x \|_2 = \left[\sum_{i=0}^{\infty} |x(i)|^2 \right]^{1/2}$,

(iii) the l_∞ norm: $\| x \|_\infty = \sup_{i \geq 0} |x(i)|$.

The corresponding formed spaces are l_1, l_2, and l_∞, respectively. One may show easily that (Exercise 5.3, Problem 6)

$$l_1 \subset l_2 \subset l_\infty.$$

Definition 5.13 A complex function $g(z)$ is said to be *analytic* in a region in the complex plane if it is differentiable there. The next result establishes an important property of l_1 sequences.

Theorem 5.14 If $x(n) \in l_1$, then

(i) $\tilde{x}(z)$ is an analytic function for $|z| \geq 1$;

(ii) $|\tilde{x}(z)| \geq \| x \|$ for $|z| \geq 1$.

Proof

(i) Since $x(n) \in l_1$, the radius of convergence of $\tilde{x}(z) = \sum_{n=0}^{\infty} x(n) z^{-n}$ is $R = 1$. Hence $\tilde{x}(z)$ can be differentiated term by term in its region of convergence $|z| > 1$. Thus $\tilde{x}(z)$ is analytic on $|z| > 1$. Furthermore, since $x(n) \in l_1, \tilde{x}(z)$ is analytic for $|z| = 1$.

(ii) This is left as Exercise 5.3, Problem 9.

We now turn our attention to the function $g(z) = z - A - \tilde{B}(z)$ in Formula (5.3.4). This function plays the role of the characteristic polynomial of linear difference equations. (See Chapter 2.) In contrast to polynomials, the function $g(z)$ may have infinitely many zeros in the complex plane. The following lemma sheds some light on the location of the zeros of $g(z)$.

Lemma 5.15 [2] The zeros of

$$g(z) = z - A - \tilde{B}(z)$$

all lie in the region $|z| < c$, for some real positive constant c. Moreover, $g(z)$ has finitely many zeros z with $|z| \geq 1$.

Proof Suppose that the zeros of $g(z)$ do not lie in any region $|z| < c$ for any positive real number c. Then there exists a sequence $\{z_i\}$ of zeros of $g(z)$ with $|z_i| \to \infty$ as $i \to \infty$. Now

$$|z_i - A| = |\tilde{B}(z_i)| \leq \sum_{n=0}^{\infty} |B(n)| |z_i|^{-n}. \tag{5.3.5}$$

Notice that the right-hand side of Inequality (5.3.5) goes to $B(0)$ as $i \to \infty$, while the left-hand side goes to ∞ as $i \to \infty$, which is a contradiction. This proves the first part of the lemma.

To prove the second part of the lemma, we first observe from the first part of the lemma that all zeros z of $g(z)$ with $|z| \geq 1$ lie in the annulus $1 \leq |z| \leq c$ for some real number c. From Theorem 5.14 we may conclude that $g(z)$ is analytic in this annulus $(1 \leq |z| \leq c)$. Therefore $g(z)$ has only finitely many zeros in the region $|z| \geq 1$ [1].

Next we embark on a program that will reveal the qualitative behavior of solutions of Eq. (5.3.1). In this program we utilize Eq. (5.3.3) which may be written as

$$\tilde{x}(z) = x(0)zg^{-1}(z). \tag{5.3.6}$$

Let γ be a circle that includes all the zeros of $g(z)$. The circle γ is guaranteed to exist by virtue of Lemma (5.15). By Formula (5.2.7) we obtain

$$x(n) = \frac{1}{2\pi i} \oint_\gamma x(0)z^n g^{-1}(z)dz \tag{5.3.7}$$

and by Formula (5.2.8) we get

$$x(n) = \text{sum of residues of } [x(0)z^n g^{-1}(z)]. \tag{5.3.8}$$

This suggests that

$$x(n) = \sum p_r(n)z_r^n \tag{5.3.9}$$

where the sum is taken over all the zeros of $g(z)$ and where $p_r(n)$ is a polynomial in n of degree less than $k-1$ if z_r is a multiple root of order k. To show the validity of Formula (5.3.9), let z_r be a zero of $g(z)$ of order k. We write the following Laurent's series expansion [1]

$$g^{-1}(z) = \sum_{n=-k}^{\infty} g_n(z - z_r)^n, \text{ for some constants } g_n,$$

$$z^n = [z_r - (z_r - z)]^n = \sum_{i=0}^{n} \binom{n}{i} z_r^{n-i}(z - z_r)^i.$$

The residue of $x(0)z^n g^{-1}$ at z_r is $x(0)$ times the coefficient of $(z-z_r)^{-1}$ in $g^{-1}(z)z^n$. The coefficient of $(z - z_r)^{-1}$ in $g^{-1}(z)z^n$ is given by

$$g_{-k}\binom{n}{k-1}z_r^{n-k+1} + g_{-k+1}\binom{n}{k-2}z_r^{n-k+2} + \ldots + g_{-1}\binom{n}{0}z_r^n. \tag{5.3.10}$$

It follows from Formula (5.3.8) that $x(n)$ may be given by Formula (5.3.9).

Formula (5.3.9) has the following important consequences.

Theorem 5.16 [3] The zero solution of Eq. (5.3.1) is uniformly stable if and only if

(a) $z - A - \tilde{B}(z) \neq 0$ for all $|z| > 1$, and

(b) If z_r is a zero of $g(z)$ with $|z_r| = 1$, then the residue of $z^n g_{-1}(z)$ at z_r is bounded as $n \to \infty$.

Proof Suppose that conditions (a) and (b) hold. If z_r is a zero of $g(z)$ with $|z_r| < 1$, then from Formula (5.3.9) its contribution to the solution $x(n)$ is bounded. On the other hand, if z_r is a zero of $g(z)$ with $|z_r| = 1$ at which the residue of $x(0)z^n g^{-1}(z)$ is bounded as $n \to \infty$, then from Formula (5.3.8), its contribution to the solution $x(n)$ is also bounded. This shows that $|x(n)| \leq L|x(0)|$ for some $L > 0$, and thus we have uniform stability. The converse is left to the reader as Exercise 5.3, Problem 10.

We observe here that a necessary and sufficient condition for condition (b) is that each zero z of $g(z)$ with $|z| = 1$ must be simple (Exercise 3.5, Problem 11).

The next result addresses the question of asymptotic stability.

Theorem 5.17 [3] The zero solution of Eq. (5.3.1) is uniformly asymptotically stable if and only if

$$z - A - \tilde{B}(z) \neq 0, \quad \text{for all } |z| \geq 1. \qquad (3.5.11)$$

Proof The proof follows easily from Formula (5.3.9) and is left to the reader as Exercise 5.3, Problem 13.

Exercise 5.3

1. Solve the Volterra difference equation $x(n + 1) = 2x(n) + \sum_{r=0}^{n} 2^{n-r} x(r)$, and then determine the stability of its zero solution.

2. Solve the Volterra difference equation $x(n + 1) = -1/2x(n) + \sum_{r=0}^{n} 3^{r-n} x(r)$, and then determine the stability of its zero solution.

3. Use Theorems 5.16 and 5.17 to determine the stability of the zero solutions of the difference equations in problems 1 and 2.

4. Without finding the solution of the equation

$$x(n + 1) = -\frac{1}{4}x(n) + \sum_{r=0}^{n} \left(\frac{1}{2}\right)^{r-n} x(r),$$

determine the stability of its zero solution.

5. Determine the stability of the zero solution of $x(n + 1) = 2x(n) - 12 \sum_{r=0}^{n} (n - r)x(r)$, using Theorem 5.16 or 5.17.

6. Prove that $l_1 \subset l_2 \subset l_\infty$.

7. Let $x = \{x_n\}$ and $y = \{y_n\}$ be two l_1 sequences. Prove that $x * y \in l_1$ by following the following steps:

(i) If $\sum_{i=0}^{\infty} x(i) = a$, $\sum_{i=0}^{\infty} y(i) = b$, and $c(n) = \sum_{i=1}^{n} x(n-i)$ $y(i)$, show that $\sum_{i=0}^{\infty} c(i) = ab$.

(ii) Prove that $\sum_{n=0}^{\infty} |c(n)| \leq \left(\sum_{i=0}^{\infty} |x(i)|\right) \left(\sum_{j=0}^{\infty} |y(j)|\right)$.

8. Prove the uniqueness of solutions of Eq. (5.3.1), that is, if $x(n)$ and $y(n)$ are solutions of Eq. (5.3.1) with $x(0) = y(0)$, then $x(n) = y(n)$ for all $n \in Z^+$.

9. If $x(n) \in l_1$ show that $|\tilde{x}(z)| \leq \| x \|_1$ for $|z| \geq 1$.

10.* Suppose that the zero solution of Eq. (5.3.1) is uniformly stable. Prove that

    (a) $g(z) = z - A - \hat{B}(z) \neq 0$ for all $|z| > 1$, and

    (b) If z_r is a zero of $g(z)$ with $|z_r| = 1$, then the residue of $z^{n-1} g^{-1}(z)$ at z_r is bounded.

11. Prove that a necessary and sufficient condition for condition (b) in Theorem 5.16 is that z_r is a simple root of $g(z)$.

12.* Prove Theorem 5.17.

5.4 Explicit Criteria for Stability of Volterra Equations

The stability results in Section 5.3 are not very practical since locating the zeros of $g(z)$ is more or less impossible in most problems. In this section we provide explicit conditions for the stability of Eq. (5.3.1). The main tools in this study are Theorems 5.17 and Rouche's Theorem (Theorem 4.16).

Theorem 5.18 [2] The zero solution of Eq. (5.3.1) is asymptotically stable if

$$\left| |A| + \left| \sum_{n=0}^{\infty} B(n) \right| \right| < 1. \tag{5.4.1}$$

Proof Let $\beta = \sum_{n=0}^{\infty} B(n)$ and $D(n) = \beta^{-1} B(n)$. Then $\sum_{n=0}^{\infty} D(n) = 1$. Furthermore, $\tilde{D}(1) = 1$ and $|\tilde{D}(z)| \leq 1$ for all $|z| \geq 1$. Let us write $g(z)$ in the form

$$g(z) = z - A - \beta \tilde{D}(z). \tag{5.4.2}$$

To prove uniform asymptotic stability of the zero solution of Eq. (5.3.1), it suffices to show that $g(z)$ has no zero z with $|z| \geq 1$. So assume there exists a zero z_r of $g(z)$ with $|z_r| \geq 1$. Then by Eq. (5.4.2) we obtain $|z_r - A| = |\beta \tilde{D}(z)| \leq |\beta|$. Using Condition (5.4.1) one concludes that $|z_r| \leq |A| + |\beta| < 1$, which is a contradiction. This concludes the proof of the theorem.

Unfortunately, we are not able to show that Condition (5.4.1) is a necessary condition for asymptotic stability. However in the next result we give a partial converse to the above theorems.

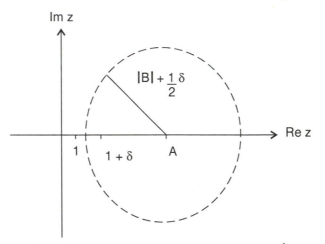

Figure 5.6. A circle with center A and radius $|B| + \frac{\delta}{2}$.

Theorem 5.19 [2] Suppose that $B(n)$ does not change sign for $n \in Z^+$. Then the zero solution of Eq. (5.3.1) is not asymptotically stable if any one of the following conditions hold:

(i) $A + \sum_{n=0}^{\infty} B(n) \geq 1$,

(ii) $A + \sum_{n=0}^{\infty} B(n) \leq -1$ and $B(n) > 0$ for some $n \in Z^+$,

(iii) $A + \sum_{n=0}^{\infty} B(n) < -1$ and $B(n) < 0$ for some $n \in Z^+$, and $\sum_{n=0}^{\infty} B(n)$ is sufficiently small.

Proof Let β and $D(n)$ be as defined in the proof of Theorem 5.18.

(i) Assume Condition (i). If $A + \beta = 1$, then clearly $z = 1$ is a root of $g(z)$ defined in Eq. (5.4.2). Hence by Theorem 5.16 the zero solution of Eq. (5.4.1) is not asymptotically stable. If $A + \beta > 1$, say $A + \beta = 1 + \delta$, then there are two areas to consider.

   (a) If $\beta < 0$, then we let γ be the circle in the complex plane with center at A and radius equal to $|\beta| + \frac{1}{2}\delta$. Then on γ (Figure 5.6) we have $|z| > 1$ and thus
   $$|\beta \tilde{D}(z)| \leq |\beta| < |z - A|. \tag{5.4.3}$$

   Let $h(z) = -\beta \tilde{D}(z)$, $f(z) = z - A$. Then from Inequality (5.4.3) $|h(z)| < |f(z)|$ on γ. Hence by Rouche's Theorem (Theorem 4.16), $g(z) = f(z) + h(z)$ and $f(z)$ have the same number of zeros inside γ. Since A is the only zero of $f(z)$ inside γ, then $g(z)$ has exactly one zero z_0 inside γ with $|z_0| > 1$. Again by using Theorem 5.16, the zero solution of Eq. (5.3.1) is not asymptotically stable.

(b) Suppose that $\beta > 0$. Since $A+\beta > 1$, it follows that $g(z) = 1-A-\beta < 0$. Moreover $|\tilde{D}(A + \beta)| \leq 1$. Thus $g(A + \beta) = \beta[1 - \tilde{D}(A + \beta)] \geq 0$. Therefore $g(z)$ has a zero between 1 and $A + \beta$ and consequently, by Theorem 5.17, the zero solution of Eq. (5.3.1) is not asymptotically stable. This completes the proof of Condition (i).

Parts (ii) and (iii) are left to the reader as Exercises 5.4, Problems 7 and 8.

The above techniques are not extendible to uniform stability. This is mainly due to the lack of easily verifiable criteria for Condition (b) of Theorem 5.16. Therefore new techniques are needed to tackle the problem of uniform stability. These techniques involve the use of Liapunov functionals (functions) which we have encountered in Chapter 4.

Let E by the space of all infinite sequences of complex numbers as defined in Definition 5.12. Then a function $V : E \to R$ is said to a *Liapunov functional* if for $x = \{x(n)\} \in E$,

(i) $V(x)$ is positive definite (Chapter 4),

(ii) $\Delta V(x) \leq 0$,

where $\Delta V(x) = V(\hat{x}) - V(x)$ and $\hat{x}(n) = x(n + 1)$ for all $n \in Z^+$.

The next result illustrates the use of Liapunov functionals in stability theory.

Theorem 5.20 [2] The zero solution of Eq. (5.3.1) is uniformly stable if

$$|A| + \sum_{j=0}^{n} |B(j)| \leq 1, \quad \text{for all } n \in Z^+ \tag{5.4.4}$$

Proof Let for $x \in E$,

$$V(x) = |x(n)| + \sum_{r=0}^{n-1} \sum_{s=n}^{\infty} |B(s - r)||x(r)| \tag{5.4.5}$$

Then

$$\Delta V(x) \;=\; \left| Ax(n) + \sum_{j=0}^{n} B(n - j)x(j) \right| + \sum_{r=0}^{n} \sum_{s=n+1}^{\infty} |B(s - r)||x(r)|$$

$$-\; |x(n)| - \sum_{r=0}^{n-1} \sum_{s=n}^{\infty} |B(s - r)||x(r)| \tag{5.4.6}$$

$$\leq \left(|A| + \sum_{j=0}^{\infty} |B(j)| - 1 \right) |x(n)|. \tag{5.4.7}$$

By Assumption (5.4.4) we thus have

$$\Delta V(x) \leq 0. \tag{5.4.8}$$

From Eq. (5.4.5) we obtain $|x(n)| \leq V(x)$. Using Inequality (5.4.8) and Expression (5.4.5) again we obtain

$$|x(n)| \leq V(x) \leq |x(0)|.$$

Consequently, the zero solution is uniformly stable (Chapter 4).

Exercise 5.4

Use Theorem 5.19 to determine the stability and instability of the zero solution of the equations in Problems 1, 2, and 3.

1. $x(n + 1) = -\frac{1}{4}x(n) + \sum_{r=0}^{n} \left(\frac{1}{3}\right)^{n+1-r} x(r)$.

2. $x(n + 1) = \frac{1}{2}x(n) + \sum_{r=0}^{n} (n - r)x(r)$.

3. $x(n + 1) = \frac{1}{3}x(n) + \sum_{r=0}^{n} e^{r-n}x(r)$.

4. [4]Find the values of a for which the zero solution of the equation $x(n) = \sum_{r=0}^{n-1} (n - r - 1)a^{n-r-1}x(r)$ is

   (i) uniformly stable

   (ii) asymptotically stable, and

   (iii) not asymptotically stable.

5. [4]Determine the values of a for which the zero solution of the equation $\Delta x(n) = -\frac{2}{3}x(n) + \sum_{r=0}^{n} (n - r)^2 a^{n-r}x(r)$ is asymptotically stable.

6. Prove Theorem 5.18 using the method of Liapunov functional used in the proof of Theorem 5.20.

7. Prove part (ii) of Theorem 5.19.

8. Prove part (iii) of Theorem 5.19.

9. Provide details of how Inequality (5.4.7) is obtained from Inequality (5.4.6).

10. (Open problem). Discuss the stability of the zero solution of Eq. (3.5.1) under the condition $A + \sum_{n=0}^{\infty} B(n) = -1$ and $\sum_{n=0}^{\infty} B(n) < 0$.

11. (Open problem). Can we omit the assumption that $\sum_{n=0}^{\infty} B(n)$ is sufficiently small in Theorem 5.19 part (iii)?

12. (Open problem). Develop a necessary and sufficient condition for the asymptotic stability of the zero solution of Eq. (5.3.1).

---

[4]$\sum_{n=0}^{\infty} na^n = \frac{a}{(1-a)^2}$ for $a < 1$, $\sum_{n=0}^{\infty} n^2 a^n = \frac{a^2+a}{(1-a)^3}$.

5.5   Volterra Systems

In this section we are mainly interested in the following Volterra system of convolution type

$$x(n + 1) = Ax(n) + \sum_{j=0}^{n} B(n - j)x(j) \tag{5.5.1}$$

where $A = (a_{ij})$ is a $k \times k$ real matrix and $B(n)$ is a $k \times k$ real matrix defined on Z^{+}. It is always assumed that $B(n) \in l_1$ i.e., $\sum_{j=0}^{\infty} |B(j)| < \infty$. The Z transform for sequences in R^k and matrices $R^{k \times k}$ is defined in the natural way, that is,

$$
\begin{aligned}
Z[x(n)] &= \left(Z(x_1(n)), Z(x_2(n)), \dots, Z(x_k(n)) \right)^T, \\
Z[B(n)] &= (Z(b_{ij}(n)).
\end{aligned}
$$

Thus all the rules and formulae for the Z transform of scalar sequences hold for vector sequences and matrices.

Taking the Z transform of both sides of Eq. (5.5.1) one obtains

$$z\, \tilde{x}(z) - z\, x(0) = A\, \tilde{x}(z) + \tilde{B}(z)\, \tilde{x}(z), |z| > R$$

which yields

$$\tilde{x}(z) = [zI - A - \tilde{B}(z)]^{-1}zx(0), |z| > R. \tag{5.5.2}$$

Theorem 5.17 for scalar equations has the following counterpart for systems.

Theorem 5.21 A necessary and sufficient condition for the uniform asymptotic stability is

$$\det(zI - A - \tilde{B}(z)) \neq 0, \quad \text{for all } |z| \geq 1. \tag{5.5.3}$$

Proof See [3]

An application of the preceding theorem will be introduced next. This will provide explicit criteria for asymptotic stability. But before introducing our result we need the following lemma concerning eigenvalues of matrices [4].

Lemma 5.22 [4] Let $G = (g_{ij})$ be a $k \times k$ matrix. If z_0 is an eigenvalue of G, then,

(i) $|z_0 - g_{ii}|\, |z_0 - g_{jj}| \leq \sum_r' |g_{ir}| \sum_r' |g_{jr}|$, for some $i, j, i \neq j$, and

(ii) $|z_0 - g_{tt}|\, |z_0 - g_{ss}| \leq \sum_r' |g_{rt}| \sum_r' |g_{rs}|$, for some $t, s, t \neq s$,

where $\sum_r' g_{ir}$ means $\left(\sum_{r=1}^{k} g_{ir} \right) - g_{ii}$.

Using the above lemma we can prove the next result. Let

$$\beta_{ij} = \sum_{n=0}^{\infty} |b_{ij}(n)|, \quad 1 \leq i, j \leq k,$$

Theorem 5.23 [2] The zero solution of Eq. (5.5.1) is uniformly asymptotically stable if either one of the following conditions hold.

(i) $\sum_{j=1}^{k} (|a_{ij}| + \beta_{ij}) < 1,$ for each $i, 1 \leq i \leq k,$ or

(ii) $\sum_{i=1}^{k} (|a_{ij}| + \beta_{ij}) < 1,$ for each $j, 1 \leq j \leq k.$

Proof (i) To prove uniform asymptotic stability under Condition (i) we need to show that Condition (5.5.3) holds. So assume the contrary, that is

$$\det(z_0 I - A - \tilde{B}(z_0)) = 0, \quad \text{for some } z_0 \text{ with } |z_0| \geq 1.$$

Then z_0 is an eigenvalue of the matrix $A + \tilde{B}(z_0)$. Hence by Condition (i) in Lemma 5.22, we have

$$|z_0 - a_{ii} - \tilde{b}_{ii}(z_0)| \, |z_0 - a_{jj} - \tilde{b}_{jj}(z_0)| \leq \sum_r^{\prime} |a_{ir} + \tilde{b}_{ir}(z_0)| \sum_r^{\prime} |a_{jr} + \tilde{b}_{jr}(z_0)|.$$

$$(5.5.4)$$

But

$$
\begin{aligned}
|z_0 - a_{ii} - \tilde{b}_{ii}(z_0)| &\geq |z_0| - |a_{ii}| - |\tilde{b}_{ii}(z_0)| \\
&\geq 1 - |a_{ii}| - |\tilde{b}_{ii}(z_0)| \\
&> \sum_r^{\prime} (|a_{ir}| + |\beta_{ir}|) \quad \text{(by Condition (i))}.
\end{aligned}
$$

Similarly,

$$|z_0 - a_{jj} - \tilde{b}_{jj}(z_0)| > \sum_r^{\prime} (|a_{jr}| + \beta_{jr}).$$

Combining both inequalities, we get

$$|z_0 - a_{ii} - \tilde{b}_{ii}(z_0)| \, |z_0 - a_{jj} - \tilde{b}_{jj}(z_0)| > \sum_r^{\prime} (|a_{ir}| + \beta_{ir}) \sum_r^{\prime} (|a_{jr}| + \beta_{jr}).$$

It is clear that this contradicts Inequality (5.5.4) if one notes that for any $1 \leq s, m \leq k$

$$|a_{st}| + \beta_{st} \geq |a_{st}| + |\tilde{b}_{st}(z_0)| \geq |a_{st} + \tilde{b}_{st}(z_0)|.$$

As in the scalar case, the above method may be extended to provide criteria for uniform stability. Again, the method of Liapunov functionals will come to the rescue.

Theorem 5.24 [2] The zero solution of Eq. (5.5.1) is uniformly stable if

$$\sum_{i=1}^{k} |a_{ij}| + \beta_{ij} \leq 1 \qquad (5.5.5)$$

for all $j = 1, 2, \ldots, k.$

Proof Define the Liapunov functional

$$V(x) = \sum_{i=1}^{k} \left[|x_i(n)| + \sum_{j=1}^{k} \sum_{r=0}^{n-1} \sum_{s=n}^{\infty} |b_{ij}(s-r)| \; |x_j(r)| \right].$$

Then

$$\Delta V_{(5.5.1)}(x) \le \sum_{i=1}^{k} \left[\sum_{j=1}^{k} |a_{ij}| \; |x_j(n)| - |x_i(n)| \right.$$
$$\left. + \sum_{j=1}^{k} \sum_{s=n}^{\infty} |b_{ij}(s-n)| \; |x_j(n)| \right]. \qquad (5.5.6)$$

A crucial but simple step is now in order. Observe that

$$\sum_{i=1}^{k} \sum_{j=1}^{k} |a_{ij}| \; |x_j(n)| = \sum_{i=1}^{k} \sum_{j=1}^{k} |a_{ji}| \; |x_i(n)|, \text{ and}$$

$$\sum_{i=1}^{k} \sum_{j=1}^{k} \sum_{s=n}^{\infty} |b_{ij}(s-n)| \; |x_j(n)| = \sum_{i=1}^{k} \sum_{j=1}^{k} \sum_{s=n}^{\infty} |b_{ij}(s-n)| \; |x_i(n)|$$

(Exercise 5.5, Problem 1).

Hence Inequality (5.5.6) now becomes

$$\Delta V_{(5.5.1)}(x) \le \sum_{i=1}^{k} \left[\sum_{j=1}^{k} |a_{ji}| + b_{ji} - 1 \right] |x_i(n)|$$
$$\le 0 \quad \text{(by Condition (5.5.5)).}$$

This implies that

$$|x(n)| \le V(x) \le \sum_{i=1}^{k} |x_i(0)| = \| x(0) \|$$

which proves uniform stability.

Exercise 5.5

1. Prove that $\sum_{i=1}^{k} \sum_{j=1}^{k} \sum_{s=n}^{\infty} |b_{ij}(s-n)| \; |x_j(n)| = \sum_{i=1}^{k} \sum_{j=1}^{k} \sum_{s=n}^{\infty} |b_{ji}(s-n)| \; |x_i(n)|.$

In Problems 2 through 6 determine whether the zero solution of the given equation is uniformly stable or uniformly asymptotically stable.

2. $x(n+1) = \sum_{j=0}^{n} B(n-j)x(j)$, where $B(n) = \begin{pmatrix} e^{-n} & 1 \\ 0 & e^{-n} \end{pmatrix}$.

3. $x(n + 1) = Ax(n) + \sum_{j=0}^{n} B(n - j)x(j)$,   where $A = \begin{pmatrix} 0 & \frac{1}{5} \\ \frac{1}{3} & \frac{1}{4} \end{pmatrix}$, $B(n)$

$= \begin{pmatrix} 4^{-n-1} & 0 \\ 0 & 3^{-n-1} \end{pmatrix}$.

4. $x(n + 1) = \sum_{j=0}^{n} B(n - j)x(j)$,   where $B(n) = \begin{pmatrix} -1 & 4^{-n-1} \\ 0 & 3^{-n-1} \end{pmatrix}$.

5. $x(n + 1) = \sum_{j=0}^{n} B(n - j)x(j)$,   where $B(n) = \begin{pmatrix} 2^{-n-1} & e^{-n-1} \\ 0 & 5^{-n-1} \end{pmatrix}$.

6.* Theorem (A. Brauer[4]). Let $G = (g_{ij})$ be a real $k \times k$ matrix. Then $\det G > 0$ if $g_{ii} > 0$, $g_{ii}g_{jj} > \sum_{r}' |g_{ir}| \sum_{r}' |g_{jr}|$,   for all $1 \leq i$, $j \leq k$, $i \neq j$. Assume that $v_{ij} = \sum_{n=0}^{\infty} b_{ij}(n) < \infty$ for $1 \leq i, j \leq k$. Suppose that the following two conditions hold

   (i) $a_{ii} + v_{ii} > 1$, $1 \leq i \leq k$,

   (ii) $(a_{ii} + v_{ii} - 1)(a_{jj} + v_{jj} - 1) > \sum_{r}' |a_{ir} + v_{ir}|$ for $1 \leq i$, $j \leq k, i \neq j$.
   Prove that

   (a) If k is odd, then the zero solution of Eq. (5.5.1) is not asymptotically stable,

   (b) If k is even, then the zero solution of Eq. (5.5.1) may or may not be asymptotically stable.

7.* (Open problem) Discuss the stability of the zero solution of Eq. (5.5.1) under the condition $a_{ii} + v_{ii} \leq 1$, $1 \leq i \leq k$. Consider the Volterra system with infinite delay

$$x(n + 1) = Ax(n) + \sum_{j=0}^{\infty} B(n - j)x(j). \qquad (5.5.7)$$

8. Mimic the proof of Theorem 5.23 to find criteria for the asymptotic stability of the zero solution of Eq. (5.5.7).

9. Mimic the proof of Theorem 5.24 to find criteria for the uniform stability of the zero solution of Eq. (5.5.7).

10. Prove Theorem 5.23 using the method of Liapunov functionals (as in Theorem 5.24).

5.6 A Variation of Constants Formula

Associated with the homogeneous system (5.5.1) we contemplate the following nonhomogeneous system

$$y(n+1) = Ay(n) + \sum_{j=0}^{n} B(n-j)y(j) + g(n) \qquad (5.6.1)$$

where $g(n) \in R^k$.

The existence and uniqueness of solutions of System (5.5.1) may be established by a straightforward argument (Exercise 5.6, Problem 14). Let $e_i = (0, \ldots, 1, \ldots, 0)^T$ be the standard ith unit vector in R^k, $1 \le i \le k$. Then there are k vector solutions $x_1(n), x_2(n), x_3(n), \ldots, x_k(n)$ of System (5.5.1) with $x_i(n) = e_i$, $1 \le i \le k$. The set solutions are linearly independent on Z^+. For if there is a nontrivial linear relation $c_1 x_1(n) + c_2 x_2(n) + \ldots + c_k x_k(n) = 0$ on Z^+ then at $n = 0$ we have $c_1 e_1 + c_2 e_2 + \ldots + c_k e_k = 0$. This proves that $c_1 = c_2 = \ldots = c_k = 0$, which is a contradiction. The $k \times k$ matrix $X(n)$, whose ith column is $x_i(n)$, is called the *fundamental matrix* of System (5.5.1). Notice that $X(n)$ is a nonsingular matrix with $X(0) = I$. Moreover, $x(n) = X(n)x_0$ is a solution of system equation (5.5.1) with $x(0) = x_0$ (Exercise 5.6, Problem 1). Furthermore, the fundamental matrix $X(n)$ satisfies the matrix equation (Exercise 5.6, Problem 2).

$$X(n+1) = AX(n) + \sum_{j=0}^{n} B(n-j)X(j). \qquad (5.6.2)$$

It should be pointed out that the fundamental matrix $X(n)$ enjoys all the properties possessed by its counterpart in ordinary difference equations (Chapter 3).

Next we give the variation of constants formula.

Theorem 5.25 Suppose that the Z transforms of $B(n)$ and $g(n)$ exist, then the solution $y(n)$ of System (5.6.1) with $y(n_0) = y_0$ is given by

$$y(n, 0, y_0) = X(n)y_0 + \sum_{r=0}^{n-1} X(n-r-1)g(r). \qquad (5.6.3)$$

Proof We first observe that (why?)

$$X(n+1) = AX(n) + \sum_{r=0}^{n} B(n-r)X(r) \qquad (5.6.4)$$

Taking the Z transform of both sides of Eq. (5.6.4), we obtain for some $R > 0$,

$$z\tilde{X}(z) - zX(0) = A\tilde{X}(z) + \tilde{B}(z)\tilde{X}(z), \quad |z| > R.$$

This yields

$$[zI - A - \tilde{B}(z)]\tilde{X}(z) = zI, \quad |z| > R. \qquad (5.6.5)$$

Since the right-hand side of Eq. (5.6.5) is nonsingular, it follows that the matrix $zI - A - \tilde{B}(z)$ is also nonsingular (why?). This implies that

$$\tilde{X}(z) = z[zI - A - \tilde{B}(z)]^{-1}, |z| > R. \qquad (5.6.6)$$

In the next step of the proof we take the Z transform of both sides of System (5.6.1) to obtain

$$\tilde{y}(z) = [zI - A - \tilde{B}(z)]^{-1}[zy_0 + \tilde{g}(z)], \quad |z| > R_1$$

for some $R_1 \geq R$, and by using Formula (5.6.6) this gives

$$\tilde{y}(z) = \tilde{X}(z)y_0 + \frac{1}{2}\tilde{X}(z)\tilde{g}(z), \quad |z| > R_1.$$

Thus

$$y(n) = Z^{-1}[\tilde{X}(z)y_0] + Z^{-1}\left[\frac{1}{2}\tilde{X}(z)\tilde{g}(z)\right]$$

$$= X(n)y_0 + \sum_{r=0}^{n-1} X(n - r - 1)g(r) \text{ (using Formulas (5.1.4) and (5.1.8))}.$$

Exercise 5.6

1. Let $X(n)$ be the fundamental matrix of system equation (5.5.1). Prove that $x(n) = X(n)x_0$ is a solution of Eq. (5.5.1) for any vector $x_0 \in R^k$.

2. Prove that the fundamental matrix $X(n)$ satisfies Eq. (5.6.2).

3. Prove that the zero solution of Eq. (5.5.1) is uniformly stable if and only if $|x(n, n_0, x_0)| \leq M|x_0|$, for some $M > 0$.

4. Prove that the zero solution of Eq. (5.5.1) is uniformly asymptotically stable if and only if there exist $M > 0, v \in (0, 1)$ such that $|x(n, n_0, x_0)| \leq Mv^{n-n}0$. (Hint: See Theorem 4.4.)

5. Solve the equation $x(n + 1) = -2\sqrt{3}\,x(n) + \sum_{r=0}^{n} 2^{n-r}(3^{1/2(n-r+1)})x(r) + 2^n(3^{n/2}), \quad x(0) = 0$.

   (a) by the Z-transform method, and

   (b) by using Theorem 5.25.

6. Solve the equation $x(n + 1) = \frac{1}{2}x(n) + \sum_{r=0}^{n}(n - r)x(r) + n$

   (a) by the Z-transform method, and

   (b) using Problem 5 and Theorem 5.25.

7. Consider the planar system $x(n + 1) = Ax(n) + \sum_{j=0}^{n} B(n - j)x(j) + g(n), \quad x(0) = 0$ where

$$A = \begin{pmatrix} -\sqrt{2} & 0 \\ 0 & -\sqrt{6} \end{pmatrix}, \quad B(n) = \begin{pmatrix} 2^{-n/2} & 0 \\ 0 & 6^{-n/2} \end{pmatrix}.$$

    (a) Find the fundamental matrix $X(n)$ of the homogeneous equation.

    (b) Use Theorem 5.25 to solve the equation when $g(n) = \begin{pmatrix} n \\ 0 \end{pmatrix}$.

8. Consider the system $\Delta x(n) = \sum_{j=0}^{n} B(n-j)x(j) + g(n)$, where $B(n) = \begin{pmatrix} 1 & 0 \\ 0 & 2^n \end{pmatrix}$.

    (a) Solve the homogeneous part when $g(n) = 0$,

    (b) Use Theorem 5.25 to find the solution of the nonhomogeneous equation
when $g(n) = \begin{pmatrix} a \\ 0 \end{pmatrix}$, where a is a constant.

9. Consider the system

$$y(n+1) = Ay(n) + g(n), \quad y(0) = y_0(*).$$

Use the Z transform to show that

    (a) $A^n = Z^{-1}[z(zI - A)^{-1}]$.

    (b) $\sum_{r=0}^{n-1} A^{n-r-1} g(r) = Z^{-1}[(zI - A)^{-1}\tilde{g}(z)]$.

    (c) Conclude that the solution of the given equation is given by $y(n) = Z^{-1}[z(zI - A)^{-1}]y_0 + Z^{-1}[(zI - A)^{-1}\tilde{g}(z)]$.

10. Use Eq. (5.6.5) to show that for some $R > 0$, $\det(zI - A - \tilde{B}(z)) \neq 0$ for $|z| > R$.

Apply the method of Problem 9 to solve equation (*) in Problem 9 if A and $g(n)$ are given as follows:

11. $A = \begin{pmatrix} 3 & -2 \\ 1 & 0 \end{pmatrix}$, $g(n) = \begin{pmatrix} n \\ 3 \\ 0 \end{pmatrix}$, $y(0) = 0$.

12. $A = \begin{pmatrix} 0.5 & 1 \\ 0 & 0.5 \end{pmatrix}$, $g(n) = 0$.

13. Prove the existence and uniqueness of the solutions of Eq. (5.6.1).

5.7   The Z Transform Versus Laplace Transform[5]

The Laplace transform plays the same role in differential equation as does the Z transform in difference equations. For a continuous function $f(t)$, the Laplace transform is defined by

$$\hat{f}(s) = L\big(f(t)\big) = \int_0^\infty e^{-st}\, f(t)\,dt.$$

If one discretizes this integral we get $\sum_{n=0}^\infty e^{-sn} f(n)$. If further we let $z = e^s$ we get the Z transform of $f(n)$, namely $\sum_{n=0}^\infty f(n)z^{-n}$. Hence given $s = \alpha + i\beta$ in the s plane (commonly called the frequency domain in engineering), then

$$z = e^{\alpha+i\beta} = e^\alpha\, e^{i\beta} = e^\alpha \cdot e^{i(\beta+2n\pi)}, \quad n \in \mathbb{Z}.$$

Hence a point in the z plane corresponds to infinitely many points in the s plane. Observe that the left half of the s plane corresponds to the interior of the unit circle $|z| < 1$ in the z plane. Thus asymptotic stability of a differential equation is obtained if all the roots of its characteristic equation have negative real parts. In difference equations this corresponds to the condition that all the roots of the characteristic equation lie inside the unit circle.

There is another method that enables us to carry the stability analysis from the s plane to the z plane, i.e., from differential equations to difference equation. For suppose that the characteristic equation of a difference equation is given by

$$P(z) = a_0 z^n + a_1 Z^{n-1} + \ldots + a_n = 0.$$

The bilinear transformation defined by

$$z = \frac{s+1}{s-1}$$

maps the interior of the unit circle to the left half plane in the complex plane (Figure 5.7). To show this we let $s = \alpha + i\beta$. Then

$$|z| = \left| \frac{\alpha + i\beta + 1}{\alpha + i\beta - 1} \right| < 1$$

or

$$\frac{(\alpha + 1)^2 + \beta^2}{(\alpha - 1)^2 + \beta^2} < 1$$

which gives $\alpha < 0$. Now substituting $z = \frac{s+1}{s-1}$ into $P(z)$ we obtain

$$a_0 \left(\frac{s+1}{s-1} \right)^n + a_1 \left(\frac{s+1}{s-1} \right)^{n-1} + \ldots + a_n = 0$$

---

[5]This section may be skipped by readers who are not familiar with the Laplace transform.

Table 5.1. Z-transform pairs

No.	$x(n)$ for $n = 0, 1, 2, 3, \ldots$ $x(n) = 0$ for $n = -1, -2, -3, \ldots$	$\tilde{x}(z) = \sum_{n=0}^{\infty} x(n) z^{-n}$
1.	1	$z/z - 1$
2.	a^n	$z/z - a$
3.	a^{n-1}	$\frac{1}{z-a}$
4.	n	$z/(z-1)^2$
5.	n^2	$z(z+1)/(z-1)^3$
6.	n^3	$z(z^2 + 4z + 1)/(z-1)^4$
7.	n^k	$(-1)^k D^k \left(\frac{z}{z-1}\right); D = z\frac{d}{dx}$
8.	na^n	$az/(z-a)^2$
9.	$n^2 a^n$	$az(z+a)/(z-a)^3$
10.	$n^3 a^n$	$az(z^2 + 4az + a^2)/(z-a)^4$
11.	$n^k a^n$	$(-1)^k D^k \left(\frac{z}{z-a}\right); D = z\frac{d}{dz}$
12.	$\sin n\omega$	$z \sin \omega/$ $(z^2 - 2z \cos \omega + 1)$
13.	$\cos n\omega$	$z(z - \cos \omega)/$ $(z^2 - 2z \cos \omega + 1)$
14.	$a^n \sin n\omega$	$az \sin n\omega/$ $(z^2 - 2az \cos \omega + a^2)$
15.	$a^n \cos n\omega$	$z(z - a \cos \omega)/$ $(z^2 - 2az \cos \omega + a^2)$
16.	$\delta_0(n)$	1
17.	$\delta_m(n)$	z^{-m}
18.	$a^n/n!$	$e^{a/z}$
19.	$\cosh n\omega$	$z(z - \cosh \omega)/$ $(z^2 - 2z \cosh \omega + 1)$
20.	$\sinh n\omega$	$z \sinh \omega/$ $(z^2 - 2z \cosh \omega + 1)$
21.	$\frac{1}{n}, n > 0$	$\ln (z/z - 1)$
22.	$e^{-\omega n} x(n)$	$\tilde{x}(e^{\omega} z)$
23.	$n^{(2)} = n(n-1)$	$2z/(z-1)^3$
24.	$n^{(3)} = n(n-1)(n-2)$	$3!z/(z-1)^4$
25.	$n^{(k)} = n(n-1) \ldots (n-k+1)$	$k!z/(z-1)^{k+1}$
26.	$x(n-k)$	$z^{-k} \tilde{x}(z)$
27.	$x(n+k)$	$z^k \tilde{x}(z) - \sum_{r=0}^{k-1} x(r) z^{k-r}$

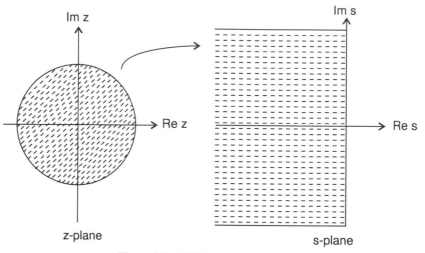

Figure 5.7. A bilinear transformation.

or

$$Q(s) = b_0 s^n + b_1 s^{n-1} + \ldots + b_n = 0.$$

We now can apply the Routh stability criterion [5] on $Q(s)$ to check if all the zeros of $Q(s)$ are in the left half plane. If this is the case then we know for sure that the zeros of $P(z)$ all lie inside the unit circle. We are not going to pursue this approach here since the computation involved is horrendous.

References

[1] R.V. Churchill and J.W. Brown, *Complex Variables and Applications*, Mc-Graw Hill, New York, 1990.

[2] S. Elaydi, "Stability of Volterra Difference Equations of Convolution Type," *Proceedings of the Special Program at Nankai Institute of Mathematics*, (Ed. Liao Shan-Tao et. al.), World Scientific, Singapore, 1993, pp. 66–73.

[3] S. Elaydi, "Global Stability of Difference Equations," *Proceedings of the First World Congress of Nonlinear Analysis*, Florida, 1992, Kluwer (forthcoming).

[4] A. Brauer, "Limits for the Characteristic Roots of a Matrix, II," *Duke Math. J.*, **14** (1947), 21–26.

[5] R.K. Miller and A.N. Michael, *Ordinary Differential Equations*, Academic, New York, 1982.

Bibliography

A main reference for the Z transform is the book by E.I. Jury, *Theory and Applications of the Z-Transform Method*, Robert E. Kreiger, Florida, 1982.

Two very readable books that include extensive treatment of the Z transform method are:

J.A.Cadzow, *Discrete Time Systems*, Prentice Hall, New Jersey, 1973.

K. Ogata, *Discrete-Time Control Systems*, Prentice Hall, New Jersey, 1987.

6

Control Theory

6.1 Introduction

In the last three decades, control theory has gained importance as a discipline for engineers, mathematicians, scientists, and other researchers. Examples of control problems include landing a vehicle on the moon, controlling the economy of a nation, manufacturing robots, controlling the spread of an epidemic, etc. Though a plethora of other books discuss continuous control theory [1, 2, 3], we will present here an introduction to discrete control theory.

We may represent a physical system we intend to control by the homogeneous difference system

$$x(n + 1) = Ax(n) \qquad (6.1.1)$$

where A is a $(k \times k)$ matrix. We extensively studied this equation in Chapters 3 and 4; here we will refer to it as an *uncontrolled system*.

To control this system, or to induce it to behave in a predetermined fashion, we introduce into the system a forcing term, or a *control $u(n)$*. Thus, the *controlled system* is the nonhomogeneous system

$$x(n + 1) = Ax(n) + u(n). \qquad (6.1.2)$$

In realizing System (6.1.2), it is assumed that the control can be applied to affect directly each of the state variables $x_1(n), x_2(n), \ldots, x_k(n)$ of the system. In most applications, however, this assumption is unrealistic. For example, in controlling an epidemic, we cannot expect to be able to affect directly all of the state variables of the system.

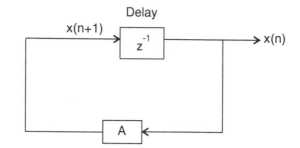

Figure 6.1a. (Uncontrolled system).

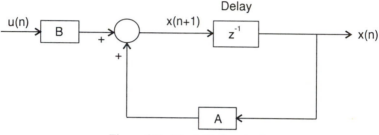

Figure 6.1b. (Controlled system).

We find another example in the realm of economics. Economists, and certain politicians even, would pay dearly to know how the rate of inflation can be controlled, especially by altering some or all of the following variables: taxes, the money supply, bank lending rates, and so on. There probably is no equation like Eq. (6.1.2) that accurately describes the rate of inflation. Thus, a more reasonable model for the controlled system may be developed: we denote it by

$$x(n + 1) = Ax(n) + Bu(n) \tag{6.1.3}$$

where B is a $(k \times m)$ matrix sometimes called the input matrix, and $u(n)$ is a $(m \times 1)$ vector. In this system, we have m control variables or components $u_1(n), u_2(n), \ldots, u_m(n)$, where $m \leq k$.

In engineering design and implementation, the system is often represented by a block diagram, as in Fig. 6.1a–b.

The delay is represented traditionally by z^{-1} since $\frac{1}{z} Z[x(n + 1)] = Z[x(n)]$. (See Fig. 6.2.)

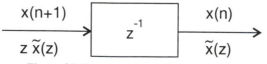

Figure 6.2. Representation of system delay.

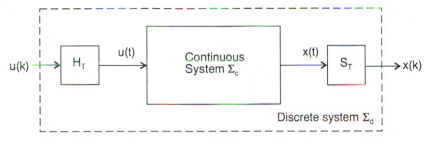

Figure 6.3. A continuous system with ideal sampler and zero-order hold.

6.1.1 *Discrete Equivalents for Continuous Systems*

One of the main areas of application for the discrete control methods developed in this chapter is the control of continuous systems, i.e., those modeled by differential and not difference equations. The reason for this is that, while most physical systems are modeled by differential equations, control laws are often implemented on a digital computer whose inputs and outputs are sequences. A common approach to control design in this case is to obtain an equivalent difference equation model for the continuous system to be controlled.

The block diagram of Fig. 6.3 shows a common method of interfacing a continuous system to a computer for control. The system $\sum_c$ has state vector $x(t)$ and input $u(t)$ and is modeled by the differential equation

$$\dot{x}(t) = \hat{A}(t) + \hat{B}u(t). \tag{6.1.4}$$

The system S_T is an *ideal sampler* which produces, given a continuous signal $x(t)$, a sequence $x(k)$ defined

$$x(k) = x(kT). \tag{6.1.5}$$

The system H_T is a *zero-order hold* which produces, given a sequence $u(k)$, a piecewise-constant continuous signal $u_c(t)$ defined

$$u(t) = u(k), \quad t \in [kT, (k+1)T). \tag{6.1.6}$$

It's not hard to check that the solution to Eq. (6.1.4) for $t \in [kT, (k+1)T)$ is given by

$$x(t) = e^{\hat{A}t}x(kT) + \int_{kT}^{t} e^{\hat{A}(t-\tau)}\hat{B}u(\tau)d\tau. \tag{6.1.7}$$

Thus a difference equation model for the overall system $\sum_d$ (indicated by the dotted box in Fig. 6.3 can be obtained by evaluating Formula (6.1.7) at $t = (k+1)T$ and using Eqs. (6.1.5) and (6.1.6).

$$x(k+1) = Ax(k) + Bu(k), \tag{6.1.8}$$

where

$$A = e^{\hat{A}T} \text{ and } B = Te^{\hat{A}T}\hat{B}. \tag{6.1.9}$$

Example 6.1 A current–controlled DC motor can be modeled by the differential equation

$$\dot{x}(t) = -\frac{1}{\tau} x(t) + \frac{K}{\tau} u(t),$$

where x is the motor's angular velocity, u is the applied armature current, and K and τ are constants. A difference equation model suitable for the design of a discrete control system for this motor can be found using Eqs. (6.1.8) and (6.1.9).

$$x(k+1) = Ax(k) + Bu(k),$$

where

$$A = e^{\hat{A}T} = e^{-T/\tau} \text{ and } B = Te^{\hat{A}T}\hat{B} = \frac{KT}{\tau} e^{-T/\tau}.$$

6.2   Controllability

In this section we are mainly interested in the problem of whether it is possible to steer a system from a given initial state to any arbitrary state in a finite time period. In other words, we would like to determine whether a desired objective can be achieved by manipulating the chosen control variables. Until 1960, transform methods were the main tools in the analysis and design of controlled systems. Such methods are referred to now as classical Control Theory. In 1960, the Swiss mathematician/engineer R.E. Kalman [4] laid down the foundation of modern control theory by introducing state space methods. Consequently, matrices have gradually replaced transforms (e.g., Z transform, Laplace transform), as the principle mathematical machinery in modern control theory [5, 6, 7].

Definition 6.2 System (6.1.3) is said to be *completely controllable* (or simply controllable) if for any $n_0 \in Z^+$, any initial state $x(n_0) = x_0$, and any given final state (the desired state) x_f, there exists a finite time $N > n_0$ and a control $u(n)$, $n_0 < n \le N$ such that $x(N) = x_f$.[1]

Remark Since System (6.1.3) is completely determined by the matrices A and B, we may speak of the controllability of the pair $\{A, B\}$.

In other words, there exists a sequence of inputs $u(0), u(1), \ldots, u(N-1)$ such that this input sequence, applied to System (6.1.3), yields $x(N) = x_f$.

Example 6.3 Consider the system governed by the equations

$$\begin{aligned}
x_1(n+1) &= a_{11}x_1(n) + a_{12}x_2(n) + bu(n), \\
x_2(n+1) &= a_{22}x_2(n).
\end{aligned}$$

Here

$$A = \begin{pmatrix} a_{11} & a_{12} \\ 0 & a_{22} \end{pmatrix}, \quad B = \begin{pmatrix} b \\ 0 \end{pmatrix}.$$

---

[1] In some books this may be referred to as completely reachable.

It will not take much time before we realize that this system is not completely controllable, since $u(n)$ has no influence on $x_2(n)$. Moreover, $x_2(n)$ is entirely determined by the second equation and is given by $x_2(n) = a_{22}^n x_2(0)$.

The above example was easy enough that we were able to determine controllability by inspection. For more complicated systems, we are going to develop some simple criteria for controllability.

The *controllability matrix* W of System (6.1.3) is defined as the $k \times km$ matrix

$$W = [B, AB, A^2 B, \ldots, A^{k-1} B]. \tag{6.2.1}$$

The controllability matrix plays a major role in control theory, as may be seen in the following important basic result.

Theorem 6.4 System (6.1.3) is *completely controllable* if and only if rank $W = k$.

Before proving the theorem, we make some few observations about it and then prove a preliminary result.

First consider the simple case when the system has only a single input, and thus the input matrix B reduces to an $m \times 1$ vector b. Hence the controllability matrix becomes the $k \times k$ matrix

$$W = [b, Ab, \ldots, A^{k-1} b].$$

The controllability condition that W has rank k means that the matrix W is nonsingular or its columns are linearly independent. For the general case, the controllability condition is that from among the km columns there are k linearly independent columns. Let us now illustrate the theorem by an example.

Example 6.5 Contemplate the system

$$
\begin{aligned}
y_1(n + 1) &= a y_1(n) + b y_2(n), \\
y_2(n + 1) &= c y_1(n) + d y_2(n) + u(n),
\end{aligned}
$$

where $ad - bc \neq 0$. Here

$$A = \begin{pmatrix} a & b \\ c & d \end{pmatrix}, \quad B = \begin{pmatrix} 0 \\ 1 \end{pmatrix},$$

and $u(n)$ is a scalar control. Now

$$W(2) = (B, AB) = \begin{pmatrix} 0 & b \\ 1 & d \end{pmatrix}$$

has rank 2 if $b \neq 0$. Thus the system is completely controllable by Theorem 6.4 if and only if $b \neq 0$.

Lemma 6.6 For any $N \geq k$, the rank of the matrix

$$[B, AB, A^2B, \ldots, A^{N-1}B]$$

is equal to the rank of the controllability matrix W.

Proof (I) Consider the matrix $W(n) = [B, AB, \ldots, A^{n-1}B]$, $n = 1, 2, 3, \ldots$. As n increases by 1, either the rank of $W(n)$ remains constant or increases by at least 1. Suppose that for some $r > 1$, rank $W(r+1) =$ rank $W(r)$. Then every column in the matrix $A^r B$ is linearly dependent on the columns of $W(r) = [B, AB, \ldots, A^{r-1}B]$. Hence

$$A^r B = B M_0 + ABM_1 + \ldots + A^{r-1}BM_{r-1} \tag{6.2.2}$$

where each M_i is an $m \times m$ matrix. By premultiplying both sides of Eq. (6.2.2) by A we obtain

$$A^{r+1}B = ABM_0 + A^2BM_1 + \ldots + A^rBM_{r-1}.$$

Thus the columns of $A^{r+1}B$ are linearly dependent on the columns of $W(r+1)$. This implies that rank $W(r+2) =$ rank $W(r+1) =$ rank $W(r)$. By repeating this process, one may conclude that

$$\text{rank } W(n) = \text{rank } W(r) \text{ for all } n > r.$$

We conclude from the above argument that rank $W(n)$ increases by at least 1 as n increases by 1 until it attains its maximum k. Hence the rank maximum of $W(n)$ is attained in at most k steps. Therefore, the maximum rank is attained at $n \leq k$, and consequently rank $W(\equiv \text{rank } W(k)) = \text{rank } W(N)$ for all $N \geq k$.

Proof (II) In the second proof we apply the Cayley–Hamilton Theorem (Chapter 3). So if $p(\lambda) = \lambda^k + p_1\lambda^{k-1} + \ldots + p_k$ is the characteristic polynomial of A, then $p(A) = 0$, i.e.,

$$A^k + p_1 A^{k-1} + \ldots + p_k I = 0$$

or

$$A^k = \sum_{i=1}^{k} q_i A^{k-1} \tag{6.2.3}$$

where $q_i = -p_i$. Multiplying Expression (6.2.5) by B, we obtain

$$A^k B = \sum_{i=1}^{k} q_i A^{k-i} B. \tag{6.2.4}$$

Thus the columns of $A^k B$ are linearly dependent on the columns of $W(k) \equiv W$. Therefore rank $W(k+1) = \text{rank } W$. By multiplying Expression (6.2.4) by A we have

$$A^{k+1}B = q_1 A^k + q_2 A^{k-1} + \ldots + q_k A.$$

Consequently, rank $W(k+2) = \text{rank } W(k+1) = \text{rank } W$. By repeating the process, one concludes that rank $W(N) = \text{rank } W$ for all $N \geq k$.

We are now ready to prove the theorem.

Proof of Theorem 6.4

Sufficiency Suppose that rank $W = k$. Let x_0 and x_f be any two arbitrary vectors in R^k. Recall that by the variation of Constants Formula (3.2.13) we have

$$x(k) - A^k x(0) = \sum_{r=0}^{k-1} A^{k-r-1} B u(r)$$

or

$$x(k) - A^k x(0) = W \bar{u}(k) \qquad (6.2.5)$$

where

$$\bar{u}(k) = \begin{pmatrix} u(k-1) \\ u(k-2) \\ \vdots \\ u(0) \end{pmatrix}.$$

Since rank $W = k$, Range $W = R^k$. Hence if we let $x(0) = x_0$ and $x(k) = x_f$, then $x_f - A^k x_o \in$ Range W. Thus $x_f - A^k x_0 = W\bar{u}$ for some vector $\bar{u} \in R^k$. Consequently, system equation (6.1.3) is completely controllable.

Necessity Assume that System (6.1.3) is completely controllable and rank $W < k$. From the proof of Lemma 6.6 (Proof I) we conclude that there exists $r \in Z^+$ such that

rank $W(1) <$ rank $W(2) \ldots <$ rank $W(r) =$ rank $W(r+1) \ldots =$ rank W.

Moreover rank $W(n) =$ rank W for all $n > k$. Furthermore, since $W(j+1) = (W(j), A^j B)$, it follows that

Range $W(1) \quad \subset \quad$ Range $W(2) \ldots \subset$ Range $W(r) =$ Range $W(r+1) \ldots$

$\qquad \qquad = \quad$ Range $W \ldots =$ Range $W(n)$

for any $n > k$.

Since rank $W < k$, Range $W \neq R^k$. Thus there exists $\xi \notin$ Range W. This implies that $\xi \notin$ Range $W(n)$ for all $n \in Z^+$. If we let $x_0 = 0$ in Formula 6.2.5 with k replaced by n we have $x(n) = W(n)\bar{u}(n)$. Hence for ξ to be equal $x(n)$ for some n, ξ must be in the range of $W(n)$. But $\xi \notin$ Range $W(n)$ for all $n \in Z^+$ implies that ξ may not be reached at any time from the origin, which is a contradiction. Therefore rank $W = k$.

Remark 6.7 There is another definition of complete controllability in the literature which I will call here "*Controllability of the origin.*" A system is *controllable to the origin* if for any $n_0 \in Z^+$ and $x_0 \in R^k$, there exists a finite time $N > n_0$ and a control $u(n)$, $n_0 < n \leq N$ such that $x(N) = 0$.

Clearly complete controllability is a stronger property than controllability to the origin. The two notions coincide in continuous–time systems. [See Ref. 2 [2].]

However, for the discrete-time system (6.1.3), controllability to the origin does not imply complete controllability unless A is nonsingular (Exercises 6.1 and 6.2, Problem 13). The following example illustrates our remark.

Example 6.8 Consider the control system $x(n + 1) = Ax(n) + Bu(n)$ with

$$A = \begin{pmatrix} 0 & 1 \\ 0 & 0 \end{pmatrix}, \quad B = \begin{pmatrix} 1 \\ 0 \end{pmatrix}. \tag{6.2.6}$$

Now for

$$x(0) = x_0 = \begin{pmatrix} x_{01} \\ x_{02} \end{pmatrix},$$

we have from Eq. (6.2.6)

$$
\begin{aligned}
x(1) &= Ax_0 + Bu(0) \\
&= \begin{pmatrix} 0 & 1 \\ 0 & 0 \end{pmatrix} \begin{pmatrix} x_{01} \\ x_{02} \end{pmatrix} + \begin{pmatrix} 1 \\ 0 \end{pmatrix} u(0) \\
&= \begin{pmatrix} x_{02} \\ 0 \end{pmatrix} + \begin{pmatrix} u(0) \\ 0 \end{pmatrix}.
\end{aligned}
$$

So if we pick $u(0) = -x_{02}$, then we will have $x(1) = 0$. Therefore system equation (6.2.6) is controllable to zero. Observe, however, that

$$\text{rank } (B, AB) = \text{rank } \begin{pmatrix} 1 & 0 \\ 0 & 0 \end{pmatrix} = 1 < 2.$$

Thus by Theorem 6.4, system equation (6.2.6) is not completely controllable.

Example 6.9 Contemplate the system $y(n + 1) = Ay(n) + Bu(n)$, where

$$A = \begin{pmatrix} 0 & 1 \\ 2 & -1 \end{pmatrix}, \quad B = \begin{pmatrix} 1 \\ 1 \end{pmatrix}.$$

Now $W(1) = \begin{pmatrix} 1 \\ 1 \end{pmatrix}$ is of rank 1,

$W(2) = \begin{pmatrix} 1 & 1 \\ 1 & 1 \end{pmatrix}$ is also of rank 1 since it is now equivalent to $= \begin{pmatrix} 1 & 1 \\ 0 & 0 \end{pmatrix}$.

Hence according to Theorem 6.4 the system is not controllable. Notice, however, that the point $\begin{pmatrix} -4 \\ 0 \end{pmatrix}$ is reachable from $\begin{pmatrix} 1 \\ 2 \end{pmatrix}$ under the control $u(n) = -2$ in time $n = 2$.

Example 6.10 Fig. 6.4 shows a card of mass m attached to a wall via a flexible linkage. The equation of motion for this system is

$$m\ddot{x} + b\dot{x} + kx = u \tag{6.2.7}$$

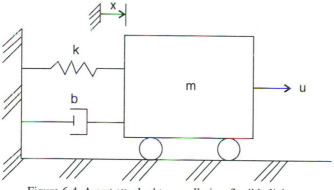

Figure 6.4. A cart attached to a wall via a flexible linkage.

where k and b are the stiffness and damping, respectively, of the linkage and u is an applied force. Equation (6.2.7) can be written in state variable form as

$$\begin{bmatrix} \dot{x} \\ \dot{v} \end{bmatrix} = \begin{bmatrix} 0 & 1 \\ -k/m & -b/m \end{bmatrix}\begin{bmatrix} x \\ v \end{bmatrix} + \begin{bmatrix} 0 \\ 1/m \end{bmatrix}u. \qquad (6.2.8)$$

Thus

$$\hat{A} = \begin{bmatrix} 0 & 1 \\ -k/m & -b/m \end{bmatrix}, \quad \hat{B} = \begin{bmatrix} 0 \\ 1/m \end{bmatrix}.$$

Recall that, given a sample period T, the matrices A and B of the equivalent discrete system are given by

$$A = e^{\hat{A}T}, \quad B = Te^{\hat{A}T}\hat{B}$$

so that their computation requires finding the exponential of the matrix $\hat{A}$. This is not so difficult as it sounds, at least when $\hat{A}$ can be diagonalized, for then we can find a matrix P such that

$$\hat{A} = P\Lambda P^{-1} \qquad (6.2.9)$$

where Λ is a diagonal matrix

$$\Lambda = \begin{bmatrix} \lambda_1 & \cdots & 0 \\ \vdots & \ddots & \vdots \\ 0 & \cdots & \lambda_k \end{bmatrix}. \qquad (6.2.10)$$

By definition

$$e^{\hat{A}T} = I + \hat{A}T + \frac{1}{2!}\hat{A}^2T^2 + \frac{1}{3!}\hat{A}^3T^3 + \cdots$$

so that substituting using Eq. (6.2.9) gives

$$e^{\hat{A}T} = Pe^{\Lambda T}P^{-1},$$

and the diagonal form (6.2.10) of Λ gives

$$e^{\hat{A}T} = P \begin{bmatrix} e^{\lambda_1 T} & \cdots & 0 \\ \vdots & \ddots & \vdots \\ 0 & \cdots & e^{\lambda_k T} \end{bmatrix} P^{-1}.$$

Returning to our example, note that if $m = 1$, $k = 2$, and $b = 3$ then

$$\hat{A} = \begin{bmatrix} 0 & 1 \\ -2 & -3 \end{bmatrix}, \quad \hat{B} = \begin{bmatrix} 0 \\ 1 \end{bmatrix}.$$

This $\hat{A}$ can be written in the form of Eq. (6.2.9), where

$$\Lambda = \begin{bmatrix} -1 & 0 \\ 0 & -2 \end{bmatrix}, \quad P = \begin{bmatrix} 1 & 1 \\ -1 & -2 \end{bmatrix}.$$

Hence

$$\begin{aligned} A &= e^{\hat{A}T} = P \begin{bmatrix} e^{-T} & 0 \\ 0 & e^{-2T} \end{bmatrix} P^{-1} \\ &= \begin{bmatrix} 2e^{-T} - e^{-2T} & e^{-T} - e^{-2T} \\ -2e^{-T} + 2e^{-2T} & -e^{-T} + 2e^{-2T} \end{bmatrix}, \\ B = Te^{\hat{A}T}\hat{B} &= T \begin{bmatrix} e^{-T} - e^{-2T} \\ -e^{-T} + 2e^{-2T} \end{bmatrix}. \end{aligned}$$

The controllability of the discrete equivalent system can then be checked by computing

$$W = [B \ AB] = T \begin{bmatrix} e^{-T} - e^{-2T} & e^{-2T} - e^{-4T} \\ -e^{-T} + 2e^{-2T} & -e^{-2T} + 2e^{-4T} \end{bmatrix}.$$

Checking the determinant gives

$$\det W = -T^2 e^{-4T}(1 - e^{-T} + e^{-2T}),$$

which is zero only if $T = 0$. Thus the cart system is controllable for any nonzero sample period.

6.2.1   Controllability Canonical Forms

Consider the second order difference equation

$$z(n + 2) + p_1 z(n + 1) + p_2 z(n) = u(n).$$

Recall from Section 3.2 that this equation is equivalent to the system

$$x(n + 1) = Ax(u) + Bu(n)$$

where

$$A = \begin{pmatrix} 0 & 1 \\ -p_2 & -p_1 \end{pmatrix}, \quad B = \begin{pmatrix} 0 \\ 1 \end{pmatrix}, \quad x = \begin{pmatrix} z(n) \\ z(n+1) \end{pmatrix}.$$

Clearly

$$W(2) = \begin{pmatrix} 0 & 1 \\ 1 & -p_1 \end{pmatrix}$$

has rank 2 for all values of p_1 and p_2. Consequently, this equation is always completely controllable.

The preceding example may be generalized to the kth order equation.

$$z(n+k) + p_1 z(n+k+1) + \ldots + p_k z(n) = u(n) \tag{6.2.11}$$

which is equivalent to the system

$$x(n+1) = Ax(n) + bu(n),$$

where

$$A = \begin{pmatrix} 0 & 1 & 0 & \cdots & 0 \\ \vdots & 0 & 1 & & \vdots \\ & & & & 1 \\ -p_k & -p_{k-1} & \cdots & & -p_1 \end{pmatrix}, \quad B = e_k = \begin{pmatrix} 0 \\ \vdots \\ 1 \end{pmatrix},$$

$$x(n) = \begin{pmatrix} z(n) \\ z(n+1) \\ \vdots \\ z(n+k-1) \end{pmatrix}.$$

Notice that

$$AB = \begin{pmatrix} 0 \\ 0 \\ \vdots \\ 1 \\ -p_1 \end{pmatrix}, \quad A^2 B = \begin{pmatrix} 0 \\ 0 \\ \vdots \\ 1 \\ * \\ * \end{pmatrix}, \quad \ldots, \quad A^{k-1} B = \begin{pmatrix} 1 \\ * \\ \vdots \\ * \end{pmatrix},$$

where the *'s are some combinations of the products of the p_i's. It follows that

$$W = \begin{pmatrix} 0 & 0 & \cdots & 1 \\ \vdots & \vdots & & \vdots \\ 0 & 1 & \cdots & * \\ 1 & * & \cdots & * \end{pmatrix}$$

is of rank k and the equation and thus the system is completely controllable. The converse of the above statement is also valid. That is to say if System (6.1.3), with

$k \times 1$ vector $B \equiv b$, is completely controllable, then it can be put in the form of a kth scalar equation (6.2.11) by a similarity transformation. To accomplish this task we start with the $k \times k$ controllability matrix $W = (b, Ab, \ldots, A^{k-1}B)$. Since System (6.1.3) is completely controllable, it follows from Theorem 6.4 that W is nonsingular. Let us write W^{-1} in terms of its rows as

$$W^{-1} = \begin{pmatrix} w_1 \\ w_2 \\ \vdots \\ w_k \end{pmatrix}.$$

We claim that the set $\{w_k, w_k A, \ldots, w_k A^{k-1}\}$ generated by the last row of W^{-1} is linearly independent. To show this, suppose that for some constants $a_1, a_2, \ldots, a_k$ we have

$$a_1 w_k + a_2 w_k A + \ldots + a_k w_k A^{k-1} = 0. \tag{6.2.12}$$

Multiplying Eq. (6.2.12) from the right by b yields

$$a_1 w_k b + a_2 w_k Ab + \ldots + a_k w_k A^{k-1} b = 0. \tag{6.2.13}$$

Since $W^{-1}W = I$, $w_k b = w_k Ab = \ldots = w_k A^{k-2} b = 0$, and $W_k A^{k-1} b = 1$. Hence it follows from Eq. (6.2.13) that $a_k = 0$. One may repeat this procedure by multiplying Eq. (6.2.12) by Ab (and letting $a_k = 0$) to conclude that $a_{k-1} = 0$. Continuing this procedure, one may show that $a_i = 0$ for $1 \leq i \leq k$. This proves our claim that the vectors $w_k, w_k A, \ldots, w_k A^{k-1}$ are linearly independent. Hence the $k \times k$ matrix

$$P = \begin{pmatrix} w_k \\ w_k A \\ \vdots \\ w_k A^{k-1} \end{pmatrix} \quad \ldots$$

is nonsingular. Define a change of coordinates for System (6.1.3) by

$$z(n) = Px(n) \tag{6.2.14}$$

which gives

$$z(n+1) = PAP^{-1} z(n) + Pb\, u(n)$$

or

$$z(n+1) = \hat{A}\, z(n) + \hat{b}\, u(n) \tag{6.2.15}$$

where

$$\hat{A} = PAP^{-1}, \quad \hat{b} = Pb \tag{6.2.16}$$

clearly

$$\hat{b} = Pb = \begin{pmatrix} 0 \\ 0 \\ \vdots \\ 0 \\ 1 \end{pmatrix}.$$

Now

$$\hat{A} = PAP^{-1} = \begin{pmatrix} w_k A \\ w_k A^2 \\ \vdots \\ w_k A^k \end{pmatrix} P^{-1}.$$

Since $w_k A$ is the second row in P, it follows that

$$w_k A\, P^{-1} = (0\ 1\ 0\ \dots\ 0).$$

Similarly

$$w_k A^2 P^{-1} = (0\ 0\ 1\ 0\ \dots\ 0)$$

$$\vdots$$

$$w_k A^{k-1} P^{-1} = (0\ 0\ \dots\ 0\ 1)$$

However

$$w_k A^k P^{-1} = (-p_k\ -p_{k-1}\ \cdots\ -p_1)$$

where the p_i's are some constants. Thus

$$\hat{A} = \begin{pmatrix} 0 & 1 & 0 & \cdots & 0 \\ 0 & 0 & 1 & \cdots & 0 \\ \vdots & & & & \vdots \\ 0 & 0 & 0 & \cdots & 1 \\ -p_k & -p_{k-1} & -p_{k-2} & \cdots & -p_1 \end{pmatrix}$$

with the characteristic equation

$$\lambda^k + p_1 \lambda^{k-1} + p_2 \lambda^{k-2} + \dots + p_k = 0.$$

Observe that A and $\hat{A}$ have the same characteristic equation. The above discussion proves the following.

Theorem 6.11 A system $x(n + 1) = Ax(n) + bu(n)$ is completely controllable if and only if it is equivalent to a kth order equation of the form (6.2.11).

System (6.2.15) is said to be in a *controllable canonical form*.

Another controllable canonical form may be obtained by using the change of variables $x(n) = Wz(n)$, where W is the controllability matrix of the system. This is a more popular form among engineers due to its simple derivative. The reader is asked in Exercises 6.1 and 6.2, Problem 20 to show the obtained controllable canonical pair $\{\tilde{A}, \tilde{b}\}$ are given by

$$\tilde{A} = \begin{pmatrix} 0 & 0 & 0 & \cdots & -p_k \\ 1 & 0 & 0 & \cdots & -p_{k-1} \\ 0 & 1 & 0 & \cdots & -p_{k-2} \\ \vdots & \vdots & & & \\ 0 & 0 & & \cdots & -p_1 \end{pmatrix}, \quad \tilde{b} = \begin{pmatrix} 1 \\ 0 \\ \vdots \\ 0 \end{pmatrix}. \qquad (6.2.17)$$

Exercises 6.1 and 6.2

In Problems 1 through 6, determine whether or not the system $x(n + 1) = A\, x(n) + B\, u(n)$ is completely controllable.

1. $A = \begin{pmatrix} -2 & 2 \\ 1 & -1 \end{pmatrix}$, $B = \begin{pmatrix} 1 \\ 0 \end{pmatrix}$.

2. $A = \begin{pmatrix} -1 & 0 \\ 0 & -2 \end{pmatrix}$, $B = \begin{pmatrix} 2 \\ 3 \end{pmatrix}$.

3. $A = \begin{pmatrix} -1 & 0 \\ 0 & -2 \end{pmatrix}$, $B = \begin{pmatrix} 2 \\ 0 \end{pmatrix}$.

4. $A = \begin{pmatrix} -2 & 1 & 0 & 0 & 0 \\ 0 & -2 & 1 & 0 & 0 \\ 0 & 0 & -2 & 0 & 0 \\ 0 & 0 & 0 & -5 & 1 \\ 0 & 0 & 0 & 0 & -5 \end{pmatrix}$, $B = \begin{pmatrix} 0 & 1 \\ 0 & 0 \\ 3 & 0 \\ 0 & 0 \\ 2 & 1 \end{pmatrix}$.

5. $A = \begin{pmatrix} -2 & 1 & 0 & 0 & 0 \\ 0 & -2 & 1 & 0 & 0 \\ 0 & 0 & -2 & 0 & 0 \\ 0 & 0 & 0 & -5 & 1 \\ 0 & 0 & 0 & 0 & -5 \end{pmatrix}$, $B = \begin{pmatrix} 0 & 1 \\ 3 & 0 \\ 0 & 0 \\ 2 & 1 \\ 0 & 0 \end{pmatrix}$.

6. $A = \begin{pmatrix} A_{11} & A_{12} \\ 0 & A_{22} \end{pmatrix}$, $B = \begin{pmatrix} B_1 \\ 0 \end{pmatrix}$, A_{11} is a $r \times r$ matrix, A_{12} is a $r \times s$ matrix, A_{22} is a $s \times s$ matrix, B_1 is a $r \times m$ matrix, where $r + s = k$.

We say that a state x_f is *reachable* from an initial state x_0 if there exists $N \in Z^+$ and a control $u(n)$, $n = 0, 1, \ldots, N - 1$, such that $x(N, x_0) = x_f$.

7. Prove that a state x_f is reachable from x_0 in time N if and only if $x_f - A^N x_0 \in$ Range $W(N)$.

8. Consider the system $x(n + 1) = Ax(n) + Bu(n)$, where $A = \begin{pmatrix} 1 & 2 & -1 \\ 0 & 1 & 0 \\ 1 & -4 & 3 \end{pmatrix}$, $B + \begin{pmatrix} 0 \\ 0 \\ 1 \end{pmatrix}$. Find a basis for the set of vectors in R^3 which are reachable from the origin.

9. Consider the system
$$x(n + 1) = \begin{pmatrix} -1 & -1 \\ 2 & -4 \end{pmatrix} x(n) + Bu(n).$$

Find for what vectors B in R^3 the system is not completely controllable.

10. Obtain a necessary and sufficient condition for

$$x_1(n + 1) = a_{11}x_1(n) + a_{12}x_2(n) + u(n),$$
$$x_2(n + 1) = a_{21}x_1(n) + a_{22}x_2(n) - u(n)$$

to be controllable.

11. Obtain a necessary and sufficient condition for $x(n + 1) = Ax(n) + Bu(n)$ to be completely controllable, where

$$A = \begin{pmatrix} a_{11} & a_{12} \\ a_{21} & a_{22} \end{pmatrix}, \quad B = \begin{pmatrix} 1 \\ 1 \end{pmatrix}.$$

12. Consider the system

$$x_1(n + 1) = x_2(n) + u_1(n) + u_2(n),$$
$$x_2(n + 1) = x_3(n) + u_1(n) - u_2(n),$$
$$x_3(n + 1) = u_1(n).$$

    a. Prove that the system is completely controllable in two steps.

    b. If $u_2(n) \equiv 0$, show that the system is completely controllable in three steps.

    c. If $u_1(n) \equiv 0$, show that the system is not completely controllable.

13. Show that if the matrix A in Eq. (6.1.3) is nonsingular, then complete controllability and controllability to the origin are equivalent.

14. Prove that if $U = WW^T$ is positive definite, then W has rank k, where W is the controllability matrix of Eq. (6.1.3). Prove also the converse.

15. Show that system equation (6.1.3) is completely controllable if $[B, \ AB, \ \ldots, \ A^{k-r}]$ has rank k, where $r = \text{rank } B$.

16. A polynomial $\varphi(\lambda) = \lambda^m + a_1\lambda^{m-1} + \ldots + a_m$ is said to be the *minimal polynomial* of a $k \times k$ matrix A if $\varphi(\lambda)$ is the lowest degree polynomial for which $\varphi(A) = 0$. It follows that $m \leq k$. Prove that system $\{A, B\}$ is completely controllable if and only if rank $[B, \ AB, \ \ldots, A^{m-r}B] = k$, where rank $B = r$.

17. For a $k \times k$ matrix A, and a $k \times m$ matrix, prove that the following statements are true.

    (i) If $\{A, B\}$ is completely controllable, then so is $\{A + BC, \ B\}$ for any $m \times 1$ vector C.

    (ii) If $\{A + BC_o, \ B\}$ is completely controllable for some $m \times 1$ vector C_o, then $\{A + BC, \ B\}$ is completely controllable for any $m \times 1$ vector C.

18. Consider the system $x(n+1) = Ax(n) + Bu(n)$, where A is a $k \times k$ matrix, B is a $k \times m$ matrix and such that A has k linearly independent eigenvectors. Prove that $\{A, B\}$ is completely controllable if and only if no row of $P^{-1}B$ has all zero elements, where $P = (\xi_1, \ldots, \xi_k)$, ξ_i's are the eigenvectors of A.

19. Suppose that in Problem 18, A does not possess k linearly independent eigenvectors and that there exists a nonsingular matrix P, where $P^{-1} AP = J$ is the Jordan Canonical form of A. Prove that $\{A, B\}$ is completely controllable if and only if

    (i) no two Jordan blocks in J are associated wtih the same eigenvalue,

    (ii) the elements of any row of $P^{-1} B$ that correspond to the last row of each Jordan block are not all zero, and

    (iii) the elements of each row of $P^{-1} B$ that correspond to distinct eigenvalues are not all zero.

20. (Another controllability canonical form) Consider the completely controllable system $x(n + 1) = x(n) + bu(n)*$ where b is a $k \times 1$ vector. Let $x(n) = Wz(n)$, where W is the controllability matrix. Then $*$ becomes $z(n + 1) = \tilde{A} z(n) + \tilde{b} u(n)$. Show that $\tilde{A}$ and $\tilde{b}$ are given by Eq. (6.2.13).

21. Consider the system $x(n + 1) = Ax(n) + Bu(n)$, where

$$A = \begin{pmatrix} 1 & 0.6 \\ 0 & 0.4 \end{pmatrix}, \quad B = \begin{pmatrix} 0.4 \\ 0.6 \end{pmatrix}.$$

The reachability set is defined by $R(n) = \{x(0): x(0)$ is reached from the origin in N steps with $|u(i)| \le 1, 1 \le i \le N\}$.

    (a) Find $R(1)$ and plot it.

    (b) Find $R(2)$ and plot it.

6.3   Observability

In the previous section, it was assumed that (the observed) output of the control system is the same as that of the state of the system $x(n)$. In practice, however, one may not be able to observe the state of the system $x(n)$ but rather an output $y(n)$ which is related to $x(n)$ in a specific manner. The mathematical model of this type of system is given by

$$\left. \begin{array}{l} x(n + 1) = Ax(n) + Bu(n), \\ y(n) = Cx(n), \end{array} \right\} \tag{6.3.1}$$

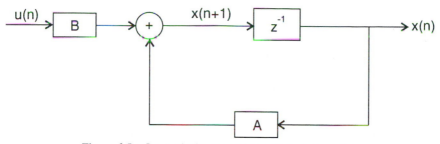

Figure 6.5a. Output is the same as the state: $y(n) = x(n)$.

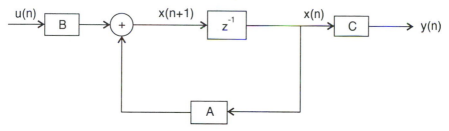

Figure 6.5b. Input-output System: $y(n) = Cx(n)$.

where $A(n)$ is a $k \times k$ matrix, B an $k \times m$ matrix, $u(n)$ and $m \times 1$ vector, and C is an $r \times k$ matrix. The control $u(n)$ is the input of the system and $y(n)$ is the output of the system as shown in Fig. 6.5a–b.

Roughly speaking, observability means that it is possible to determine the state of a system $x(n)$ by measuring only the output $y(n)$. Hence it is useful in solving the problem of reconstructing unmeasurable state variables from measurable ones. The input-output System (6.3.1) is *completely observable* if for any $n_0 \geq 0$, there exists $N > n_0$ such that the knowledge of $u(n)$ and $y(n)$ for $n_0 \leq n \leq N$ suffices to determine $x(n_0) = x_0$.

Example 6.12 Consider the system (Fig. 6.6)

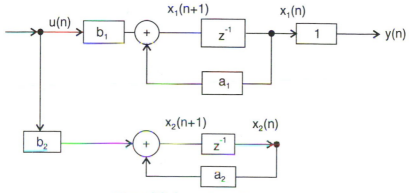

Figure 6.6. A nonobservable system.

$$
\begin{aligned}
x_1(n+1) &= a_1 x_1(n) + b_1 u(n), \\
x_2(n+1) &= a_2 x_2(n) + b_2 u(n), \\
y(n) &= x_1(n).
\end{aligned}
$$

This system is not observable, since the first equation shows that $x_1(n) = y(n)$ is completely determined by $u(n)$ and $x_1(0)$ and that there is no way to determine $x_2(0)$ from the output $y(n)$.

In discussing observability, one may assume that the control $u(n)$ is identically zero. This obviously simplifies our exposition. To explain why we can do this without loss of generality, we write $y(n)$ using the variation of constants Formula (3.2.13) for $x(n)$:

$$
y(n) = Cx(n)
$$

or

$$
y(n) = C\, A^{n-n_0} + \sum_{j=n_0}^{n-1} C\, A^{n-j-1}\, B\, u(j).
$$

Since C, A, B, and u are all known, the second term on the right-hand side of this last equation is known. Thus it may be subtracted from the observed value $y(n)$. Hence, for investigating a necessary and sufficient condition for complete observability it suffices to consider the case when $u(n) \equiv 0$.

We now present a criterion for complete observability that is analogous to that of complete controllability.

Theorem 6.13 System (6.3.1) is *completely observable* if and only if the $rk \times k$ observability matrix

$$
V = \begin{bmatrix} C \\ CA \\ CA^2 \\ \vdots \\ CA^{k-1} \end{bmatrix} \tag{6.3.2}
$$

has rank k.

Proof By applying the variation of constants Formula (3.2.13) on Eq. (6.3.1) we obtain

$$
\begin{aligned}
y(n) &= C\, x(n) \\
&= C\left[A^n x_0 + \sum_{r=0}^{n-1} A^{n-r-1} B\, u(r) \right]. \tag{6.3.3}
\end{aligned}
$$

Let

$$
\hat{y}(n) = y(n) - \sum_{r=0}^{n-1} C\, A^{n-r-1} B\, u(r) \tag{6.3.4}
$$

Using Formula (6.3.3), Eq. (6.3.4) may now be written as

$$
\hat{y}(n) = C\, A^n x_0. \tag{6.3.5}
$$

Putting $n = 0, 1, 2, \ldots, k - 1$ in Eq. (6.3.5) yields

$$
\begin{bmatrix} \hat{y}(0) \\ \hat{y}(1) \\ \vdots \\ \hat{y}(k-1) \end{bmatrix} = \begin{bmatrix} C \\ CA \\ \vdots \\ CA^{k-1} \end{bmatrix} x_0.
\tag{6.3.6}
$$

Suppose that rank $V = k$. Then Range $V = R^k$. Now if $y(n), u(n)$ are given for $0 \le n \le k - 1$, then it follows from Eq. (6.3.4) that $\hat{y}(n), 0 \le n \le k - 1$, is also known. Hence there exists $x_0 \in R^k$ such that Eq. (6.3.6) holds. Hence System (6.3.1) is completely observable. Conversely, suppose System (6.3.1) is completely observable. Let us write

$$
V(N) = \begin{bmatrix} C \\ CA \\ CA^2 \\ \vdots \\ CA^{N-1} \end{bmatrix} = (C^T, A^T C^T, (A^T)^2 C^T, \ldots, (A^T)^{N-1} C^T)^T.
\tag{6.3.7}
$$

Then from Theorem 6.13, $V(N)$ is of rank k if and only if the observability matrix $V \equiv V(k)$ has rank k. Therefore, if x_0 can be uniquely determined from N observations $y(0), y(1), \ldots, y(N-1)$, it can be so determined from $y(0), y(1), \ldots, y(k-1)$. Thus rank $V = k$.

Notice that the matrix B does not play any role in determining observability. This confirms our earlier remark that in studying observability, one may assume that $u(n) \equiv 0$. Henceforth, we may speak of the observability of the pair $\{A, C\}$.

Example 6.3 revisited

Consider again Example 6.3. The system may be written as

$$
\begin{pmatrix} x_1(n+1) \\ x_2(n+1) \end{pmatrix} = \begin{pmatrix} a_1 & 0 \\ 0 & a_2 \end{pmatrix} \begin{pmatrix} x_1(n) \\ x_2(n) \end{pmatrix} + \begin{pmatrix} b_1 \\ b_2 \end{pmatrix} u(n),
$$

$$
y(n) = (1 \ 0) \begin{pmatrix} x_1(n) \\ 0 \end{pmatrix}.
$$

Thus $A = \begin{pmatrix} a_1 & 0 \\ 0 & a_2 \end{pmatrix}$ and $C = (1 \ 0)$. It follows that the observability matrix is given by $V = \begin{pmatrix} 1 & 0 \\ a_1 & 0 \end{pmatrix}$. Since rank $V = 1 < 2$, the system is not completely observable by virtue of Theorem 6.13.

Finally, we give an example to illustrate the above results.

Example 6.14 Consider the input-output system (Fig. 6.7)

$$
\begin{aligned}
x_1(n+1) &= x_2(n), \\
x_2(n+1) &= -x_1(n) + 2x_2(n) + u(n), \\
y(n) &= c_1 x_1(n) + c_2 x_2(n).
\end{aligned}
$$

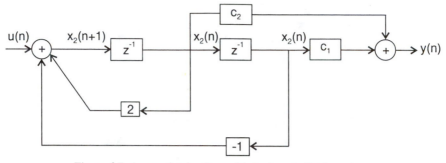

Figure 6.7. A completely observable and controllable system.

Then $A = \begin{pmatrix} 0 & 1 \\ -1 & 2 \end{pmatrix}$, $B = \begin{pmatrix} 0 \\ 1 \end{pmatrix}$, and $C = (c_1, c_2)$. The observability matrix is given by

$$V = \begin{pmatrix} C \\ CA \end{pmatrix} = \begin{pmatrix} c_1 & c_2 \\ -c_2 & c_1 + 2c_2 \end{pmatrix}.$$

By adding the first column to the second column in V we obtain the matrix

$$\hat{V} = \begin{pmatrix} c_1 & c_1 + c_2 \\ -c_2 & c_1 + c_2 \end{pmatrix}.$$

Observe that rank $\hat{V} = 2$ if and only if $c_1 + c_2 \neq 0$. Since rank $V = $ rank $\hat{V}$, it follows that the system is completely observable if and only if $c_1 + c_2 \neq 0$ (or $c_2 \neq -c_1$).

We may also note the system is completely controllable.

Example 6.15 Example 6.10 looked at the controllability of a cart attached to a fixed wall via a flexible linkage using an applied force u. A dual question can be posed: If the force on the cart is a constant, can its magnitude be observed by measuring cart position? In order to answer this question, the state equation (6.2.8) must be *augmented* with one additional equation, representing the assumption that the applied force is constant:

$$\begin{bmatrix} \dot{x} \\ \dot{v} \\ \dot{u} \end{bmatrix} = \begin{bmatrix} 0 & 1 & 0 \\ -k/m & -b/m & 1/m \\ 0 & 0 & 0 \end{bmatrix} \begin{bmatrix} x \\ v \\ u \end{bmatrix},$$

$$y = \begin{bmatrix} 1 & 0 & 0 \end{bmatrix} \begin{bmatrix} x \\ v \\ u \end{bmatrix}.$$

Using the same values $m = 1$, $k = 2$, and $b = 3$ as in Example 6.10,

$$\hat{A} = \begin{bmatrix} 0 & 1 & 0 \\ -2 & -3 & 1 \\ 0 & 0 & 0 \end{bmatrix},$$

can be written as

$$\hat{A} = P\Lambda P^{-1},$$

where

$$\Lambda = \begin{bmatrix} -1 & 0 & 0 \\ 0 & -2 & 0 \\ 0 & 0 & 0 \end{bmatrix}, \quad P = \begin{bmatrix} 1 & 1 & 1 \\ -1 & -2 & 0 \\ 0 & 0 & 2 \end{bmatrix}.$$

Thus

$$A = e^{\hat{A}T} = Pe^{\Lambda T}P^{-1} =$$
$$\begin{bmatrix} 2e^{-T} - e^{-2T} & e^{-T} - e^{-2T} & \frac{1}{2} + e^{-T} + \frac{1}{2}e^{-2T} \\ -2e^{-T} + 2e^{-2T} & -e^{-T} + 2e^{-2T} & -e^{-T} - e^{-2T} \\ 0 & 0 & 1 \end{bmatrix}.$$

To check observability, we must compute

$$V = \begin{bmatrix} C \\ CA \\ CA^2 \end{bmatrix} =$$
$$\begin{bmatrix} 1 & 0 & 0 \\ 2e^{-T} - e^{-2T} & e^{-T} - e^{-2T} & \frac{1}{2} + e^{-T} + \frac{1}{2}e^{-2T} \\ 2e^{-2T} - e^{-4T} & e^{-2T} - e^{-4T} & \frac{1}{2} + 2e^{-T} + e^{-2T} + \frac{1}{2}e^{-4T} \end{bmatrix}$$

and its determinant

$$\det V = e^{-T} + 2e^{-2T} - 4e^{-3T} - 2e^{-4T} + 3e^{-5T}$$
$$= e^{-T}(1 + e^{-T})(1 - e^{-T})^2(1 + 3e^{-T}).$$

The factored form above shows that, since T is real, $\det V = 0$ only if $T = 0$. The system is therefore observable for all nonzero T.

Theorem 6.13 establishes a duality between the notions of controllability and observability. The following definition formalizes the notion of duality.

Definition 6.16 The dual system of Eq. (6.3.1) is given by

$$\left. \begin{array}{l} x(n+1) = A^T x(n) + C^T u(n), \\ y(n) = B^T x(n). \end{array} \right\} \tag{6.3.8}$$

Notice that the controllability matrix $\bar{W}$ of System (6.3.8) may be given by

$$\bar{W} = [C^T, A^T C^T, (A^T)^2 C^T, \ldots, (A^T)^{k-1} C^T].$$

Furthermore, the observability matrix V of System (6.3.1) is the transpose of $\bar{W}$, i.e.,

$$V = \bar{W}^T.$$

But since rank $\bar{W}$ = rank $\bar{W}^T$ = rank V, we have the following conclusion

Theorem 6.17 *(Duality Principal)* System (6.3.1) is completely controllable if and only if its dual System (6.3.8) is completely observable.

Remark 6.18 In Remark 6.7 we introduced a weaker notion of controllability, namely, controllability to the origin. In this section we have established a duality between complete controllability and complete observability. To complete our analysis we need to find a dual notion for controllability to the origin. Fortunately, such a notion does exist and it is called *constructability*. System (6.3.1) is said to be *constructible* if there exists a positive integer N such that for given $u(0), u(1), \ldots, u(N-1)$, and $y(0), y(1), \ldots, y(N-1)$, it is possible to find the state vector $x(N)$ of the system. Since the knowledge of $x(0)$ yields $x(N)$ by the variation of constants formula, it follows that complete observability implies constructibility. The two notions are in fact equivalent if the matrix A is nonsingular. The diagram below illustrates the relations among various notions of controllability and observability.

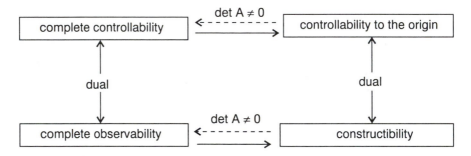

Finally we give an example to demonstrate that constructibility does not imply complete observability.

Example 6.19 Contemplate a dual of the system in Example 6.8:

$$\begin{pmatrix} x_1(n+1) \\ x_2(n+1) \end{pmatrix} = \begin{pmatrix} 0 & 0 \\ 1 & 0 \end{pmatrix} \begin{pmatrix} x_1(n) \\ x_2(n) \end{pmatrix} + \begin{pmatrix} 1 \\ 0 \end{pmatrix} u(n),$$

$$y(n) = (1 \quad 0) \begin{pmatrix} x_1(n) \\ x_2(n) \end{pmatrix}.$$

The observability matrix is given by

$$V = \begin{pmatrix} 1 & 0 \\ 0 & 0 \end{pmatrix}$$

whose rank is 1. It follows from Theorem 6.13 that the system is not completely observable. However, if we know $u(0)$, $u(1)$, and $y(0)$, $y(1)$, then form the second equation we find $x_1(1) = y(1)$. The first equation yields $x_1(2) = u(1)$ and $x_2(2) = x_1(1)$. Thus

$$x(2) = \begin{pmatrix} x_1(2) \\ x_2(2) \end{pmatrix}$$

is now obtained and consequently, the system is constructible.

6.3.1   Observability Canonical Forms

Consider again the completely observable system

$$x(n + 1) = A\,x(n) + b\,u(n), \atop y(n) = Cx(n) \Biggr\} \qquad (6.3.9)$$

where b is a $k \times 1$ vector and C is a $1 \times k$ vector. Recall that in Section 6.2 we constructed two controllability canonical forms of System (6.3.1). By exactly parallel procedures we can obtain two *observability canonical forms* corresponding to System (6.3.9). Both procedures are based on the nonsingularity of the observability matrix

$$V = \begin{pmatrix} C \\ CA \\ \vdots \\ CA^{k-1} \end{pmatrix}.$$

If we let $z(n) = V\,x(n)$ in Eq. (6.3.11) we obtain the first observability canonical form (Exercise 6.3, Problem 10)

$$z(n + 1) = \bar{A}\,z(n) + \bar{b}\,u(n), \atop y(n) = \bar{c}\,z(n) \Biggr\} \qquad (6.3.10)$$

where

$$\bar{A} = \begin{pmatrix} 0 & 1 & 0 & \cdots & 0 \\ 0 & 0 & 1 & \cdots & 0 \\ \vdots & \vdots & \vdots & \cdots & \vdots \\ 0 & 0 & 0 & \cdots & 1 \\ -p_k & -p_{k-1} & -p_{k-2} & \cdots & -p_1 \end{pmatrix}, \quad \bar{c} = (1\ 0\ 0\ \cdots\ 0),$$

$$\bar{b} = Vb. \qquad (6.3.11)$$

In Exercise 6.3, Problem 11 the reader is asked to find a change of variable that yields the other observability canonical $\{\tilde{A},\ \tilde{c}\}$, with

$$\tilde{A} = \begin{pmatrix} 0 & 0 & \cdots & 0 & -p_k \\ 1 & 0 & \cdots & 0 & -p_{k-1} \\ 0 & 1 & \cdots & 0 & -p_{k-2} \\ \vdots & \vdots & & \vdots & \\ 0 & 0 & \cdots & 1 & -p_1 \end{pmatrix}, \quad \tilde{c} = (0\ 0\ \cdots\ 1). \qquad (6.3.12)$$

Exercise 6.3

1. Consider the input-output system

$$x(n + 1) = A\,x(n) + B\,u(n), \atop y(n) = C\,x(n)$$

where $A = \begin{pmatrix} 0 & 1 \\ 2 & -1 \end{pmatrix}$.

(a) If $C = (0, 2)$, show that the pair$\{A, C\}$ is observable. Then find $x(0)$ if $y(0) = a$ and $y(1) = b$.

(b) If $C = (2, 1)$, show that the pair $\{A, C\}$ is unobservable.

2. Determine the observability of the pair $\{A, C\}$, where

$$ A = \begin{pmatrix} 0 & 1 & 0 \\ 0 & 0 & 1 \\ \frac{-1}{4} & \frac{1}{4} & 1 \end{pmatrix}, \quad C = \begin{pmatrix} 2 & -3 & -2 \\ 2 & 3 & 1 \end{pmatrix}. $$

3. Consider the system defined by

$$ \begin{pmatrix} x_1(n + 1) \\ x_2(n + 1) \end{pmatrix} = \begin{pmatrix} a & b \\ c & d \end{pmatrix} \begin{pmatrix} x_1(n) \\ x_2(n) \end{pmatrix} + \begin{pmatrix} 1 \\ 1 \end{pmatrix} u(n), $$

$$ y(n) = (1 \ 0) \begin{pmatrix} x_1(n) \\ x_2(n) \end{pmatrix}. $$

Determine the conditions on $a, b, c,$ and d for complete state controllability and complete observability. In Problems 4 and 5, determine the observability of the pair $\{A, C\}$.

4. $A = \begin{pmatrix} 2 & 1 & 0 & 0 & 0 \\ 0 & 2 & 1 & 0 & 0 \\ 0 & 0 & 2 & 0 & 0 \\ 0 & 0 & 0 & -3 & 1 \\ 0 & 0 & 0 & 0 & -3 \end{pmatrix}, \quad C = \begin{pmatrix} 1 & 1 & 1 & 0 & 1 \\ 0 & 1 & 1 & 1 & 0 \end{pmatrix}.$

5. $A = \begin{pmatrix} 2 & 1 & 0 & 0 & 0 \\ 0 & 2 & 1 & 0 & 0 \\ 0 & 0 & 2 & 0 & 0 \\ 0 & 0 & 0 & -3 & 1 \\ 0 & 0 & 0 & 0 & -3 \end{pmatrix}, \quad C = \begin{pmatrix} 1 & 1 & 1 & 0 & 1 \\ 0 & 1 & 1 & 0 & 0 \end{pmatrix}.$

6. Show that the pair $[A, C]$, where

$$ \begin{pmatrix} A_{11} & 0 \\ A_{21} & A_{22} \end{pmatrix}, \quad C = (C_1 \ 0), A_{11}: r \times r; \quad A_{21}: m \times r; $$

$$ A_{22}: m \times m; C_1: p \times r $$

is not completely observable for any submatrices $A_1, A_{21}, A_{22},$ and C_1.

7. Prove that system equation (6.3.2) is completely observable if and only if rank $[C, \ CA, \ \ldots, \ C \ A^{m-r}]^T = k$, where m is the degree of the minimal polynomial of A, and $r = \text{rank } C$.

8. Prove that system equation (6.3.2) is completely observable if and only if the matrix $V^T V$ is positive definite, where V is the observability matrix of $\{A, C\}$.

9. Show that the kth order scalar equation

$$
\begin{aligned}
z(n+k) + p_1 z(n+k-1) + \ldots + p_k z(n) &= u(n), \\
y(n) &= c\, z(n)
\end{aligned}
$$

is completely observable if $c \neq 0$.

10. Verify that the change of variable $z(n) = V\, x(n)$ produces the observability canonical pair $\{\bar{A}, \bar{c}\}$ defined in Expression (6.3.13).

11. Devise a change of variable that will transform system equation (6.3.11) to the observability canonical pair $\{\tilde{A}, \tilde{c}\}$ given by Expression (6.3.14). (Hint: See Section 6.2.)

12. Consider System (6.3.2), where $P^{-1} AP$ is a diagonal matrix. Show that a necessary and sufficient condition for complete observability is that none of the columns of the $r \times k$ matrix CP consists of all zero elements.

13. Consider system equation (6.3.2), where $P^{-1} AP$ is in the Jordan form J. Show that necessary and sufficient conditions for complete observability of the system are

   (i) no two Jordan blocks in J corresponds to the same eigenvalue of A,

   (ii) none of the columns of CP that correspond to the first row of each Jordan block consists of all zero elements, and

   (iii) no columns of CP that correspond to distinct eigenvalues consist of all zero elements.

14. Let P be a nonsingular matrix. Show that if the pair $\{A, C\}$ is completely observable, then so is the pair $\{P^{-1}AP, CP\}$.

15. Show that if the matrix A in system equation (6.3.2) is nonsingular, the complete observability and constructibility are equivalent.

16. Consider the completely observable system

$$
\begin{aligned}
x(n+1) &= A\, x(n) + B\, u(n), \\
y(n) &= C\, x(n)
\end{aligned}
$$

where a is a $k \times k$ matrix and C is a $1 \times k$ vector. Let $M = (C^T, A^T C^T, \ldots, (A^T)^{k-1} C^T)$.

(a) Show that $M^T A (M^T)^{-1} = \begin{pmatrix} 0 & 1 & 0 & \cdots & 0 \\ 0 & 0 & 1 & \cdots & 0 \\ \vdots & \vdots & \vdots & & \vdots \\ 0 & 0 & 0 & \cdots & 1 \\ -a_k & -a_{k-1} & -a_{k-2} & \cdots & -a_1 \end{pmatrix}$

where $a_i, 1 \le i \le k$ are the coefficients of the characteristic polynomial

$$|A - \lambda I| = \lambda^k + a_1 \lambda^{k-1} + \ldots + a_k.$$

(b) Write down the corresponding *canonical controllable system* by letting $\tilde{x}(n) = M^T x(n)$. Then deduce a necessary condition on C for complete observability of the original system.

17. Consider the system

$$\begin{aligned} x(n+1) &= A\,x(n), \\ y(n) &= C\,x(n) \end{aligned}$$

where A is a $k \times k$ matrix and C is a $1 \times k$ vector. Prove that the system is completely observable if and only if the matrix

$$G = (C, CA^{-1}, CA^{-2}, \ldots, CA^{-k+1}) \text{ is nonsingular.}$$

6.4   Stabilization by State Feedback (Design via Pole Placement)

Feedback controls are used in may aspects of our lives, from the brakes system of a car to central air conditioning. The method has been used by engineers since the beginning of time. However, the systematic study of stabilization by state feedback control is of a more recent origin (See [2, 3]) and dates back to the 1960's. The idea of state feedback is simple: it is assumed the state vector $x(n)$ can be directly measured, and the control $u(n)$ is adjusted based on this information. Consider the (open–loop) time-invariant control system shown in Fig. 6.8a whose equation is

$$x(n+1) = A\,x(n) + B\,u(n) \tag{6.4.1}$$

where as before A is a $k \times k$ matrix, and B a $k \times m$ matrix.

Suppose we apply linear feedback $u(n) = -Kx(n)$, where K is a real $m \times k$ matrix called the *state feedback* or *gain state* matrix. The resulting (closed-loop) system (Fig. 6.8b) obtained by substituting $u = -Kx$ into Eq. (6.4.1) is

$$x(n+1) = Ax(n) - BKx(n)$$

or

$$x(n+1) = (A - BK)x(n). \tag{6.4.2}$$

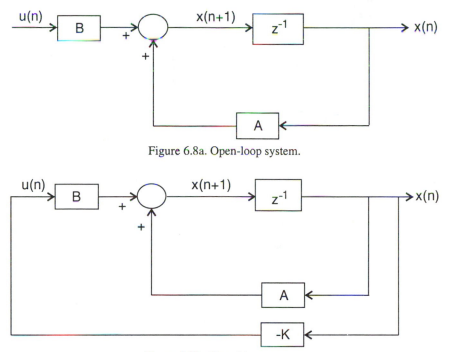

Figure 6.8a. Open-loop system.

Figure 6.8b. Closed-loop system.

The objective of feedback control is to choose K in a way such that the resulting
System (6.4.2) would behave in a prespecified manner. For example, if one wishes
to stabilize System (6.4.1), that is to make its zero solution asymptotically stable,
K must be chosen so that all the eigenvalues of $A - BK$ lie inside the unit disk.
    We now give the main result in this section.

Theorem 6.20 Let $\Lambda = \{\mu_1, \mu_2, \ldots, \mu_k\}$ be an arbitrary set of k complex
numbers such that $\bar{\Lambda} = \{\bar{\mu}_1, \bar{\mu}_2, \ldots, \bar{\mu}_k\} = \Lambda$. Then the pair $\{A, B\}$ is completely
controllable, if and only if there exists a matrix K such that eigenvalues of $A - BK$
are the set Λ.
    Since the proof of the theorem is rather lengthy we first present the proof for the
case $m = 1$, i.e., when B is a $(k \times 1)$ vector and $u(n)$ is a scalar. We start the proof by
writing the characteristic polynomial of A, $|A - \lambda I| = \lambda^k + a_1\lambda^{k-1} + a_2\lambda^{k-2} + \ldots + a_k$.
Suppose also that

$$\prod_{i=1}^{k} (\lambda - \mu_i) = \lambda^k + b_1\lambda^{k-1} + b_2\lambda^{k-2} + \ldots + b_k.$$

Define $T = WM$ where W is the controllability matrix of rank k defined in Eq.
(6.2.1) as

$$W = (B, AB, \ldots, A^{k-1}B)$$

and

$$M = \begin{pmatrix} a_{k-1} & a_{k-2} & \cdots & a_1 & 1 \\ a_{k-2} & a_{k-3} & \cdots & 1 & 0 \\ \vdots & \vdots & & \vdots & \vdots \\ a_1 & 1 & \cdots & 0 & 0 \\ 1 & 0 & & 0 & 0 \end{pmatrix}.$$

Then (Exercise 6.4, Problem 12)

$$\bar{A} = T^{-1}AT = \begin{pmatrix} 0 & 1 & 0 & \cdots & 0 \\ 0 & 0 & 1 & \cdots & 0 \\ \vdots & \vdots & \vdots & & \\ 0 & 0 & 0 & \cdots & 1 \\ -a_k & -a_{k-1} & -a_{k-2} & \cdots & -a_1 \end{pmatrix} \qquad (6.4.3)$$

and

$$\bar{B} = T^{-1}B = (0 \ \ 0 \ \ \cdots \ \ 0 \ \ 1)^T.$$

Letting $x(n) = T\bar{x}(n)$ in System (6.4.2) we get the equivalent system

$$\bar{x}(n+1) = (\bar{A} - \bar{B}\bar{K})\bar{x}(n) \qquad (6.4.4)$$

where

$$\bar{K} = KT. \qquad (6.4.5)$$

Choose

$$\bar{K} = (b_k - a_k, b_{k-1} - a_{k-1}, \ldots, b_1 - a_1). \qquad (6.4.6)$$

Then

$$\bar{B}\bar{K} = \begin{pmatrix} 0 & 0 & \cdots & 0 \\ 0 & 0 & \cdots & 0 \\ \vdots & \vdots & & \vdots \\ 0 & 0 & \cdots & 0 \\ b_k - a_k & b_{k-1} - a_{k-1} & & b_1 - a_1 \end{pmatrix}.$$

Observe that $A - BK$ is similar to $\bar{A} - \bar{B}\bar{K}$ since $\bar{A} - \bar{B}\bar{K} = T^{-1}AT - T^{-1}BKT = T^{-1}(A - BK)T$. Thus

$$|A - BK - \lambda I| = |\bar{A} - \bar{B}\bar{K} - \lambda I| = \begin{vmatrix} -\lambda & 1 & \cdots & 0 \\ 0 & -\lambda & \cdots & 0 \\ \vdots & \vdots & & \vdots \\ 0 & 0 & \cdots & 1 \\ -b_k & -b_{k-1} & \cdots & -b_1 - \lambda \end{vmatrix}$$

$$= \lambda^n + b_1\lambda^{n-1} + \ldots + b_k$$

which has Λ as its set of eigenvalues. Hence the required feedback (gain) matrix is given by

$$K = \bar{K}T^{-1}$$
$$= (b_k - a_k, b_{k-1} - a_{k-1}, \ldots, b_1 - a_1)T^{-1}.$$

Example 6.21 Consider the control system $x(n + 1) = Ax(n) + Bx(n)$ with

$$A = \begin{pmatrix} 1 & -3 \\ 4 & 2 \end{pmatrix}, \quad B = \begin{pmatrix} 1 \\ 1 \end{pmatrix}.$$

Find a state feedback gain matrix K such that the eigenvalues of the closed loop system are $\frac{1}{2}$, and $\frac{1}{4}$.

Solution
Method 1

$$|A - \lambda I| = \begin{vmatrix} 1 - \lambda & -3 \\ 4 & 2 - \lambda \end{vmatrix} = \lambda^2 - 3\lambda + 14.$$

So

$$a_1 = -3, \quad a_2 = 14.$$

Also

$$\left(\lambda - \frac{1}{2}\right)\left(\lambda - \frac{1}{4}\right) = \lambda^2 - \frac{3}{4}\lambda + \frac{1}{8} \quad (*)$$

So

$$b_1 = -\frac{3}{4} \text{ and } b_2 = \frac{1}{8}.$$

Now

$$W = \begin{pmatrix} 1 & -2 \\ 1 & 6 \end{pmatrix}, \quad M = \begin{pmatrix} -3 & 1 \\ 1 & 0 \end{pmatrix}.$$

Hence

$$T = WM = \begin{pmatrix} 1 & -2 \\ 1 & 6 \end{pmatrix}\begin{pmatrix} -3 & 1 \\ 1 & 0 \end{pmatrix} = \begin{pmatrix} -5 & 1 \\ 3 & 1 \end{pmatrix}$$

and

$$T^{-1} = -\frac{1}{8}\begin{pmatrix} 1 & -1 \\ -3 & -5 \end{pmatrix}.$$

Therefore

$$K = (b_2 - a_2, b_1 - a_1)T^{-1}$$

or

$$K = \left(-13\frac{7}{8}, \ 2\frac{1}{4}\right) \cdot \left(-\frac{1}{8}\right)\begin{pmatrix} 1 & -1 \\ -3 & -5 \end{pmatrix}$$
$$= \begin{pmatrix} \frac{165}{64} & \frac{-21}{64} \end{pmatrix}.$$

Method 2 In this method we substitute $K = (k_1 k_2)$ into the characteristic polynomial $|A - BK - \lambda I|$ and then match the coefficients of powers in λ with the desired characteristic polynomial (*).

$$\begin{vmatrix} 1 - k_1 - \lambda & -3 - k_2 \\ 4 - k_1 & 2 - k_2 - \lambda \end{vmatrix} = \lambda^2 - \lambda (3 - k_1 - k_2) + 14 - 5k_1 + 3k_2. \quad (**)$$

Comparing the coefficients of powers in λ in (*) and (**) we obtain

$$3 - k_1 - k_2 = \frac{3}{4},$$

$$14 - 5k_1 + 3k_2 = \frac{1}{8}.$$

This gives us $k_1 = \frac{165}{64}$, and $k_2 = \frac{-21}{64}$.

Hence

$$K = \begin{pmatrix} \frac{165}{64} & \frac{-21}{64} \end{pmatrix}.$$

To prove the general case $m > 1$ of Theorem 6.20 we need the following preliminary result.

Lemma 6.22 If the pair $\{A, B\}$ is completely controllable and the columns of B, assumed nonzero, are $b_1, b_2, \ldots, b_m$, then there exist matrices $K_i, 1 \leq i \leq m$ such that the pairs $\{A - BK_i, b_i\}$ are completely controllable.

Proof Let us consider the case $i = 1$. Since the controllability matrix W has rank k (full rank), one may select a basis of R^k consisting of k columns of W. One such selection would be the $k \times k$ matrix

$$M = \begin{pmatrix} b_1, Ab_1, \ldots, A^{r_1-1}b_1, b_2, Ab_2, \ldots, a^{r_2-1}b_2, \ldots \end{pmatrix}$$

where r_i is the smallest integer such that $A^{r_i} b_i$ is linearly dependent on all the preceding vectors. Define an $m \times k$ matrix L having its r_1 column equal to $e_2 = (0, 1, \ldots, 0)^T$, its $(r_1 + r_2)$th column equal to $e_3 = (0, 0, 1, \ldots, 0)^T$, and so on, all its other columns being zero. We claim that the desired matrix K_1 is given by $K_1 = LM^{-1}$. To verify the claim, we compare the corresponding columns in both sides of $K_1 M = L$. It follows immediately that

$$K_1 b_1 = 0, \quad K_1 A b_1 = 0, \quad \ldots, \quad K_1 A^{r_1-1} b_1 = 0, \quad K_1 b_2 = e_2,$$

$$K_1 A b_2 = 0, \quad \ldots, \quad K_1 A^{r_2-1} b_1 = 0, \quad K_1 b_3 = e_2, \text{ etc.}$$

Thus we have

$$\begin{pmatrix} b_1, (A - BK_1)b_1, (A - BK_2)^2 b_1, \ldots, (A - BK_2)^{k-1} b_1 \end{pmatrix} = W(k)$$

which has rank k by assumption. This proves our claim. We are now ready to give the proof of the general case $m > 1$ in Theorem 6.20.

Proof of Theorem 6.20 Let K_1 be the matrix in Lemma 6.22. Then by Lemma 6.22, it follows that the pair $\{A - BK_1, b_1\}$ is completely controllable. And by the proof of the theorem for $m = 1$, there exists a $1 \times$ vector ξ such that the eigenvalues of $A + BK_1 + b_1\xi$ are the set Λ. Let $\bar{K}$ be the $m \times k$ matrix whose first row is ξ and all other rows are zero. Then the desired feedback (gain) matrix is given by $K = K_1 + \bar{K}$. Since $u = -Kx$ this gives

$$
\begin{aligned}
x(n + 1) &= (A - BK)x(n) \\
&= (A - BK_1 - b_1\xi)x(n).
\end{aligned}
$$

To prove the converse, select K_o such that $(A - BK_o)^n \to 0$ as $n \to \infty$, that is the spectral radius $\rho(A - BK_o) < 1$, and K_1 such that $\rho(A - BK_1) = \left\{\exp\left(\frac{2\pi n}{k}\right): n = 0, 1, \ldots, k - 1\right\}$, the kth roots of unity. Then clearly $(A - BK_1)^k = 1$. Suppose that for some vector $\xi \in R^k$, $\xi^T A^n B = 0$ for all $n \in Z^+$. Then for any matrix K,

$$
\begin{aligned}
\xi^T(A - BK)^n &= \xi^T(A - BK)(A - BK)^{n-1} \\
&= (\xi^T A - \xi^T BK)(A - BK)^{n-1} \\
&= \xi^T A(A - BK)^{n-1}, \quad \text{since } \xi^T B = 0 \\
&= \xi^T A(A - BK)(A - BK)^{n-2} \\
&= \xi^T A^2(A - BK)^{n-2}, \quad \text{since } \xi^T AB = 0.
\end{aligned}
$$

Continuing this procedure we obtain

$$
\xi^T(A - BK)^n = \xi^T A^n, \quad \text{for all } n \in Z^+.
$$

This implies that

$$
\xi^T[(A - BK_o)^n - (A - BK_1)^n] = 0, \quad \text{for all } n \in Z^+,
$$

or

$$
\xi^T[(A - BK_o)^{kr} - 1] = 0, \quad \text{for all } r \in Z^+.
$$

Letting $r \to \infty$, we have $(A - BK_o)^{kr} \to 0$ and consequently, $\xi^T = 0$. This implies that the pair $\{A, B\}$ is completely controllable.

An immediate consequence of Theorem 6.20 is a simple sufficient condition for *stabilizability*. A system $x(n + 1) = Ax(n) + Bu(n)$ is *stabilizable* if one can find a feedback control $u(n) = -Kx(n)$ such that the zero solution of the resulting closed-loop system $x(n + 1) = (A - BK)x(n)$ is asymptotically stable. In other words the pair $\{A, B\}$ is stabilizable if for some matrix K, $A - BK$ is a stable matrix (i.e., all its eigenvalues lie inside the unit disk).

Corollary 6.23 System (6.4.1) is stabilizable if it is completely controllable.

The question still remains whether or not we can stabilize an uncontrollable system. Well, the answer is yes and no, as it may be seen by the following example.

Example 6.24 Consider the control system

$$x(n + 1) = Ax(n) + Bu(n), \text{ where}$$

$$A = \begin{pmatrix} 0 & a & b \\ 1 & d & e \\ 0 & 0 & h \end{pmatrix}, \quad B = \begin{pmatrix} 1 & \beta_1 \\ 0 & \beta_2 \\ 0 & 0 \end{pmatrix},$$

Let us write

$$A = \begin{pmatrix} A_{11} & A_{12} \\ 0 & A_{22} \end{pmatrix}, \quad B = \begin{pmatrix} B_1 \\ 0 \end{pmatrix},$$

where

$$A_{11} = \begin{pmatrix} 0 & a \\ 1 & d \end{pmatrix}, \quad A_{12} = \begin{pmatrix} b \\ e \end{pmatrix}, \quad A_{22} = (h), \quad B_1 = \begin{pmatrix} 1 & \beta_1 \\ 0 & \beta_2 \end{pmatrix}.$$

If $x = \begin{pmatrix} y \\ z \end{pmatrix}$, then our system may be written as

$$\begin{aligned} y(n + 1) &= A_{11}y(n) + A_{12}z(n) + B_1 u(n), \\ z(n + 1) &= A_{22}z(n). \end{aligned}$$

It is easy to verify that the pair $\{A_{11}, B_1\}$ is completely controllable. Hence by Theorem 6.20, there is a (2×2) gain matrix $\bar{K}$ such that $A_{11} + B_1\bar{K}$ is a stable matrix. Letting $K = (\bar{K})(0)$, then

$$A - BK = \begin{pmatrix} A_{11} - B_1\bar{K} & * \\ 0 & h \end{pmatrix}.$$

Hence the matrix $A - BK$ is stable if and only if $|h| < 1$.

In the general case, a system is stabilizable if and only if the uncontrollable part is asymptotically stable (Exercise 6.4, Problem 8). In this instance, from the columns of the controllability matrix W, we select a basis for the controllable part of the system and extend it to a basis S for R^k. The change of variables $x = Py$, where P is the matrix whose columns are the elements of S, transforms our system to

$$y(n + 1) = \bar{A}y(n) + \bar{B}u,$$

where

$$\bar{A} = \begin{pmatrix} A_{11} & A_{12} \\ 0 & A_{22} \end{pmatrix}, \quad \bar{B} = \begin{pmatrix} B_1 \\ 0 \end{pmatrix}.$$

Here the pair $\{A_{11}, B_1\}$ is controllable. Hence the system is stabilizable if and only if the matrix A_{22} is stable.

6.4.1   Stabilization of Nonlinear Systems by Feedback

Before ending this section, let us turn our attention to the problem of stabilizing a nonlinear system

$$x(n + 1) = f(x(n), u(n)), \tag{6.4.7}$$

where $f : R^k \times R^m \to R^k$. The objective is to find a feedback control

$$u(n) = h(x(n)) \tag{6.4.8}$$

in such a way that the equilibrium point $x^* = 0$ of the closed loop system

$$x(n + 1) = f(x(n)), h(x(n)) \tag{6.4.9}$$

is asymptotically stable (locally!). We make the following assumptions:

(i) $f(0, 0) = 0$ and

(ii) f is continuously differentiable, $A = \frac{\partial f}{\partial x}(0, 0)$ is a $k \times k$ matrix, $B = \frac{\partial f}{\partial u}(0, 0)$ is a $k \times m$ matrix.

Under the above conditions, we have the following surprising result.

Theorem 6.25 If the pair $\{A, B\}$ is controllable, then the nonlinear System (6.4.7) is stabilizable. Moreover, if K is the gain matrix for the pair $\{A, B\}$, then the control $u(n) = -Kx(n)$ may be used to stabilize System (6.4.7).

Proof Since the pair $\{A, B\}$, is controllable, there exists a feedback control $u(n) = -Kx(n)$ which stabilizes the linear part of the system, namely,

$$y(n + 1) = Ay(n) + Bv(n).$$

We are going to use the same control on the nonlinear system. So let $g: R^k \to R^k$ be a function defined by $g(x) = f(x, -Kx)$. Then system equation (6.4.7) becomes

$$x(n + 1) = g(x(n)) \tag{6.4.10}$$

with

$$\left. \frac{\partial g}{\partial x} \right|_{x=0} = A - BK.$$

Since, by assumption, the zero solution of the linearized system

$$y(n + 1) = (A - BK)y(n) \tag{6.4.11}$$

is asymptotically stable, it follows by Theorem 4.22 that the zero solution of System (6.4.10) is also asymptotically stable. This completes the proof of the theorem.

Example 6.26 Consider the nonlinear difference system

$$
\begin{aligned}
x_1(n + 1) &= 2\sin(x_1(n)) + x_2 + u_1(n), \\
x_2(n + 1) &= x_1^2(n) - x_2(n) - u_2(n).
\end{aligned}
$$

Find a control that stabilizes the system.

Solution One may check easily the controllability of the linearized system $\{A, B\}$, where

$$A = \begin{pmatrix} 2 & 1 \\ 0 & -1 \end{pmatrix}, \quad B = \begin{pmatrix} 1 \\ -1 \end{pmatrix}$$

after some computation, a gain matrix for the linearized system is $K = (2.015, 0.975)$ where the eigenvalues of $A - BK$ are $\frac{1}{2}$ and 0.1. As implied by Theorem 6.22, the control $u(n) = -Kx(n)$ would stabilize the nonlinear system, where $K = (2.015, 0.975)$.

Exercise 6.4

In Problems 1 through 3 determine the gain matrix K that stabilizes the system $\{A, B\}$.

1. $A = \begin{pmatrix} 0 & 1 \\ -0.16 & -1 \end{pmatrix}, \quad B = \begin{pmatrix} 0 \\ 1 \end{pmatrix}.$

2. $A = \begin{pmatrix} 2 & 1 & 1 \\ -2 & 1 & 0 \\ -2 & -1 & 0 \end{pmatrix}, \quad B = \begin{pmatrix} 0 & 1 \\ 1 & 0 \\ 0 & -2 \end{pmatrix}.$

3. $A = \begin{pmatrix} 0 & 1 & 0 \\ 0 & 0 & 1 \\ -2 & 1 & 3 \end{pmatrix}, \quad B = \begin{pmatrix} 0 \\ 0 \\ 1 \end{pmatrix}.$

4. Determine the matrices B for which the system $\{A, B\}$, $A = \begin{pmatrix} 1 & -1 & 2 \\ 0 & \frac{1}{2} & 1 \\ \frac{1}{2} & \frac{-1}{2} & 1 \end{pmatrix}$, is (a) controllable and (b) stabilizable.

5. Consider the second order equation

$$x(n + 2) + a_1 x(n + 1) + a_2 x(n) = u(n).$$

Determine a gain control $u(n) = c_1 x(n) + c_2 x(n + 1)$ which stabilizes the equation. (Hint: Put the equation into a system form.)

6. Describe an algorithm for decomposing the system $x(n+1) = Ax(n) + Bu(n)$ into its controllable and uncontrollable parts when A is a (3×3) matrix, and B is a (3×2) matrix.

7. Generalize the result of Problem 6 to the case where A is a $k \times k$ matrix and B is a $k \times r$ matrix.

8. *Show that the pair $\{A, B\}$ is stabilizable if and only if the uncontrollable part of the system is asymptotically stable.

9. *Deadbeat Response.* If the eigenvalues of the matrix $A - BK$ are all zero, then the solutions of the system $x(n+1) = (A - BK)x(n)$ will read 0 in finite time. It is then said that the gain matrix K produces a *deadbeat response.* Suppose that A is a (3×3) matrix and B a (3×1) vector.

   (a) Show that the desired feedback matrix K for the deadbeat response is given by

   $$K = [1 \ 0 \ 0][\xi_1 \ \xi_2 \ \xi_3]^{-1}$$

   where

   $$\xi_1 = A^{-1}B, \quad \xi_2 = (A^{-1})^2 B, \quad \xi_3 = (A^{-1})^3 B.$$

   (b) Show that the vectors ξ_1, ξ_2, and ξ_3 are generalized eigenvectors of the matrix $A - BK$. [i.e., $(A - BK)\xi_1 = 0$, $(A - BK)\xi_2 = \xi_1$, $(A - BK)\xi_3 = \xi_2$].

10. Ackermann's Formula:

    Let $\Lambda = \{\mu_1, \mu_2, \ldots, \mu_k\}$ be the desired eigenvalues for the completely controllable pair $\{A, B\}$, with $\Lambda = \bar{\Lambda}$. Show that the feedback (gain) matrix K can be given by

    $$K = (0 \ 0 \ \ldots \ 0)(B \ AB \ \ldots \ A^{k-1}B)^{-1}p(A),$$

    where

    $$p(\lambda) = \prod_{i=1}^{k} (\lambda - \mu_i) = \lambda^k + \alpha_1\lambda^{k-1} + \ldots + \alpha_k.$$

11. Let $\Lambda = \{\mu_1, \mu_2, \ldots, \mu_k\}$ be a set of complex numbers with $\Lambda = \bar{\Lambda}$. Show that if the pair $\{A, C\}$ is completely observable, then there exists a matrix L such that the eigenvalues of $A - LC$ are the set Λ.

12. Verify Formula (6.4.3).

13. Find a stabilizing control for the system

    $$\begin{aligned} x_1(n + 1) &= 3x_1(n) + x_2^2(n) - \text{sat}(2x_2(n) + u(n)), \\ x_2(n + 1) &= \sin x_1(n) - x_2(n) + u(n), \end{aligned}$$

    where

    $$\text{sat } y = \begin{cases} y & \text{if } |y| \leq 1, \\ \text{sign } y & \text{if } |y| > 1 \end{cases}.$$

14. Find a stabilizing control for the system

    $$\begin{aligned} x_1(n + 1) &= 2x_1(n) + x_2(n) + x_3^3(n) + u_1(n) + 2u_2(n), \\ x_2(n + 1) &= x_1^2(n) + \sin x_2(n) + x_2^2(n) + u_1^2(n) + u_2(n), \\ x_3(n + 1) &= x_1^4(n) + x_2^3(n) + \frac{1}{2}x_3(n) + u_1(n). \end{aligned}$$

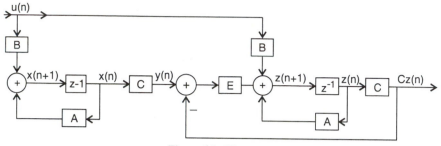

Figure 6.9. Observer.

15. (Research problem). Find sufficient conditions for the stabilizability of a time–variant system

$$x(n + 1) = A(n)x(n) + B(n)u(n).$$

16. (Research problem). Extend the result of Problem 15 to nonlinear time–variant systems.

6.5   Observers

Theorem 6.20 provides a method of finding a control $u(n) = -Kx(n)$ that stabilizes a given system. This method clearly requires the knowledge of all state variables $x(n)$. Unfortunately, in many systems of practical importance, the entire state vector is not available for measurement. Faced with this diffficulty, we are led to construct an estimate of the full state vector based on the available measurements. Let us consider again the system

$$\left. \begin{array}{l} x(n + 1) = Ax(n) + Bu(n), \\ y(n) = Cx(n) \end{array} \right\}. \tag{6.5.1}$$

To estimate the state vector $x(n)$ we construct the k-dimensional *observer* (Fig. 6.9)

$$z(n + 1) = Az(n) + E[y(n) - Cz(n)] + Bu(n) \tag{6.5.2}$$

where E is a $k \times r$ matrix to be determined later. Notice that unlike $x(n)$, the state observer $z(n)$ can be obtained from available data. To see this, let us write the observer (6.5.2) in the form

$$z(n + 1) = (A - EC)z(n) + Ey(n) + Bu(n). \tag{6.5.3}$$

We observe here that the inputs to the observer involve $y(n)$ and $u(n)$ which are available to us.

The question remains whether the observer state $z(n)$ is a good estimate of the original state $x(n)$. One way to check the goodness of this estimator is to insure that the error $e(n) = z(n) - x(n)$ goes to zero as $n \to \infty$. To achieve this objective we write the error equation in $e(n)$ by subtracting Eq. (6.5.2) from Eq. (6.5.1) and using $y(n) = Cx(n)$. Hence

$$z(n + 1) - x(n + 1) = [A - EC][z(n) - x(n)]$$

or

$$e(n + 1) = [A - EC]e(n). \tag{6.5.4}$$

Clearly if the zero solution of System (6.5.4) is asymptotically stable (i.e., the matrix $A - EC$ is stable), then the error vector $e(n)$ tends to zero. Thus the problem reduces to finding a matrix E such that the matrix $A - EC$ has all its eigenvalues inside the unit disk. The following result gives a condition under which this can be done.

Theorem 6.27   If System (6.5.1) is completely observable, then an observer (6.5.2) can be constructed such that the eigenvalues of the matrix $A - EC$ are arbitrarily chosen. In particular, one can choose a matrix E such that the error $e(n) = z(n) - x(n)$ in the estimate of the state $x(n)$ by the state observer $z(n)$ tends to zero.

Proof   Since the pair $\{A, C\}$ is completely observable, it follows from Section 4.3 that the pair $\{A^T, C^T\}$ is completely controllable. Hence by Theorem 6.20 the matrix E can be chosen such that $A^T - C^T E^T$ has an arbitrary set of eigenvalues, which is the same as the set of eigenvalues of the matrix $A - EC$.

Moreover, if we choose the matrix E such that all the eigenvalues of the matrix $A - EC$ are inside the unit disk, then $e(n) \to 0$ (see Corollary 3.24).

6.5.1   *Eigenvalue Separation Theorem*

Suppose that the system

$$
\begin{aligned}
x(n + 1) &= Ax(n) + Bu(n), \\
y(n) &= Cx(n)
\end{aligned}
$$

is both completely observable and completely controllable. Assuming the the state vector $x(n)$ is available, we can use Theorem 6.25 to find a feedback control $u(n) = -Kx(n)$ such that in the closed-loop system

$$x(n + 1) = (A - BK)x(n)$$

the eigenvalues of $A - BK$ can be chosen arbitrarily. Next we use Theorem 6.27 to choose a state observer $z(n)$ to estimate the state $x(n)$ in such a way that the eigenvalues of $A - EC$ in the observer

$$z(n + 1) = (A - EC)z(n) + Ey(n) + Bu(n)$$

can also be chosen arbitrarily.

In practice a feedback control may be obtained using the state observer $z(n)$ instead of the original state $x(n)$ (whose components are not all available for measurement). In other words, we use the feedback control

$$u(n) = -Kz(n). \qquad (6.5.5)$$

The resulting composite system is given by

$$\begin{aligned} x(n+1) &= Ax(n) - BKz(n), \\ z(n+1) &= (A - EC)z(n) + ECx(n) - BKz(n). \end{aligned}$$

It follows that

$$e(n+1) = z(n+1) - x(n+1) = (A - EC)e(n).$$

Hence we have the following composite system

$$\left. \begin{aligned} x(n+1) &= (A - BK)x(n) + BKe(n), \\ e(n+1) &= (A - EC)e(n) \end{aligned} \right\}.$$

The system matrix is given by

$$A = \begin{pmatrix} A - BK & BK \\ 0 & A - EC \end{pmatrix}$$

whose characteristic polynomial is the product of the characteristic polynomials of $(A - BK)$ and $(A - EC)$. Hence the eigenvalues of A are either eigenvalues of $A - BK$ or eigenvalues of $A - EC$ which we can choose arbitrarily. Thus we have proved the following result.

Theorem 6.28 (Eigenvalue Separation Theorem) Consider the system

$$\begin{aligned} x(n+1) &= Ax(n) + Bu(n), \\ y(n) &= Cx(n) \end{aligned}$$

with the observer

$$z(n+1) = (A - EC)z(n) + Ey(n) + Bu(n)$$

and the feedback control

$$u(n) = -Kz(n).$$

Then the characteristic polynomial of this composite system is the product of the characteristic polynomials of $A - BK$ and $A - EC$. Furthermore, the eigenvalues of the composite system can be chosen arbitrarily.

Example 6.29 Consider the system

$$\begin{aligned} x(n+1) &= Ax(n) + Bu(n), \\ y(n) &= Cx(n) \end{aligned}$$

where

$$A = \begin{pmatrix} 0 & -\frac{1}{4} \\ 1 & -1 \end{pmatrix}, \quad B = \begin{pmatrix} 0 \\ 1 \end{pmatrix}, \quad C = (0 \quad 1).$$

Design a state observer so that the eigenvalues of the observer matrix $A - EC$ are $\frac{1}{2} + \frac{1}{2}i$ and $\frac{1}{2} - \frac{1}{2}i$.

Solution The observability matrix is given by

$$\begin{pmatrix} C \\ CA \end{pmatrix} = \begin{pmatrix} 0 & 1 \\ 1 & -1 \end{pmatrix}$$

which has full rank 2. Thus the system is completely observable and the desired observer feedback gain matrix E may be now determined. The characteristic equation of the observer is given by $\det(A - EC - \lambda I) = 0$. If

$$E = \begin{pmatrix} E_1 \\ E_2 \end{pmatrix},$$

then we have

$$\left| \begin{pmatrix} 0 & -\frac{1}{4} \\ 1 & -1 \end{pmatrix} - \begin{pmatrix} E_1 \\ E_2 \end{pmatrix} (0 \quad 1) - \begin{pmatrix} \lambda & 0 \\ 0 & \lambda \end{pmatrix} \right| = 0$$

which reduces to

$$\lambda^2 + (1 + E_2)\lambda + E_1 + \frac{1}{4} = 0. \tag{6.5.6}$$

From assumption the desired characteristic equation is given by

$$\left(\lambda - \frac{1}{2} - \frac{1}{2}i \right) \left(\lambda - \frac{1}{2} + \frac{1}{2}i \right) = 0$$

or

$$\lambda^2 - \lambda + \frac{1}{2} = 0. \tag{6.5.7}$$

Comparing Eqs. (6.5.6) and (6.5.7) yields

$$E_1 = \frac{1}{4}, \quad E_2 = -2$$

Thus $E = \begin{pmatrix} \frac{1}{4} \\ -2 \end{pmatrix}$.

Example 6.30 Fig. 6.10 shows a metallic sphere of mass m, suspended in the magnetic field generated by an electromagnet. The equation of motion for this system is

$$m\ddot{x}_t = mg - k\frac{u_t^2}{x_t}, \tag{6.5.8}$$

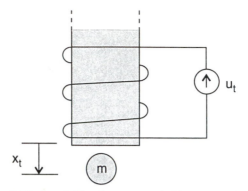

Figure 6.10. A metallic sphere suspended in a magnetic field.

where x_t is the distance of the sphere from the magnet, u_t is the current driving the electromagnet, g is the acceleration of gravity, and k is a constant determined by the properties of the magnet.

It's easy to check that Eq. (6.5.9) has an equilibrium at

$$x_t = x_0 = 1,$$

$$u_t = u_0 = \sqrt{mg/k}.$$

Linearizing Eq. (6.5.9) about this equilibrium gives the following approximate model in terms of the deviations $x = x_t - x_0$ and $u = u_t - u_0$:

$$\ddot{x} - \frac{g}{k}x = -2\sqrt{kg/m}\, u,$$

or, in state variable form:

$$\begin{bmatrix} \dot{x} \\ \dot{v} \end{bmatrix} = \begin{bmatrix} 0 & 1 \\ \frac{g}{k} & 0 \end{bmatrix} \begin{bmatrix} x \\ v \end{bmatrix} + \begin{bmatrix} 0 \\ -2\sqrt{kg/m} \end{bmatrix} u.$$

Thus

$$\hat{A} = \begin{bmatrix} 0 & 1 \\ \frac{g}{k} & 0 \end{bmatrix}, \quad \hat{B} = \begin{bmatrix} 0 \\ -2\sqrt{kg/m} \end{bmatrix}.$$

$\hat{A}$ can be written in the form

$$\hat{A} = P\Lambda P^{-1}$$

where

$$\Lambda = \begin{bmatrix} \sqrt{\frac{g}{k}} & 0 \\ 0 & -\sqrt{\frac{g}{k}} \end{bmatrix}, \quad P = \frac{1}{\sqrt{2}} \begin{bmatrix} 1 & 1 \\ \sqrt{\frac{g}{k}} & -\sqrt{\frac{g}{k}} \end{bmatrix},$$

so

$$A = e^{\hat{A}T} = \begin{bmatrix} \cosh\sqrt{\tfrac{g}{k}}\,T & \sqrt{\tfrac{k}{g}}\,\sinh\sqrt{\tfrac{g}{k}}\,T \\ \sqrt{\tfrac{g}{k}}\,\sinh\sqrt{\tfrac{g}{k}}\,T & \cosh\sqrt{\tfrac{g}{k}}\,T \end{bmatrix},$$

$$B = T e^{\hat{A}T}\hat{B} = -2T \begin{bmatrix} \sqrt{k/m}\,\sinh\sqrt{\tfrac{g}{k}}\,T \\ \sqrt{g/m}\,\cosh\sqrt{\tfrac{g}{k}}\,T \end{bmatrix}.$$

The discrete equivalent system is thus controllable since

$$\det W = |[B \; AB]|$$
$$= \left| -2T\sqrt{k/m} \begin{bmatrix} \sinh\sqrt{\tfrac{g}{k}}\,T & 2\sinh\sqrt{\tfrac{g}{k}}\,T\cosh\sqrt{\tfrac{g}{k}}\,T \\ \sqrt{\tfrac{g}{k}}\,\sinh\sqrt{\tfrac{g}{k}}\,T & 2\sqrt{\tfrac{g}{k}}\,\sinh\sqrt{\tfrac{g}{k}}\,T\cosh\sqrt{\tfrac{g}{k}}\,T \end{bmatrix} \right|$$
$$= ce\sqrt{\tfrac{g}{k}}\,T\,\sinh\sqrt{\tfrac{g}{k}}\,T,$$

where $c = 0$ only if $T = 0$.

If the position deviation x of the ball from equilibrium can be measured then the system is also observable, since then we have the measurement equation

$$y = [1 \; 0]\begin{bmatrix} x \\ v \end{bmatrix}$$

and hence

$$C = [1 \; 0].$$

Observability is easily verified by computing

$$\det V = \left| \begin{bmatrix} C \\ CA \end{bmatrix} \right| = \left| \begin{bmatrix} 1 & 0 \\ \cosh\sqrt{\tfrac{g}{k}}\,T & \sqrt{\tfrac{k}{g}}\,\sinh\sqrt{\tfrac{g}{k}}\,T \end{bmatrix} \right|$$
$$= \sqrt{\tfrac{k}{g}}\,\sinh\sqrt{\tfrac{g}{k}}\,T,$$

which is zero only if $T = 0$. Before continuing, fix $m = k = 0.1, g = 10$, and $T = 0.01$. Thus

$$A = \begin{bmatrix} 1.0050 & 0.0100 \\ 1.0017 & 1.0050 \end{bmatrix}, \quad B = \begin{bmatrix} -0.0020 \\ -0.2010 \end{bmatrix}.$$

Note that A is unstable, with eigenvalues $\lambda_1 = 1.1052$ and $\lambda_2 = 0.9048$.

Controllability of $\{A, B\}$ implies that a stabilizing state feedback gain $K = [k_1, k_2]$ can be found. Moreover the eigenvalues of the resulting system matrix $A - BK$ can be assigned arbitrarily. In our example

$$A - BK = \begin{bmatrix} 1.0050 + 0.0020k_1 & 0.0100 + 0.0020k_2 \\ 1.00017 + 0.2010k_1 & 1.0050 + 0.2010k_2 \end{bmatrix}$$

so that

$$|\lambda I - A + BK| = \lambda^2 - (2.0100 + 0.002k_1 + 0.201k_2)\lambda + 0.2000k_2 + 1$$

and eigenvalues of $\lambda_1 = \frac{1}{2}$ and $\lambda_2 = -\frac{1}{2}$ (both inside the unit circle) can be obtained by choosing

$$K = [k_1, k_2] = [-376.2492 - 6.2500].$$

Observability of $\{A, C\}$ implies that an asymptotic observer can be constructed to produce an estimate of the system state vector from measurements of x. The observer gain $L = [l_1, l_2]^T$ can be chosen not only to ensure that the state estimate converges, but to place the observer eigenvalues arbitrarily. In our example

$$A - LC = \begin{bmatrix} 1.0050 - l_1 & 0.0100 \\ 1.0017 - l_2 & 1.0050 \end{bmatrix}$$

so that

$$|\lambda I - A + LC| = \lambda^2 + (l_1 - 2.0100)\lambda - 1.0050l_1 + 0.0100l_2 + 1$$

and eigenvalues of $\lambda_1 = \frac{1}{4}$ and $\lambda_2 = -\frac{1}{4}$ can be obtained by choosing

$$L = \begin{bmatrix} l_1 \\ l_2 \end{bmatrix} \begin{bmatrix} 2.0100 \\ 95.5973 \end{bmatrix}.$$

The eigenvalue separation theorem ensures that combining this observer with the state feedback controller designed above will produce a stable closed loop system with eigenvalues at $\pm\frac{1}{2}$ and $\pm\frac{1}{4}$.

Exercise 6.5
In Problems 1 through 4 design an observer so that the eigenvalues of the matrix $E - AC$ are as given.

1. $A = \begin{pmatrix} 1 & 1 \\ 0 & -1 \end{pmatrix}$, $C = (1 \quad 1)$,

   $\lambda_1 = \frac{1}{2}$, $\lambda_2 = -\frac{1}{4}$.

2. $A = \begin{pmatrix} 0 & 1 \\ 1 & 0 \end{pmatrix}$, $C = (0 \quad 1)$,

   $\lambda_1 = \frac{1}{2} - \frac{1}{4}i$, $\lambda_2 = \frac{1}{2} + \frac{1}{4}i$.

3. $A = \begin{pmatrix} 1 & 0 & 0 \\ 0 & 0 & 1 \\ 0 & 1 & 0 \end{pmatrix}$, $C = (0 \quad 1 \quad 0)$,

   $\lambda_1 = \frac{1}{2}$, $\lambda_2 = \frac{1}{4} - \frac{1}{4}i$, $\lambda_3 = \frac{1}{4} + \frac{1}{4}i$.

4. $A = \begin{pmatrix} 0 & 1 & 0 \\ 0 & -1 & 1 \\ 0 & 0 & -1 \end{pmatrix}$, $\quad C = \begin{pmatrix} 0 & 1 & 0 \\ 1 & 0 & 1 \end{pmatrix}$,

$\lambda_1 = \frac{1}{2}$, $\quad \lambda_2 = -\frac{1}{4}$, $\quad \lambda_3 - \frac{1}{2}$.

5. Reduced Order Observers:

Consider the completely observable system

$$\begin{aligned} x(n+1) &= Ax(n) + Bu(n), \\ y(n) &= Cx(n) \end{aligned} \tag{6.5.9}$$

where it is assumed that the $r \times k$ matrix C has rank r (i.e., the r measurements are linearly independent). Let H be a $(k-r) \times k$ matrix such that the matrix

$$P = \begin{pmatrix} H \\ C \end{pmatrix}$$

is nonsingular. Let

$$\bar{x}(n) = Px(n). \tag{6.5.10}$$

Then $\bar{x}$ may be written as

$$\bar{x} = \begin{pmatrix} w(n) \\ y(n) \end{pmatrix}$$

where $w(n)$ is $k-r$ dimensional and $y(n)$ is the r-dimensional of outputs.

(a) Use Eq. (6.5.10) to show that system equation (6.5.9) may be put in the form

$$\begin{pmatrix} w(n+1) \\ y(n+1) \end{pmatrix} = \begin{pmatrix} A_{11} & A_{12} \\ A_{21} & A_{22} \end{pmatrix} \begin{pmatrix} w(n) \\ y(n) \end{pmatrix} + \begin{pmatrix} B_1 \\ B_2 \end{pmatrix} u(n). \tag{6.5.11}$$

(b) Multiply the bottom part of Eq. (6.5.11) by any $(k-r)xr$ matrix E to show that

$$\begin{aligned} W(n+1) - Ey(n+1) &= (A_{11} - EA_{21})[W(n) - Ey(n)] \\ &+ [A_{11}E - EA_{21}E + A_{12} - EA_{22}]y(n) \\ &+ (B_1 - EB_2)u(n) \end{aligned} \tag{6.5.12}$$

(c) If $v(n) = w(n) - Fy(n)$, show that

$$\begin{aligned} v(n+1) &= (A_{11} - EA_{21})v(n) + [A_{11}E - EA_{21}E + A_{12} \\ &\quad - EA_{22}]y(n) \\ &+ (B_1 - FB_2)u(n). \end{aligned} \tag{6.5.13}$$

(d) Explain why we can take an observer of system equation (6.5.9) as the $(k - r)$ dimensional system

$$
\begin{aligned}
z(n+1) \ &= \ (A_{11} - EA_{21})z(n) + [A_{11}E - EA_{21}E + A_{12} \\
&\quad -EA_{22}]y(n) \\
&+ \ (B_1 - FB_2)u(n).
\end{aligned}
\tag{6.5.14}
$$

(e) Let $e(n) = z(n) - v(n)$. Show that

$$
e(n+1) = (A_{11} - EA_{21})e(n). \tag{6.5.15}
$$

6. Prove that if the system equation (6.5.9) is completely observable, then the pair $\{A_{11}, A_{21}\}$ in Eq. (6.5.11) is completely observable.

7. Prove the Eigenvalue Separation Theorem 6.28 for reduced order observers.

8. Consider the system

$$
\begin{aligned}
x_1(n+1) \ &= \ x_2(n), \\
x_2(n+1) \ &= \ -x_1(n) + 2x_2(n) + u(n), \\
y(n) \ &= \ x_1(n).
\end{aligned}
$$

Construct a one dimensional observer with a zero eigenvalue.

References

[1] S. Barnett, *Introduction to Mathematical Control Theory*, Claredon, Oxford 1975.

[2] T. Kailath, *Linear Systems*, Prentice Hall, New Jersey, 1980.

[3] D.G. Luenberger, *Introduction to Dynamic Systems, Theory, Models & Applications*, John Wiley & Sons, New York 1979.

[4] R.E. Kalman, and J.E. Bertram, "Control System Analysis and Design via the Second Method of Liapunov: I. Continuous-Time Systems; II. Discrete-Time Systems," *ASME J. Basic Eng., Ser. D*, **82** (1960), 371–93, 394–400.

[5] J.P. La Salle, *The Stability and Control of Discrete Processes: Applied Mathematical Sciences*, Vol. 82, Springer, New York, 1986.

[6] K. Ogata, *Discrete-Time Control Systems*, Prentice-Hall, New Jersey, 1987.

[7] L. Weiss, "Controllability, Realization and Stability of Discrete-Time Systems," *Siam J. Control*, **10**, (1972), 230–251.

7

Asymptotic Behavior of Difference Equations

In Chapters 4 and 5 we were mainly interested in stability questions. In other words, we wanted to know whether solutions of a difference equation converge to zero or to an equilibrium point. In asymptotic theory, we are concerned rather with obtaining asymptotic formulae for the manner in which solutions tend to zero or a constant. We begin this chapter by introducing the reader to the tools of the trade.

7.1 Tools of Approximation

The symbols $\sim$, o, and O are the main tools of approximating functions and are widely used in all branches of science. For the benefit of our readers, we shall give our definitions for functions defined on the real or complex numbers. Hence sequences will be treated as a special case of the general theory.

We start with the symbol O (big oh).

Definition 7.1. Let $f(t)$ and $g(t)$ be two functions defined on R or C. Then we say that $f(t) = O(g(t))$, $(t \to \infty)$, if there is a positive constant M such that

$$|f(t)| \leq M|g(t)| \text{ for all } t \geq t_0.$$

Equivalently, $f(t) = O(g(t))$ if $\left|\frac{f(t)}{g(t)}\right|$ is bounded for $t \geq t_0$. In other words $f = O(\varphi)$ if f is of order not exceeding φ.

Example 7.2.

(a) Show that

$$\left(\frac{n}{t^2 + n^2}\right)^n = O\left(\frac{1}{t^n}\right), \qquad (n \to \infty), \text{ for } n \in Z^+.$$

Solution Without loss of generality we assume $t > 1$. We have $t^2 + n^2 = (t - n)^2 + 2nt \geq 2nt$. Hence

$$\left(\frac{n}{t^2 + n^2}\right)^n \leq \frac{1}{(2t)^n} = \frac{1}{2^n}\left(\frac{1}{t^n}\right) \leq \frac{1}{t^n}, \qquad \text{for } n \in Z^+, t > 1.$$

It follows that

$$\left(\frac{n}{t^2 + n^2}\right)^n = O\left(\frac{1}{t^n}\right)$$

with the constant $M = 1$ being independent of n.

(b) Show that

$$\sin\left(n\pi + \frac{1}{n}\right) = O\left(\frac{1}{n}\right), (n \to \infty).$$

Solution Recall that $\sin\left(n\pi + \frac{1}{n}\right) = (-1)^n \sin\frac{1}{n}$. Thus

$$\left|\frac{\sin\left(n\pi + \frac{1}{n}\right)}{1/n}\right| = \left|\frac{\sin\frac{1}{n}}{1/n}\right|.$$

If we let $u = \frac{1}{n}$, then $\lim_{n\to\infty}\left|\frac{\sin\frac{1}{n}}{1/n}\right| = \lim_{u\to 0}\left|\frac{\sin u}{u}\right| = 1.$

Hence we conclude that $\left|\left(\sin\frac{1}{n}\right)/1/n\right|$ is bounded, which gives the required result.

(c) Show that $t^2 \log t + t^3 = O(t^3), (t \to \infty)$.

Solution $\left|\frac{t^2 \log t + t^3}{t^3}\right| = 1 + \left|\frac{\log t}{t}\right|.$

Using the first derivative test one may show that the function $y = \log t/t$ attains its maximum value $\frac{1}{e}$ as $t = e$. Hence $|\log t/t| \leq \frac{1}{e} < 1$ and thus $\left|(t^2 \log t + t^3)/t^3\right| \leq 2$. This proves the required result.

Remark We would like to point out here that the relation defined by O is not symmetric, i.e., if $f = O(g)$, then it is not necessarily true that $g = O(f)$. To illustrate this point we cite some simple examples such as $x = O(x^2), (x \to \infty)$, but $x^2 \neq O(x), (x \to \infty)$ or $\bar{e}^x = O(1), (x \to \infty)$, but $1 \neq O(\bar{e}^x), (x \to \infty)$ since $\frac{1}{\bar{e}^x} \to \infty, (x \to \infty)$.

However it is true that the relation O is transitive, that is to say if $f = O(g)$ and $g = O(h)$, then $f = O(h)$ (Exercise 7.1, Problem 1). In this case we say that $f = O(h)$ is a better approximation of f than $f = O(g)$.

Next we give the definition of the symbol o (small oh).

Definition 7.3. If $\lim_{t\to\infty} \frac{f(t)}{g(t)} = 0$, then we say that

$$f(t) = o(g(t)), \qquad (t \to \infty).$$

Example 7.4.

(a) Show that $t^2 \log t + t^3 = o(t^4)$, $(t \to \infty)$.

Solution $\lim_{t\to\infty} \dfrac{t^2 \log t + t^3}{t^4} = \lim_{t\to\infty} \dfrac{\log t}{t^2} + \lim_{t\to\infty} \dfrac{1}{t}$.

Using L'Hôpital's Rule we have

$$\lim_{t\to\infty} \frac{\log t}{t^2} = \lim_{t\to\infty} \frac{1}{2t^2} = 0.$$

Hence

$$\lim_{t\to\infty} \frac{t^2 \log t + t^3}{t^4} = 0,$$

and the required conclusion follows.

(b) Show that $o(g(t)) = g(t)o(1)$, $(t \to \infty)$.

Solution Let $f(t) = o(g(t))$, $(t \to \infty)$. Then

$$\lim_{t\to\infty} \frac{f(t)}{g(t)} = 0,$$

which implies that $\frac{f(t)}{g(t)} = o(1)$, $(t \to \infty)$. Consequently, $f(t) = g(t)o(1)$, $(t \to \infty)$.

The reader may sense correctly that the symbol o plays a much less important role than the symbol O.

Finally, we introduce the asymptotic equivalence relation $\sim$.

Definition 7.5. If $\lim_{t\to\infty} \frac{f(t)}{g(t)} = 1$, then we say that f is asymptotic to g, $(t \to \infty)$, and we write $f \sim g$, $(t \to \infty)$.

Notice that if $f \sim g$ as $t \to \infty$, then

$$\lim_{t\to\infty} \frac{f(t) - g(t)}{g(t)} = 0.$$

This implies from Definition 7.3 that $f(t) - g(t) = o(g(t)) = g(t)o(1)$ (Example 7.4). Hence we have

$$f(t) = g(t)[1 + o(1)].$$

Thus, it appears that the symbol $\sim$ is superfluous, since, as it has been demonstrated above, $f \sim g$ can be conveniently written as $f = g(1 + o(1))$.

Example 7.6.

(a) Show that $\sinh t \sim \frac{1}{2}e^t, (t \to \infty)$.

Solution $\lim_{t \to \infty} \frac{\sinh t}{\frac{1}{2}e^t} = \lim_{t \to \infty} \frac{\frac{1}{2}(e^t - e^{-t})}{\frac{1}{2}e^t} = 1.$

(b) Show that $t^2 \log t + t^3 \sim t^3, (t \to \infty)$.

Solution $\lim_{t \to \infty} \frac{t^2 \log t + t^3}{t^3} = 1 + \lim_{t \to \infty} \frac{\log t}{t}$

$= 1 + 0$ (using L'Hôpital's Rule)

$= 1.$

Notice that from Examples 7.2(c) and 7.6(b) we have $t^3 \sim t^2 \log t + t^3$ and $t^2 \log t + t^3 = O(t^3)$. It is also true that $t^2 \log t + 2t^3 = O(t^3)$, but $t^2 \log t + 2t^3$ is not asymptotic to t^3, since

$$\lim_{t \to \infty} \frac{t^2 \log t + 2t^3}{t^3} = 2.$$

Hence this remark leads to the following definitions:

$$O(g) := \{f : f = O(g)\},$$

and

$$o(g) := \{f : f = o(g)\}.$$

Example 7.7. Show that $o(f) = O(f)$ but $O(f) \ne o(f)$.

Proof Let $g = o(f)$. Then $\lim_{t \to \infty} \frac{g(t)}{f(t)} = 0$. Hence $\left|\frac{g(t)}{f(t)}\right|$ is bounded, which implies that $o(f) = g = O(f)$. To prove that $O(f) \ne o(f)$, let $f(t) = t^2 + 1, g(t) = 2t^2$. Then clearly $g = O(f)$. Since $\lim_{t \to \infty} \frac{g(t)}{f(t)} = 2 \ne 0, g \ne o(f)$.

Before ending this section we will entertain the curious reader by introducing the Prime Number Theorem, well known in the discipline of Number Theory. It says that the number of primes $\prod(t)$ which are less than the real number t is asymptotic to $t/(\log t), (t \to \infty)$, that is

$$\prod(t) \sim \frac{t}{\log t}, \qquad (t \to \infty).$$

For a proof of this result the reader may consult [1].

Another interesting asymptotic result is the so-called Stirling's Formula

$$n! \sim n^n \sqrt{2\pi n} \, e^{-n}, (n \to \infty).$$

A proof of this formula may be found in [18].

Exercise 7.1

1. Show that $\dfrac{t^2}{1+t^3} + \log(1+t^2) = O(\log t), (t \to \infty)$.

2. Show that $\sinh t = O(e^t), \qquad (t \to \infty)$.

3. Show that $O(g(t)) = g(t)O(1), \qquad (t \to \infty)$.

4. Show that

   (i) $\dfrac{1}{t-1} = \dfrac{1}{t}\left[1 + \dfrac{1}{t} + O\left(\dfrac{1}{t^2}\right)\right], \qquad (t \to \infty)$,

   (ii) $\dfrac{1}{t-1} = \dfrac{1}{t}\left[1 + \dfrac{1}{t} + o\left(\dfrac{1}{t}\right)\right], \qquad (t \to \infty)$.

5. Show that $\sinh\left(\dfrac{1}{t}\right) = o(1), \qquad (t \to \infty)$.

6. Show that

   (i) $[O(t)]^2 = O(t^2) = o(t^3)$,

   (ii) $t + o(t) = O(t)$.

7. Show that

   (i) $\sin(O(t^{-1})) = O(t^{-1})$,

   (ii) $\cos(t + \alpha + o(1)) = \cos(t + \alpha) + o(1)$, for any real number α.

8. Prove that $\sim$ is an equivalence relation.

9. Prove that both relation o and O are transitive.

10. Let $f_n(t) = O(g_n(t)), n = 1, \ldots, N$. Prove that

   (i) $\displaystyle\sum_{n=1}^{N} a_n f_n(t) = O\left(\sum_{n=1}^{N} a_n g_n(t)\right), \qquad (t \to \infty), \qquad a_n \in G$,

   (ii) $\displaystyle\sum_{n=1}^{N} a_n f_n(t) = O\left(\prod_{n=1}^{N} a_n g_n(t)\right), \qquad (t \to \infty), \qquad a_n \in G$.

11. Show that if O is replaced by o in Problem 10, the relations (i) and (ii) still hold.

12. Suppose that $f(t) = O(t), (t \to \infty)$ and $g(t) = O(t^2), (t \to \infty)$. Show that for any nonzero constants $a, b, a\,f(t) + b\,g(t) = O(g(t)), (t \to \infty)$.

13. If $f = O(g), (t \to \infty)$, show that

   (i) $O(o(f)) = o(O(f)) = o(g)$,

   (ii) $O(f)o(g) = o(f)o(g) = o(fg)$.

14. Let f be a positive nonincreasing function of t, and $f(t) \sim g(t), (t \to \infty)$. Prove that $\sup_{s>t} f(s) \sim g(t), (t \to \infty)$. [Hint: Use $f(t) = (1 + o(1))g(t)$]

15. Suppose that the functions f and g are continuous and have convergent integrals on $[1, \infty)$. If $f(t) \sim g(t), (t \to \infty)$, prove that

$$\int_t^\infty f(s)ds \sim \int_t^\infty g(s)ds (t \to \infty).$$

16. Consider the exponential integral $E_n(x)$ defined by

$$E_n(x) = \int_t^\infty \frac{e^{-xt}}{t^n}dt (x > 0), \quad \text{where } n \text{ is a positive integer.}$$

   (a) Show that $E_n(x)$ satisfies the difference equation

$$E_{n+1}(x) = \frac{1}{n}\left[e^{-x} - xE_n(x)\right].$$

   (b) Use integration by parts to show that

$$E_n(x) = \frac{e^{-x}}{x}\left(1 + 0\left(\frac{1}{x}\right)\right), \quad (x \to \infty).$$

   (c) Show that

$$E_n(x) = \frac{e^{-x}}{n-1}\left[1 + O\left(\frac{1}{n-2}\right)\right], (n \to \infty).$$

$$\left[\text{Hint: Show that } \int_1^\infty \frac{e^{-xt}}{t^n-1}dt \leq \frac{e^{-x}}{n-2}\right].$$

17. Show that

$$\int_0^\infty \frac{e^{-1}}{x+t} = \frac{1}{x}[1 - \frac{1}{x} + O\left(\frac{1}{x^2}\right)], (x \to \infty).$$

   [Hint: Use integration by parts].

18. Show that

$$\sum_{k=1}^n k^k = n^n\left[1 + O\left(\frac{1}{n}\right)\right], \qquad (n \to \infty).$$

$$\left[\text{Hint: Write } \sum_{k=1}^n k^k = n^n\left[1 + \frac{(n-1)^{n-1}}{n^n} + \cdots + \frac{1}{n^n}\right].\right]$$

7.2   Poincaré's Theorem

In this section we introduce to the reader the theorems of Poincaré and Perron. Both theorems deal with the asymptotic behavior of linear difference equations with nonconstant coefficients. It is widely accepted among researchers in difference equations that the theorem of Poincaré [2, 1885] marks the beginning of research in the qualitative theory of linear difference equations. Thirty-six years later, Perron [3, 1921] made some significant improvements to Poincaré's Theorem.

To motivate our study we will take the reader on a short excursion to the much simpler linear equations with constant coefficients of the form

$$x(n + k) + p_1 x(n + k - 1) + \cdots + p_k x(n) = 0, \qquad (7.2.1)$$

where the p_i's are real or complex numbers. The characteristic equation of Eq. (7.2.1) is given by

$$\lambda^k + p_1 \lambda^{k-1} + \cdots + p_k = 0. \qquad (7.2.2)$$

Let $\lambda_1, \lambda_2, \ldots, \lambda_k$ be the characteristic roots of Eq. (7.2.2). Then there are two main cases to consider.

Case 1. Suppose that distinct characteristic roots have distinct moduli, i.e., if $\lambda_i \neq \lambda_j$ then $|\lambda_i| \neq |\lambda_j|$ for all $1 \leq i, j \leq k$.

For the convenience of the reader we will divide Case 1 into two subcases.

Subcase a. Assume that all characteristic roots are distinct. So, by relabeling them, one may write the characteristic roots in descending order

$$|\lambda_1| > |\lambda_2| > \cdots > |\lambda_k|.$$

Then the general solution of Eq. (7.2.1) is given by

$$x(n) = c_1 \lambda_1^n + c_2 \lambda_2^n + \cdots + c_k \lambda_k^n. \qquad (7.2.3)$$

Hence if $c_1 \neq 0$, we have

$$\lim_{n \to \infty} \frac{x(n + 1)}{x(n)} = \lim_{n \to \infty} \frac{c_1 \lambda_1^{n+1} + c_2 \lambda_2^{n+1} + \cdots + c_k \lambda_k^{n+1}}{c_1 \lambda_1^n + c_2 \lambda_2^n + \cdots + c_k \lambda_k^n}$$

$$= \lim_{n \to \infty} \lambda_1 \left[\frac{c_1 + c_2 \left(\frac{\lambda_2}{\lambda_1}\right)^{n+1} + \cdots + c_k \left(\frac{\lambda_k}{\lambda_1}\right)^{n+1}}{c_1 + c_2 \left(\frac{\lambda_2}{\lambda_1}\right)^{n} + \cdots + c_k \left(\frac{\lambda_k}{\lambda_1}\right)^{n}} \right]$$

$$= \lambda_1, \qquad \text{since} \quad \left|\frac{\lambda_i}{\lambda_1}\right| < 1, i = 2, \ldots, k.$$

Similarly, if $c_1 = 0, c_2 \neq 0$ we obtain

$$\lim_{n \to \infty} \frac{x(n + 1)}{x(n)} = \lambda_2.$$

And in general if $c_1 = c_2 = \cdots = c_{i-1} = 0$, $c_i \neq 0$, then

$$\lim_{n\to\infty} \frac{x(n+1)}{x(n)} = \lambda_i.$$

Subcase b. Now suppose that there are some repeated characteristic roots. For simplicity assume that λ_1 is of multiplicity r, so $\lambda_1 = \lambda_2 = \cdots = \lambda_r$, $|\lambda_1| = |\lambda_2| = \cdots = |\lambda_r| > |\lambda_{r+1}| > \cdots > |\lambda_k|$. Then the general solution of Eq. (7.2.1) is given by

$$x(n) = c_1\lambda_1^n + c_2 n\lambda_1^n + \cdots + c_r n^{r-1}\lambda_1^n + c_{r+1}\lambda_{r+1}^n + \cdots + c_k\lambda_k^n.$$

Then one may show easily that this case is similar to Subcase a (Exercise 7.2, Problem 1).

Case 2. There exist two distinct characteristic roots λ_r, λ_j with $|\lambda_r| = |\lambda_j|$. This may occur if λ_r and λ_j are conjugates, i.e., $\lambda_r = \alpha + i\beta$, $\lambda_j = \alpha - i\beta$ for some real numbers α and β. For simplicity, let us assume that $r = 1$, $j = 2$, so $\lambda_r \equiv \lambda_1$ and $\lambda_j \equiv \lambda_2$. We write $\lambda_1 = \alpha + i\beta = re^{i\theta}$, $\lambda_2 = \alpha - i\beta = re^{-i\theta}$, where $r = (\alpha^2 + \beta^2)^{1/2}$, $\theta = \tan^{-1}\left(\frac{\beta}{\alpha}\right)$. Then the general solution of Eq. (7.2.1) is given by

$$x(n) = c_1 r^n e^{in\theta} + c_2 r^n e^{-in\theta} + c_3\lambda_3^n + \cdots + c_k\lambda_k^n.$$

Hence

$$\lim_{n\to\infty} \frac{x(n+1)}{x(n)} = \frac{r\left(c_1 e^{i(n+1)\theta} + c_2 e^{-i(n+1)\theta}\right) + c_3\lambda_3^{n+1} + \cdots + c_k\lambda_k^{n+1}}{c_1 e^{in\theta} + c_2 e^{-in\theta} + c_3\lambda_3^n + \cdots + c_k\lambda_k^n}. \tag{7.2.4}$$

Since $e^{in\theta} = \cos n\,\theta + i\sin\theta$, $e^{-in\theta} = \cos n\,\theta - i\sin\theta$ do not tend to definite limits as $n \to \infty$, we conclude that Limit (7.2.4) does not exist. For particular solutions the limit may exist. For example if $|\lambda_1| = |\lambda_2| > |\lambda_3| > \cdots > |\lambda_k|$, and

(a) $c_1 \neq 0$, $c_2 = 0$, then $\displaystyle\lim_{n\to\infty} \frac{x(n+1)}{x(n)} = re^{i\theta} = \lambda_1$,

(b) $c_1 = 0$, $c_2 \neq 0$, then $\displaystyle\lim_{n\to\infty} \frac{x(n+1)}{x(n)} = re^{-i\theta} = \lambda_2$.

Case (2) may also occur if $\lambda_i = -\lambda_j$. It is left to the reader as Exercise 7.2, Problem 2 to verify that in this case, too, $\lim_{n\to\infty}(x(n+1))/x(n)$ does not exist.

We now summarize the above discussion in the following theorem.

Theorem 7.8. Let $x(n)$ be any nonzero solution of Eq. (7.2.1). Then

$$\lim_{n\to\infty} \frac{x(n+1)}{x(n)} = \lambda_m \tag{7.2.5}$$

for some characteristic root λ_m, provided that distinct characteristic roots have distinct moduli. Moreover, if there are two or more distinct roots λ_r, λ_j with the same modulus ($|\lambda_r| = |\lambda_j|$), Limit (7.2.5) may not exist in general, but particular solutions can always be found for which Limit (7.2.5) exists and is equal to a given characteristic root λ_m.

Example 7.9. Consider the difference equation

$$x(n + 2) + \mu x(n) = 0.$$

(a) If $\mu = \beta^2$, then the characteristic equation is given by

$$\lambda^2 + \beta^2 = 0.$$

Hence the characteristic roots are $\lambda_1 = \beta i = \beta e^{-i}\pi/2$ and $\lambda_2 = -\beta i = \beta e^i \pi/2$. The general solution is given by

$$x(n) = c_1 \beta^n e^{i\frac{n\pi}{2}} + c_2 \beta^n e^{-i\frac{n\pi}{2}}.$$

So

$$\lim_{n\to\infty} \frac{x(n + 1)}{x(n)} = \beta \left(\frac{c_1 e^{i(n+1)\pi/2} + c_2 e^{-i(n+1)\pi/2}}{c_1 e^{in\pi/2} + c_2 e^{-in\pi/2}} \right)$$

which does not exist. However, if we pick the particular solution

$$\overline{x}(n) = c_1 \beta^n e^{i\frac{n\pi}{2}},$$

then

$$\lim_{n\to\infty} \frac{\overline{x}(n + 1)}{\overline{x}(n)} = \beta e^{i\frac{\pi}{2}} = \beta i.$$

Similarly, for the solution

$$\hat{x}(n) = c_2 \beta^n e^{-i\frac{n\pi}{2}}, \qquad \lim_{n\to\infty} \frac{\hat{x}(n + 1)}{\hat{x}(n)} = -\beta i.$$

(b) If $\mu = -\beta^2$, then the characteristic roots are $\lambda_1 = \beta$, $\lambda_2 = -\beta$. The general solution is given by $x(n) = c_1 \beta^n + c_2(-\beta)^n$.

Hence

$$\lim_{n\to\infty} \frac{x(n + 1)}{x(n)} = \lim_{n\to\infty} \frac{c_1 \beta^{n+1} + c_2(-\beta)^{n+1}}{c_1 \beta^n + c_2(-\beta)^n}$$

$$= \beta \lim_{n\to\infty} \frac{c_1 + c_2(-1)^{n+1}}{c_1 + c_2(-1)^n}. \tag{7.2.6}$$

Limit (7.2.6) does not exist and oscillates between $\beta(c_1 + c_2)/(c_1 - c_2)$ and $\beta(c_1 - c_2)/(c_1 + c_2)$. Notice that for the solution

$$\overline{x}(n) = c_1 \beta^n, \qquad \lim_{n\to\infty} \frac{\overline{x}(n + 1)}{\overline{x}(n)} = \beta,$$

and for the solution

$$\tilde{x}(n) = c_2(-\beta)^n, \qquad \lim_{n \to \infty} \frac{\tilde{x}(n+1)}{\tilde{x}(n)} = -\beta.$$

In 1885 the French mathematician Henri Poincaré [16] extended the above observations to equations with nonconstant coefficients of the form

$$x(n+k) + p_1(n)x(n+k-1) + \cdots + p_k(n)x(n) = 0 \qquad (7.2.7)$$

such that there are real numbers p_i, $1 \le i \le k$, with

$$\lim_{n \to \infty} p_i(n) = p_i, \qquad 1 \le i \le k. \qquad (7.2.8)$$

We shall call an equation of the form Eqs. (7.2.7) through (7.2.8) a difference equation of *Poincaré type*. The characteristic equation associated with Eq. (7.2.7) is

$$\lambda^k + p_1\lambda^{k-1} + \cdots + p_k = 0. \qquad (7.2.9)$$

The underlying idea behind Poincaré's Theorem is that since the coefficients of a difference equation of Poincaré's type are nearly constant for large n, one would expect solutions of Eq. (7.2.7) to exhibit some of the properties of the solutions of the corresponding constant coefficient difference Equation (7.2.1) as stated in Theorem 7.8.

Theorem 7.10 (Poincaré Theorem). Suppose that the roots $\lambda_1, \lambda_2, \ldots, \lambda_k$, of Eq. (7.2.9) have distinct moduli. Then for any nontrivial solution $x(n)$ of Eq. (7.2.7)

$$\lim_{n \to \infty} \frac{x(n+1)}{x(n)} = \lambda_i \qquad (7.2.10)$$

for some i.

Proof See [2].

Poincaré Theorem was later improved in 1921 by Perron who proved the following result.

Theorem 7.11 (Perron Theorem). Assume that $P_k(n) \ne 0$ for all $n \in Z^+$. Then, under the assumptions of Theorem 7.10, Eq. (7.2.7) has a fundamental set of solutions $\{x_1(n), x_2(n), \ldots, x_k(n)\}$ with the property

$$\lim_{n \to \infty} \frac{x_i(n+1)}{x_i(n)} = \lambda_i, \qquad 1 \le i \le k. \qquad (7.2.11)$$

Proof A proof of this theorem may be found in Meschkowski [4], p. 10.

It is questionable whether Poincaré's theorem remains valid if Eq. (7.2.7) has characteristic roots with equal moduli. Perron himself addressed this question and gave the following example which shows that Poincaré's Theorem may fail in this case.

Example 7.12. Consider the difference equations

$$x(n + 2) - \left(1 + \frac{(-1)^n}{n + 1}\right) x(n) = 0, \qquad (7.2.12)$$

$$x(n + 2) + \frac{1}{n + 4} x(n + 1) - \frac{n + 1}{n + 4} x(n) = 0. \qquad (7.2.13)$$

The associated characteristic equation for both equations is given by $\lambda^2 - 1 = 0$. Hence the characteristic roots are $\lambda_1 = 1$, $\lambda_2 = -1$ with $|\lambda_1| = |\lambda_2| = 1$. One may verify that a solution of the difference equation (7.2.12) may be given by

$$x(2n + 1) = \prod_{j=1}^{n} \left(1 - \frac{1}{2j}\right), \qquad x(2n) = 0, \qquad \text{for all } n \in Z^+.$$

It follows that $\lim_{n \to \infty} \frac{x(n+1)}{x(n)}$ does not exist, and Poincaré's Theorem fails in this case. However, for the difference equation (7.2.13), a fundamental set of solutions is given by

$$x_1(n) = \frac{1}{(n + 1)(n + 2)}, \qquad x_2(n) = \frac{(-1)^{n+1}(2n + 3)}{4(n + 1)(n + 2)}.$$

It follows that

$$\lim_{n \to \infty} \frac{x_1(n + 1)}{x_1(n)} = 1 = \lambda_1, \qquad \lim_{n \to \infty} \frac{x_2(n + 1)}{x_2(n)} = -1 = \lambda_2$$

and Perron's Theorem does not fail this time.

Example 7.13. Consider the difference equation

$$x(n + 2) - \frac{n}{n + 1} x(n + 1) + \frac{1}{n} x(n) = 0.$$

The associated characteristic equation is given by

$$\lambda^2 - \lambda = 0$$

with characteristic roots $\lambda_1 = 1$, $\lambda_2 = 0$. Hence by Perron's Theorem there exist solutions $x_1(n)$, $x_2(n)$ such that

$$\lim_{n \to \infty} \frac{x_1(n + 1)}{x_1(n)} = 1 \text{ and } \lim_{n \to \infty} \frac{x_2(n + 1)}{x_2(n)} = 0.$$

What can we conclude about the solutions $x_1(n)$ and $x_2(n)$? Well, $x_1(n)$ may be equal to a constant c, a polynomial in n such as

$$a_k n^k + a_{k-1} n^{k-1} + \cdots + a_o, \frac{1}{n}, \log n \text{ and others.}$$

The solution $x_2(n)$ may be equal to 0, e^{-2^n}, e^{-n^2}, etc.

The reader may correctly conclude from the preceding examples that Poincaré or Perron's Theorem provides only partial results about the asymptotic behavior of solutions of linear difference equations. The question remains whether we can use Perron's Theorem to write an asymptotic expression of solutions of equations of Poincaré type. Using null sequences, Wimp [18] devised an elegant and simple method to address the above question. Recall that $v(n)$ is called a *null sequence* if $\lim_{n\to\infty} v(n) = 0$.

Lemma 7.14. Suppose that $\lim_{n\to\infty} \frac{x(n+1)}{x(n)} = \lambda \neq 0$.

Then

$$x(n) = \pm\lambda^n e^{nv(n)} \qquad (7.2.14)$$

for some null sequence $v(n)$.

Proof Let

$$y(n) = \left| \frac{x(n)}{\lambda^n} \right|.$$

Then

$$\lim_{n\to\infty} \frac{y(n+1)}{y(n)} = \lim_{n\to\infty} \left| \frac{1}{\lambda} \frac{x(n+1)}{x(n)} \right| = 1.$$

If we let $z(n) = \log y(n)$, then we have

$$\begin{aligned}
\lim_{n\to\infty} z(n+1) - z(n) &= \lim_{n\to\infty} \log \left(\frac{y(n+1)}{y(n)} \right) \\
&= \log \lim_{n\to\infty} \frac{y(n+1)}{y(n)} \\
&= 0.
\end{aligned}$$

Hence for a given $\varepsilon > 0$ there exists a positive integer N such that

$$|z(n+1) - z(n)| < \varepsilon/2, \text{ for all } n \geq N.$$

Moreover, for $n \geq N$ we obtain

$$\begin{aligned}
|z(n) - z(N)| &\leq \sum_{r=N+1}^{n} |z(r) - z(r-1)| \\
&< \frac{\varepsilon}{2}(n - N).
\end{aligned}$$

Hence

$$\begin{aligned}
\left| \frac{z(n)}{n} \right| &< \frac{\varepsilon}{2}\left(1 - \frac{N}{n}\right) + \left| \frac{z(N)}{n} \right| \\
&< \frac{\varepsilon}{2} + \frac{\varepsilon}{2} = \varepsilon, \text{ for sufficiently large } n.
\end{aligned}$$

It follows that $\lim_{n\to\infty} \frac{z(n)}{n} = 0$ or $z(n) = n\nu(n)$ for some null sequence $\nu(n)$. This completes the proof of the lemma.

Example 7.15. Use Lemma 7.14 and Perron's Theorem to find asymptotic estimates of a fundamental set of solutions of the difference equation

$$y(n + 2) + \frac{n + 1}{n + 2}y(n + 1) - \frac{2n}{n + 2}y(n) = 0.$$

Solution The associated characteristic equation is given by

$$\lambda^2 + \lambda - 2 = 0$$

with roots $\lambda_1 = 1$, $\lambda_2 = -2$. By Perron's Theorem, there is a fundamental set of solutions $y_1(n)$, $y_2(n)$ with

$$\lim_{n\to\infty} \frac{y_1(n + 1)}{y_1(n)} = 1, \qquad \lim_{n\to\infty} \frac{y_2(n + 1)}{y_2(n)} = -2.$$

Thus by Lemma 7.14 we obtain

$$y_1(n) = e^{n\nu(n)}, \qquad y_2(n) = (-2)^n e^{n\mu(n)}$$

for some null sequences $\nu(n)$ and $\mu(n)$.

For the curious reader we note that an exact fundamental set of solutions is given by

$$y_1(n) = \frac{1}{n}, \qquad y_2(n) = (-2)^n/2.$$

Exercise 7.2.

1. Prove that each nontrivial solution $x(n)$ of the second order difference equation

$$x(n + 2) + p_1 x(n + 1) + p_2 x(n) = 0$$

   with double characteristic roots $\lambda_1 = \lambda_2 = \lambda$ satisfies $\lim_{n\to\infty}(x(n+1))/x(n) = \lambda$.

2. Suppose that the characteristic roots λ_1, λ_2 of

$$x(n + 2) + p_1 x(n + 1) + p_2 x(n) = 0$$

   are such that $\lambda_1 = -\lambda_2$. Prove that $\lim_{n\to\infty}(x(n + 1))/x(n)$ does not exist for some solution $x(n)$.

3. Consider the difference equation

$$x(n+3)-(\alpha+\beta+\gamma)x(n+2)+(\alpha\beta+\beta\gamma+\gamma\alpha)x(n+1)-\alpha\beta\gamma u(x) = 0 \quad (7.2.15)$$

   where α, β, γ are constants.

(a) Show that the characteristic roots of Eq. (7.2.15) are $\lambda_1 = \alpha$, $\lambda_2 = \beta$, and $\lambda_3 = \gamma$

(b) If $|\alpha| > |\beta| > |\gamma|$, find a fundamental set of solutions $x_1(n)$, $x_2(n)$, and $x_3(n)$ with

$$\lim_{n\to\infty} \frac{x_1(n+1)}{x_1(n)} = \alpha, \qquad \lim_{n\to\infty} \frac{x_2(n+1)}{x_2(n)} = \beta,$$

$$\lim_{n\to\infty} \frac{x_3(n+1)}{x_3(n)} = \gamma.$$

(c) If $|\alpha| = |\beta|$, $\alpha \neq \beta$, $|\alpha| > |\gamma|$, find a fundamental set of solutions $x_1(n)$, $x_2(n)$, and $x_3(n)$ such that $\lim_{n\to\infty}(x_1(n+1))/x_1(n) = \alpha$, $\lim_{n\to\infty}(x_2(n+1))/x_2(n) = \beta$, $\lim_{n\to\infty}(x_3(n+1))/x_3(n)\gamma$.

4. Consider the difference equation

$$x(n+2) + \frac{1}{n+4}x(n+1) - \frac{n+1}{n+4}x(n) = 0.$$

Use iteration to show that $\lim_{n\to\infty}(x(n+1))/x(n)$ does not exist for every solution $x(n)$.

5. Consider the equation $x(n+2) - ((n+2)+2(-1)^n)/(n+2)^3(n+3)x(n) = 0$.

Use iteration to show that $\lim_{n\to\infty}(x(n+1))/x(n)$ does not exist for any solution $x(n)$.

6. Consider the difference equation

$$x(n+2) - \left(3 + \frac{2n-1}{n^2 - 2n - 1}\right)x(n+1) + 2\left(1 + \frac{2n-1}{n^2 - 2n - 1}\right)x(n) = 0$$

(a) Use Lemma 7.14 and Perron's Theorem to find asymptotic estimates of a fundamental set of solutions of the equation.

(b) Verify that $x_1(n) = 2^n$ and $x_2(n) = n^2$ constitute a fundamental set of solutions.

7. Let α be a number whose modulus is greater than that of characteristic root of a difference equation of Poincaré's type (7.2.7). Prove that

$$\lim_{n\to\infty} \frac{x(n)}{\alpha^n} = 0$$

for any solution $x(n)$ of the equation.

8. Suppose that $\lim_{n\to\infty}(x(n+1))/x(n) = \lambda > 0$. Prove that for any $\delta \in (0, \lambda)$

(i) $|u(t)| = 0(\lambda + \delta)^n$, and

(ii) $(\lambda + \delta)^n = 0(u(n))$ (Hint: Use Lemma 7.15.)

9. Consider the equation $x(n + 2) - (n + 1)x(n + 1) - 2n^2x(n) = 0$.

   (a) Transform the equation into an equation of Poincaré type by letting $x(n) = (n - 1)! \, y(n)$.

   (b) Use part (a) to get an asymptotic estimate of a fundamental set of solutions.

10. Use the scheme of Problem 9 to find an asymptotic set of a fundamental set of solutions of the equation

$$x(n + 2) + 4^n x(n + 1) + 4n(n - 1)x(n) = 0. \text{ (Hint: Let } x(n) = (n - 2)! \, y(n).)$$

11. Suppose that both $\sum_{n=n_0}^{\infty} u(n)$ and $\sum_{n=n_0}^{\infty} u^2(n)$ converge and $1 + u(n) > 0$ for all $n \geq n_0$. Prove that $\lim_{n \to \infty} \prod_{i=n_0}^{n-1}(1 + u(i)) = c$ for some positive constant c.

$$\left[\text{Hint: Notice first that } \log \prod_{i=n_0}^{n-1}(1 + u(i)) = \sum_{i=n_0}^{n-1} \log(1 + u(i)). \right]$$

Now for $|u(n)| < \frac{1}{2}$, $|\log(1 + u(n)) - u(n)| \leq u^2(n)$, by Taylor's Theorem.

12. Consider the equation

$$(n + 2)x(n + 2) - (n + 3)x(n + 1) + 2x(n) = 0 \qquad (7.2.16)$$

   (a) Show that $1, 0$ are the characteristic roots of the equation.

   (b) Put

$$\frac{x(n + 1)}{x(n)} = 1 + \mu(n) \qquad (7.2.17)$$

   in Eq. (7.2.16), where $\mu(n)$ is a null sequence, and show that the equation becomes

$$(n + 2)\mu(n + 1) = 1 - \frac{2}{1 + \mu(n)}. \qquad (7.2.18)$$

   (c) Show that $\frac{2}{1+\mu(n)} = 2 + 0(\mu(n))$ (use the Mean Value Theorem).

   (d) Use part (c) to show that Eq. (7.2.18) is equivalent to

$$\mu(n + 1) = -\frac{1}{n + 1} = +0\left(\frac{1}{n^2}\right). \qquad (7.2.19)$$

   (e) Show that $x(n + 1) = n/n + 1 \left(1 + 0\left(\frac{1}{n^2}\right)\right) x(n)$. $\qquad (7.2.20)$
   [Hint: Substitute Eq. (7.2.19) into Eq. (7.2.17).]

(f) Prove that $x(n) \sim \frac{c}{n}, (n \to \infty)$.

[Hint: Solve Eq. (7.2.20) and then use Problem 11.]

13. Show that Eq. (7.2.16) has another solution $\bar{x} \sim c\frac{2^n}{n!}, (n \to \infty)$.

14. Use the scheme of Problem 12 to find asymptotic estimates of a fundamental set of solutions of the equation

$$(n + 1)x(n + 2) - (n + 4)x(n + 1) + x(n) = 0.$$

$$\left[\text{Hint: } x_1(n) \sim n^2, x_2(n) \sim \frac{1}{(n + 2)!}, (n \to \infty). \right]$$

15. Show that the equation $x(n + 2) - (n + 1)x(n + 1) + (n + 1)x(n) = 0$ has solutions $x_1(n), x_2(n)$ with asymptotic estimates

$$x_1(n) \sim c(n - 2)!, \qquad x_2(n) = an, (n \to \infty).$$

16.* (Hard) Consider the equation of Poincaré type

$$x(n + 2) - (2 + p_1(n))x(n + 1) + (1 + p_2(n))x(n) = 0$$

where $p_1(n) \geq p_2(n)$ for all $n \in Z^+$. Show that if $x(n)$ is a solution which is not constantly zero for large values of n, then $\lim_{n \to \infty}(x(n + 1))/x(n) = 1$.

17.* (Hard) Consider the equation

$$x(n + 2) + P_1(n)x(n + 1) + P_2(n)x(n) = 0$$

with $\lim_{n \to \infty} P_1(n) = p_1, \lim_{n \to \infty} P_2(n) = p_2$. Let η be a positive constant such that $|x(n + 1)/x(n)|^2 > |p_2| + \eta$ for sufficiently large n. Suppose that the characteristic roots λ_1, λ_2 of the associated equation are such that $|\lambda_1| \geq |\lambda_2|$.

Prove that $\lim_{n \to \infty}(x(n + 1))/x(n) = \lambda_1$. [Hint: Let $y(n) = (x(n + 1))/x(n)$.]

7.3   Second Order Difference Equations

In this section we focus our attention on the asymptotic behavior of certain types of second order difference equations. These types of equations were chosen due to the simplicity of the required techniques. For general difference equations the reader may consult Section 7.5. The first equation we consider in this section is given by

$$\Delta^2 y(n) + p(n)y(n) = 0. \tag{7.3.1}$$

This equation has the equivalent form

$$y(n + 2) - 2y(n + 1) + (1 + p(n))y(n) = 0. \tag{7.3.1}'$$

If $p(n) \to 0$ as $n \to \infty$, then Eq. (7.3.1)' is of Poincaré's type. However, Perron's Theorem does not apply since the characteristic roots of the associated equation are $\lambda_1 = \lambda_2 = -1$.

Thus we need to develop new techniques that are effective in dealing with Eq. (7.3.1). We will consider Eq. (7.3.1) as a perturbation of the equation

$$\Delta^2 x(n) = 0 \tag{7.3.2}$$

which has the linearly independent solutions $x_1(n) = c_1$, $x_2(n) = c_2 n$.

Our objective here is to prove that Eq. (7.3.1) has two linearly independent solutions $x_1(n)$ and $x_2(n)$ such that $x_1(n) \sim c_1$, $x_2(n) \sim c_2(n)$. But before stating the first result we need the following simple result.

Lemma 7.16 (Multiple Summation Reduced to Single Summation). For any function $f(n)$ defined on $n \geq n_0$,

$$\sum_{r=n_0}^{n-1} \sum_{j=n_0}^{r-1} f(j) = \sum_{j=n_0}^{n-1} (n - j - 1) f(j). \tag{7.3.3}$$

Proof Reversing the order of summations in $\sum_{r=n_0}^{n-1} \sum_{j=n_0}^{r} f(j)$ we obtain (Fig. 7.1)

$$\sum_{r=n_0}^{n-1} \sum_{j=n_0}^{r} f(j) = \sum_{j=n_0}^{n-1} f(j) \sum_{r=j}^{n-1} 1$$

$$= \sum_{j=n_0}^{n-1} (n - j) f(j). \tag{7.3.4}$$

Now

$$\sum_{r=n_0}^{n-1} \sum_{j=n_0}^{r-1} f(j) = \sum_{r=n_0}^{n-1} \sum_{j=n_0}^{r} f(j) - \sum_{r=n_0}^{n-1} f(r)$$

$$= \sum_{r=n_0}^{n-1} (n - j - 1) f(j)$$

using Eq. (7.3.4).

We are now ready to give the first asymptoticity result for Eq. (7.3.1)

Theorem 7.17. Suppose that

$$\sum_{j=1}^{\infty} j |p(j)| < \infty \tag{7.3.5}$$

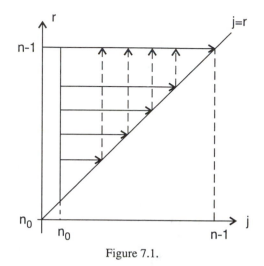

Figure 7.1.

Then Eq. (7.3.1) has two solutions $y_1(n)$, $y_2(n)$ such that $y_1(n) \sim 1$ and $y_2(n) \sim n$, $n \to \infty$.

Proof We put Eq. (7.3.1) in the form

$$\Delta^2 y(n) = -p(n)y(n). \qquad (7.3.6)$$

Taking the antidifference twice of both sides of Eq. (7.3.6) (using Formula (2.1.16)) one obtains

$$y(n) = c_1 + c_2 n - \sum_{r=1}^{n-1} \sum_{j=1}^{r-1} p(j)y(j). \qquad (7.3.7)$$

Using Eq. (7.3.3) in Eq. (7.3.7) we then have

$$y(n) = c_1 + c_2 n - \sum_{j=1}^{n-1} (n - j - 1)p(j)y(j).$$

Thus for $n \geq 1$,

$$|y(n)| \leq (|c_1| + |c_2|)n + n \sum_{j=1}^{n-1} |p(j)||y(j)|,$$

or

$$\frac{|y(n)|}{n} \leq |c_1| + |c_2| + \sum_{j=1}^{n-1} j|p(j)| \frac{|y(j)|}{j}.$$

By the discrete Gronwall's inequality (Lemma 4.2.1) we have

$$\frac{|y(n)|}{n} \leq (|c_1| + |c_2|) \exp\left(\sum_{j=1}^{n-1} j|p(j)| \right)$$

$$\leq c_3 \qquad \text{(by Assumption 7.3.4)}$$

or

$$|y(n)| \le c_3 n. \tag{7.3.8}$$

Applying the difference operator Δ on both sides of Eq. (7.3.7) we get

$$\Delta y(n) = c_2 - \sum_{j=1}^{n-1} p(j)y(j). \tag{7.3.9}$$

Now by Inequality (7.3.8) we have

$$\sum_{j=1}^{n-1} |p(j)| \, |y(j)| \le c_3 \sum_{j=1}^{n-1} j|p(j)| < \infty$$

which implies that $\sum_{j=1}^{\infty} p(j)y(j)$ converges to a limit M.

Using this information in Eq. (7.3.9) we conclude that

$$\lim_{n \to \infty} \Delta y(n) = c_2 - \sum_{j=1}^{\infty} p(j)y(j)$$
$$= c_2 - M.$$

Letting $c_2 = M$, we get the first solution $\Delta y_1(n) \to 0$ as $n \to \infty$ or $y_1(n) \sim 1$. Furthermore, if we choose $c_2 \ne M$, then we get the second solution $\Delta y_2(n) \to c_2 - M \ne 0$ or $y_2(n) \sim n$. This completes the proof of the theorem.

The preceding result has been extended to more general equations such as the so-called Emden–Fowler Equation

$$\Delta^2 y(n) + p(n)|y(n)|^{\gamma} \operatorname{sgn} y(n) = 0 \tag{7.3.10}$$

where $\gamma \ne 1$ is a positive real number and $\operatorname{sgn} y(n) \begin{cases} 1 \text{ if } y(n) > 0 \\ -1 \text{ if } y(n) < 0 \end{cases}$.

For example if $\gamma = 2m + 1$, for some $m \in Z^+$, Eq. (7.3.10) becomes $\Delta^2 y(n) + p(n)y^{2m+1}(n) = 0$, whether $y(n)$ is positive or negative (Exercise 7.3, Problem 6).

In order to study Eq. (7.3.10) we need to develop a new Gronwall's type inequality.

Lemma 7.18. Suppose that

$$u(n) \le a + b \sum_{j=n_0}^{n-1} c(j)u^{\gamma}(j),$$

where $1 \ne \gamma > 0, a \ge 0, b > 0, c(j) > 0$, and $u(j) > 0$ for $j \ge n_0$. Then

$$u(n) \le \left[a^{1-\gamma} + b(1 - \gamma) \sum_{j=n_0}^{n-1} c(j) \right]^{\frac{1}{1-\gamma}}, \tag{7.3.11}$$

provided that for $\gamma > 1$, $a^{1-\gamma} + b(1 - \gamma) \sum_{j=n_0}^{n-1} c(j) > 0$ for all $n \geq n_0$.

Proof Let

$$v(n) = a + b \sum_{j=n_0}^{n-1} c(j) u^{\gamma}(j). \tag{7.3.12}$$

Then $u(n) \leq v(n)$ for all $n \geq n_0$. Furthermore,

$$\Delta v(n) = bc(n) u^{\gamma}(n). \tag{7.3.13}$$

From Eq. (7.3.13) it follows that

$$\frac{\Delta v(n)^{1-\gamma}}{1 - \gamma} = \int_{n}^{n+1} \frac{dv(t)}{v^{\gamma}(t)} \leq \frac{\Delta v(n)}{v^{\alpha}(n)} \leq \frac{\Delta v(n)}{u^{\alpha}(n)}. \tag{7.3.14}$$

Using Eq. (7.3.12) in Eq. (7.3.14) one concludes that

$$\frac{\Delta v^{1-\gamma}(n)}{1 - \gamma} \leq b\, c(n)$$

or

$$\Delta v^{1-\gamma}(n) \leq (1 - \gamma) b\, c(n).$$

Hence

$$v^{1-\gamma}(n) \leq v^{1-\gamma}(n_0) + \sum_{j=n_0}^{n-1} (1 - \gamma) b\, c(j).$$

Thus

$$u(n) \leq \left[a^{1-\gamma} + \sum_{j=n_0}^{n-1} (1 - \gamma) bc(j) \right]^{\frac{1}{1-\gamma}}.$$

Theorem 7.19. Suppose that

$$\sum_{j=n_0}^{\infty} j^{\gamma} |p(j)| = M < \infty. \tag{7.3.15}$$

Then each solution $y(n)$ of the Eq. (7.3.10) with the initial condition $y(n_0)$ that satisfies

$$\left[|\Delta y(n_0)| + \left| \frac{y(n_0)}{n_0} - \Delta y(n_0) \right| \right]^{1-\gamma} + 2(1 - \gamma) M > 0 \tag{7.3.16}$$

is such that $y(n) \sim n$.

Proof Let $A(n) = \Delta y(n)$, $B(n) = y(n) - n\Delta y(n)$. Then

$$y(n) = n\Delta y(n) + B(n)$$

or

$$y(n) = nA(n) + B(n). \tag{7.3.17}$$

Moreover,

$$\Delta A(n) = \Delta^2 y(n)$$
$$= -p(n)|y(n)|^\gamma \operatorname{sgn} y$$

or

$$\Delta A(n) = p(n)[|nA(n) + B(n)|]^\gamma \operatorname{sgn} y. \tag{7.3.18}$$

Since $B(n) = y(n) - n\Delta y(n)$, it follows that

$$\Delta B(n) = \Delta y - (n+1)\Delta^2 y(n) - \Delta y(n)$$
$$= -(n+1)\Delta A(n). \tag{7.3.19}$$

Using Eq. (7.3.18) in Eq. (7.3.19) we obtain

$$\Delta B(n) = (n+1)p(n)[|nA(n) + B(n)|]^\gamma \operatorname{sgn} y(n). \tag{7.3.20}$$

Taking the antidifference of Eq. (7.3.18) and Eq. (7.3.19), we have

$$A(n) = A(n_0) - \sum_{j=n_0}^{n-1} p(j)[jA(j) + B(j)]^\gamma \operatorname{sgn} y(j),$$

$$B(n) = B(n_0) + \sum_{j=n_0}^{n-1} (j+1)p(j)[jA(j) + B(j)]^\gamma \operatorname{sgn} y(j),$$

or

$$\frac{B(n)}{n} = \frac{B(n_0)}{n} + \frac{1}{n}\sum_{j=n_0}^{n-1} (j+1)p(j)[jA(j) + B(j)]^\gamma \operatorname{sgn} y(j). \tag{7.3.21}$$

From Eq. (7.3.20) we conclude that

$$|A(n)| \le |A(n_0)| + \sum_{j=n_0}^{n-1} |p(j)|j^\gamma \left| A(j) + \frac{B(j)}{j} \right|^\gamma. \tag{7.3.22}$$

Realizing that $n \ge n_0$ and $n \ge j+1$ in Eq. (7.3.21) we obtain

$$\frac{|B(n)|}{n} \le \frac{|B(n_0)|}{n} + \sum_{j=n_0}^{n-1} |p(j)|j^\gamma \left| A(j) + \frac{B(j)}{j} \right|^\gamma. \tag{7.3.23}$$

Adding Eq. (7.3.22) to Eq. (7.3.23) we obtain

$$|A(n)| + \frac{|B(n)|}{n} \le \left(|A(n_0)| + \frac{|B(n_0)|}{n_0} \right) + 2\sum_{j=n_0}^{n-1} |p(j)|j^\gamma \left| A(j) + \left(\frac{B(j)}{j}\right) \right|^\gamma,$$

or

$$u(n) \le u(n_0) + 2 \sum_{j=n_0}^{n-1} c(j)u^{\gamma}(j), \tag{7.3.24}$$

where $u(n) = |A(n)| + \frac{|B(n)|}{n}$, $c(j) = |p(j)|j^{\gamma}$.
By Inequality (7.3.11) it follows that

$$u(n) \le \left[u(n_0)^{1-\gamma} + 2(1-\gamma) \sum_{j=n_0}^{n-1} c(j) \right]^{\frac{1}{1-\gamma}} \le c_1.$$

But

$$\left| A(n) + \frac{B(n)}{n} \right| = \frac{|x(n)|}{n} \le |A(n)| + \frac{|B(n)|}{n} = u(n) \le c_1.$$

This implies that $x(n) \sim n$.

Notes: All the results in this section are based on the paper of Drozdowicz [5]. In the same paper he also investigated the equation

$$\Delta^2 y(n) = p(n)y(n+1) + q(n)y^{2m+1}(n), \qquad m \in z^+.$$

In [6] Drozdowicz and Popenda extended their study to the more general equation

$$\Delta^2 y(n) + p(n)f(y(n)) = 0.$$

The so-called self-adjoint equations of the form

$$\Delta(p(n)\Delta y(n)) + q(n)y^{\gamma}(n) = 0, \qquad \gamma \in (0, 1)$$

have been studied in [7].

Exercise 7.3.

Investigate the asymptotic behavior of the solutions of the following difference equations.

1. $\Delta^2 y(n) + e^{-n} y(n) = 0$.

2. $y(n+2) - 2y(n+1) + \left(1 - \dfrac{1}{n^3}\right) y(n) = 0$.

3. $\Delta^2 y(n) + e^{-n} y^3(n) = 0$.

4. $\Delta^2 y(n) - \dfrac{1}{n^2}(|y(n)|)^{1/2} = 0$.

5. $\Delta^2 y(n) + \dfrac{1}{n^2}(y(n))^{1/3} = 0$.

6. Show that if $\gamma = 2m + 1$, for some $m \in Z^+$, then Eq. (7.3.9) becomes

$$\Delta^2 y(n) + p(n)y^{2m+1}(n) = 0.$$

7. Extend Theorem 7.17 to the third order equation

$$\Delta^3 y(n) + p(n)y(n) = 0.$$

8.* Extend Theorem 7.17 to the kth order equation

$$\Delta^k y(n) + p(n)y(n) = 0.$$

9. Consider the difference equation

$$\Delta^2 y(n) = p(n)y(n+1) \text{ with } \sum_{j=n_0}^{\infty} j|p(j)| < \infty.$$

Show that the equation has two solutions $y_1(n)$ and $y_2(n)$ such that $y_1(n) \sim 1$ and $y_2(n) \sim n$.

10. Investigate the asymptotic behavior of solutions of the equation

$$\Delta^2 y(n) = \frac{(-1)^n}{n^{\alpha+1}} y(n+1)$$

where $\alpha > 1$.

11. Investigate the asymptotic behavior of solutions of

$$\Delta^2 y(n) = \frac{p(n)}{n^{\alpha+1}} y(n+1)$$

where

$$\gamma > 1, \quad \left| \sum_{j=1}^{n} p(j) \right| \le M < \infty$$

for all $n > 1$.

12. Show that the difference equation

$$\Delta^2 x(n) = p(n)x(n+1)$$

has two linearly independent solutions $x_1(n)$ and $x_2(n)$ such that

$$\det \begin{bmatrix} x_1(n) & x_2(n) \\ \Delta x_1(n) & \Delta x_2(n) \end{bmatrix} = -1.$$

13.* [Drozdowicz] Suppose that $x_1(n)$ and $x_2(n)$ are two linearly independent solutions of the equation

$$\Delta^2 x(n) = p(n)x(n+1).$$

In addition assume that for a sequence $q(n)$ we have

$$\sum_{j=n_0}^{\infty} |q(j)||u(j)| = M < \infty$$

where, for a specific $m \in Z^+$,

$$u(n) = \max \left\{ |x_1(n+1)| |x_1(n)|^{2m+1}, \qquad |x_1(n+1)| |x_2(n)|^{2m+1}, \right.$$
$$\left. |x_2(n+1)| |x_1(n)|^{2m+1}, \qquad |x_2(n+1)| |x_2(n)|^{2m+1} \right\}.$$

Show that there exist solutions $y(n)$ of the equation

$$\Delta^2 y(n) = p(n)y(n+1) + q(n)y^{2m+1}(n)$$

such that

$$y(n) = \alpha(n)x_1(n) + \beta(n)x_2(n),$$

with

$$\lim_{n \to \infty} \alpha(n) = a, \qquad \lim_{n \to \infty} \beta(n) = b$$

for some constants a, b.

[Hint: Use Problem 12 and then let $A(n) = x_2(n)\Delta y(n) - \Delta x_2(n)y(n)$ and $B(n) = \Delta x_1(n)y(n) - x_1(n)\Delta y(n)$ then mimic the proof of Theorem 7.19.]

7.4   Asymptotically Diagonal Systems

In this section we derive conditions under which solutions of a perturbed diagonal system are asymptotic to solutions of the unperturbed diagonal system. As a byproduct we obtain asymptotic results for nonautonomous kth order scalar difference equations.

We begin our study by considering the perturbed diagonal system

$$y(n+1) = (D(n) + B(n))y(n) \qquad (7.4.1)$$

and the unperturbed diagonal system

$$x(n+1) = D(n)x(n), \qquad (7.4.2)$$

where $D(n) = \text{diag}(\lambda_1(n), \lambda_2(n), \ldots, \lambda_k(n))$, $\lambda_i(n) \neq 0$ for all $n \geq n_0 \geq 0$, $1 \leq i \leq k$, and $B(n)$ is a $(k \times k)$ matrix defined for $n \geq n_0 \geq 0$. $(k \times k)$ matrices defined for $n \geq n_0 \geq 0$. The fundamental matrix of System (7.4.2) is given by

$$\Phi(n) = \text{diag} \left(\prod_{r=n_0}^{n-1} \lambda_1(r), \prod_{r=n_0}^{n-1} \lambda_2(r), \ldots, \prod_{r=n_0}^{n-1} \lambda_k(r) \right). \qquad (7.4.3)$$

Define $\Phi_1(n) = \text{diag}(\mu_1, \mu_2, \ldots, \mu_k(n))$ by letting

$$\mu_i(n) = \begin{cases} \displaystyle\prod_{r=n_0}^{n-1} \lambda_i(r), & \text{if } \displaystyle\prod_{r=m}^{n-1} |\lambda_i(r)| \leq M \text{ for all } n \geq m \geq n_0 \\ 0, & \text{otherwise} \end{cases}$$

Define $\Phi_2(n) = \Phi(n) - \Phi_1(n)$. Then both $\Phi_1(n)$ and $\Phi_2(n)$ are fundamental matrices of System (7.4.2) (Exercise 7.4, Problem 7).

We are now ready for the definition of the important notion of *dichotomy*.

Definition 7.20. System (7.4.2) is said to possess an *ordinary dichotomy* if there exists a constant M such that

(i) $\|\Phi_1(n)\Phi^{-1}(m)\| \leq M$,   for $n \geq m \geq n_0$,

(ii) $\|\Phi_2(n)\Phi^{-1}(m)\| \leq M$,   for $m \geq n \geq n_0$.

Notice that if $D(n)$ is constant, then System (7.4.2) always possesses an ordinary dichotomy.

After wading through the complicated notation above, here is an example.

Example 7.21. Consider the difference system $x(n + 1) = D(n)x(n)$ with

$$D(n) = \begin{pmatrix} 1 + \dfrac{1}{n+1} & 0 & 0 & 0 \\ 0 & 0.5 & 0 & 0 \\ 0 & 0 & n+1 & 0 \\ 0 & 0 & 0 & \dfrac{1}{n+2} \end{pmatrix}.$$

Then a fundamental matrix of the system may be given by

$$\Phi(n) = \text{diag}\left(\prod_{j=0}^{n-1}\left(1 + \frac{1}{j+1}\right), (0.5)^n, \prod_{j=0}^{n-1}(j+1), \prod_{j=0}^{n-1}\left(\frac{1}{j+2}\right) \right)$$

$$= \text{diag}\left(n+1, (0.5)^n, n!, \frac{1}{(n+1)!} \right).$$

From this we deduce that

$$\Phi_1(n) = \text{diag}\left(0, (0.5)^n, 0, \frac{1}{(n+1)!} \right)$$

and

$$\Phi_2(n) = \text{diag}\,(n+1, 0, n!, 0).$$

In addition, one can easily verify that $\Phi_1(n)$ and $\Phi_2(n)$ are fundamental matrices of the system, that is

$$\Phi_1(n+1) = D(n)\Phi_1(n)$$

and

$$\Phi_2(n+1) = D(n)\Phi_2(n).$$

Finally,

$$\Phi_1(n)\Phi^{-1}(m) = \text{diag}\left(0, (0.5)^{n-m}, 0, \frac{1}{(n+1)(n), \ldots, (m+2)}\right).$$

Hence

$$\|\Phi_1(n)\Phi^{-1}(m)\| \leq 1, \qquad \text{for } n \geq m \geq 0.$$

Similarly,

$$\Phi_2(n)\Phi^{-1}(m) = \text{diag}\left(\frac{n+1}{m+1}, 0, \frac{n!}{m!}, 0\right), \qquad \text{for } m \geq n \geq n_0.$$

Hence

$$\|\Phi_2(n)\Phi^{-1}(m)\| \leq 1, \qquad \text{for } m \geq n \geq n_0.$$

We are now ready to establish a new variation of constants formula that is very useful in asymptotic theory.

Theorem 7.22 (Variation of constants formula). Suppose that System (7.4.2) possesses an ordinary dichotomy and the following condition holds

$$\sum_{n=n_0}^{\infty} \|B(n)\| < \infty. \qquad (7.4.4)$$

Then for each bounded solution $x(n)$ of Eq. (7.4.2), there corresponds a bounded solution $y(n)$ of Eq. (7.4.1) given by

$$y(n) = x(n) + \sum_{j=n_0}^{n-1} \Phi_1(n)\Phi^{-1}(j+1)B(j)y(j) - \sum_{j=n}^{\infty} \Phi_2(n)\Phi^{-1}(j+1)B(j)y(j).$$

$$(7.4.5)$$

The converse also holds; for each bounded solution $y(n)$ of Eq. (7.4.1) there corresponds a bounded solution $x(n)$ of Eq. (7.4.2).

Proof Let $x(n)$ be a bounded solution of Eq. (7.4.2). By using the method of successive approximation, we will produce a corresponding bounded solution $y(n)$ of Eq. (7.4.1). We define a sequence $\{y_i(n)\}(i = 1, 2, \ldots)$ by letting $y_1(n) = x(n)$ and

$$y_{i+1}(n) = x(n) + \sum_{j=n_0}^{n-1} \Phi_1(n))\Phi^{-1}(j+1)B(j)y_i(j)$$

$$- \sum_{j=n}^{\infty} \Phi_2(n)\Phi^{-1}(j+1)B(j)y_i(j). \qquad (7.4.6)$$

First we prove that $y_i(n)$ is bounded on the discrete interval $[n_0, \infty)$. This task will be accomplished by an induction on i. From our assumption we have $|y_1(n)| =$

$|x(n)| \leq c_1$, for some constant c_1. Now assume that $|y_i(n)| \leq c_i$, for some constant c_i. Then by Definition (7.20) we have

$$|y_{i+1}(n)| \leq c_1 + M c_i \sum_{j=n_0}^{\infty} |B(j)| = c_{i+1}.$$

where $M = \max\{M_1, M_2\}$. Hence $y_i(n)$ is bounded for each i.

In the next step we show that the sequence $\{y_i(n)\}$ converges uniformly on the discrete interval $[n_0, \infty)$.

Now using Definition (7.20) we have for $i = 1, 2, \ldots$

$$|y_{i+2}(n) - y_{i+1}(n)| \leq M \sum_{j=n_0}^{\infty} \|B(j)\| \, |y_{i+1}(j) - y_i(j)|.$$

Hence by induction on i (Exercise 7.4, Problem 10)

$$|y_{i+1}(n) - y_i(n)| \leq \left[M \sum_{j=n_0}^{\infty} \|B(j)\| \right]^i c_1. \tag{7.4.7}$$

We choose n_0 sufficiently large such that

$$M \sum_{j=n_0}^{\infty} \|B(j)\| = \eta < 1. \tag{7.4.8}$$

Thus $|y_{i+1}(n) - y_i(n)| \leq c_1 \eta^i$, and consequently, $\sum_{i=1}^{\infty} \{y_{i+1}(n) - y_i(n)\}$ converges uniformly on $n \geq n_0$ (by the Weierstrass M-test).[1]

We define

$$y(n) = y_1(n) + \sum_{i=1}^{\infty} \{y_{i+1}(n) - y_i(n)\} = \lim_{i \to \infty} y_i(n).$$

Hence $|y(n)| \leq L$, for some constant L. Letting $i \to \infty$ in Eq. (7.4.6) we obtain Eq. (7.4.5). The second part of the theorem is left to the reader as Exercise 7.4, Problem 12.

If the condition of ordinary dichotomy is strengthened then we obtain the following important result in asymptotic theory.

Theorem 7.23. Suppose that the following assumption holds:

Assumption (H) $\begin{cases} \text{(i) System (7.4.2) possesses an ordinary dichotomy} \\ \text{(ii) } \lim_{n \to \infty} \Phi_1(n) = 0 \text{ as } n \to \infty \end{cases}$

---

[1] *Weierstrass M-test*: Let $u_n(x)$, $n = 1, 2, \ldots$, be defined on a set A with range in R. Suppose that $|u_n(x)| \leq M_n$ for all n and for all $x \in A$. If the series of constants $\sum_{n=1}^{\infty} M_n$ converges, then $\sum_{n=1}^{\infty} u_n(x)$ and $\sum_{n=1}^{\infty} |u_n(x)|$ converges uniformly on A.

If, in addition, Condition (7.4.3) holds, then for each bounded solution $x(n)$ of Eq. (7.4.2) there corresponds a bounded solution $y(n)$ of Eq. (7.4.1) such that

$$y(n) = x(n) + o(1). \qquad (7.4.9)$$

Proof Let $x(n)$ be a bounded solution of Eq. (7.4.2). Then by using Formula (7.4.9) we obtain, for a suitable choice of m (to be determined later),

$$y(n) = x(n) + \Phi_1(n) \sum_{j=n_0}^{m-1} \Phi^{-1}(j+1)B(j)y(j) + \Psi(n) \qquad (7.4.10)$$

where

$$\Psi(n) = \Phi_1(n) \sum_{j=m}^{n-1} \Phi^{-1}(j+1)B(j)y(j) - \sum_{j=n}^{\infty} \Phi_2(n)\Phi^{-1}(j+1)B(j)y(j). \qquad (7.4.11)$$

Now recall that from Theorem 7.22, $\|y\| \leq L$, for some $L > 0$. Hence from Formula (7.4.11) it follows that

$$|\Psi(n)| \leq ML \sum_{j=m}^{\infty} \|B(j)\|.$$

Thus for $\varepsilon > 0$, there exists a sufficiently large m such that $|\Psi(n)| < \varepsilon/2$. Since $\Phi_1(n) \to 0$ as $n \to \infty$, it follows from Formula (7.4.10) that $|y(n) - x(n)| < \varepsilon$, for sufficiently large n. Therefore $y(n) = x(n) + o(1)$ or $y(n) \sim x(n)$.

Our next objective is to apply the preceding theorem to produce a discrete analogue of Levinson's Theorem [8, 9]. We start our analysis by making the change of variables

$$y(n) = \prod_{r=n_0}^{n-1} \lambda_i(r)z(n), \qquad \text{for a specific } i, 1 \leq i \leq k. \qquad (7.4.12)$$

Then Eq. (7.4.1) becomes

$$z(n+1) = (D_i(n) + B_i(n))z(n), \qquad (7.4.13)$$

where

$$D_i(n) = \text{diag}\left(\frac{\lambda_1(n)}{\lambda_i(n)}, \ldots, 1, \ldots, \frac{\lambda_k(n)}{\lambda_i(n)}\right),$$

$$B_i(n) = \frac{1}{\lambda_i(n)}B(n).$$

Associated with Eq. (7.4.13) is the unperturbed diagonal system

$$x(n+1) = D_i(n)x(n). \qquad (7.4.14)$$

To make the proof of our main theorem more transparent we introduce the following lemma.

Lemma 7.24. For system equation (7.4.14), Assumption (H) holds if either one of the following two conditions are satisfied.

Condition (L) For each pair $\lambda_i(n)$, $\lambda_j(n)$, $i \neq j$, either $|\lambda_i(n)| < |\lambda_j(n)|$, for all $n \geq n_0$, or $|\lambda_i(n)| \geq |\lambda_j(n)|$, for all $n \geq n_0$.

Condition (E) Let $\gamma_{ij}(n) := \prod_{r=n_0}^{n} (|\lambda_i(r)|)/|\lambda_j(r)|$. Then either $\gamma_{ij}(n) \to 0$ as $n \to \infty$ (independent of n_o) or no subsequence of $\{\gamma_{ij}(n)\}$ converges to zero as $n \to \infty$.

Proof The proof is omitted and left to the reader to do as Exercise 7.4, Problem 9.

The following example shows that Condition (E) is weaker than Condition (L). It is however easier to check Condition (L) in many problems.

Example 7.25. Consider the matrix

$$D(n) = \operatorname{diag}\left(2 + \sin\left(\frac{2n+1}{2}\right)\pi, 2 - \sin\left(\frac{2n+1}{2}\right)\pi, 2\right).$$

Then

$$D_1(n) = \operatorname{diag}\left(\mu_1(n), \mu_2(n), \mu_3(n)\right)$$

where

$$\mu_1(n) = 1, \qquad \mu_2(n) = \frac{2 - \sin\left(\frac{2n+1}{2}\right)\pi}{2 + \sin\left(\frac{2n+1}{2}\right)\pi}, \qquad \mu_3(n) = \frac{2}{2 + \sin\left(\frac{2n+1}{2}\right)\pi}.$$

Clearly $\mu_2(n)$ and $\mu_3(n)$ do not satisfy Condition (L) but do satisfy Condition (E), since

$$\gamma_2(n) = \prod_{j=0}^{n} \mu_2(j) = \begin{cases} 1/3, & \text{if } n \text{ is even} \\ 1, & \text{if } n \text{ is odd} \end{cases}.$$

$$\gamma_3(n) = \prod_{j=0}^{n} \mu_3(j) = \frac{2}{3} \times \frac{2}{1} \times \frac{2}{3} \times \cdots \times \begin{cases} 2/3, & \text{if } n \text{ is even} \\ 2, & \text{if } n \text{ is odd} \end{cases}.$$

Hence $\gamma_3(n) \to \infty$ as $n \to \infty$ but no subsequence of it converges to zero.

Next we give the fundamental theorem in the asymptotic theory of difference equations; the discrete analogue of Levinson's Theorem [8].

Theorem 7.26. Suppose that either Condition (L) or Condition (E) holds and for each i, $1 \leq i \leq k$,

$$\sum_{n=n_o}^{\infty} \frac{1}{|\lambda_i(n)|} \|B(n)\| < \infty. \tag{7.4.15}$$

Then System (7.4.1) has a fundamental set of k solutions $y_i(n)$ such that

$$y_i(n) = (e_i + o(1)) \prod_{n=n_o}^{n-1} \lambda_i(r), \tag{7.4.16}$$

where e_i is the standard unit vector in R^k where its components are all zero except that the ith component is 1.

Proof Notice that under Condition (L) or (E) it follows from Lemma (7.24) that Eq. (7.4.14) satisfies Assumption (H). Moreover, from Assumption (7.4.11), $B_i(n)$ satisfies Condition (7.4.3). Thus we can apply Theorem 7.23 on Eqs. (7.4.13) and (7.4.14). Observe that since the ith diagonal element in $D_i(n)$ is 1, it follows that $x(n) = e_i$ is a bounded solution of Eq. (7.4.14). By Theorem 7.23, there corresponds a solution $z(n)$ of Eq. (7.4.13) such that $z(n) = e_i + o(1)$. Now conclusion (7.4.16) follows immediately by substituting for $z(n)$ from Formula (7.4.12).

Example 7.27. Consider the difference system $y(n + 1) = A(n)y(n)$, where

$$A(n) = \begin{pmatrix} \dfrac{n^2 + 2}{2n^2} & 0 & \dfrac{1}{n^3} \\ 0 & 1 & 0 \\ \dfrac{1}{2^n} & 0 & n \end{pmatrix}.$$

To apply Theorem 7.26 we need to write $A(n)$ in the form $D(n) + B(n)$ with $D(n)$ a diagonal matrix and $B(n)$ satisfying Condition (7.4.15). To achieve that we let

$$D(n) = \begin{pmatrix} \dfrac{1}{2} & 0 & 0 \\ 0 & 1 & 0 \\ 0 & 0 & n \end{pmatrix}, \qquad B(n) = \begin{pmatrix} \dfrac{1}{n^2} & 0 & \dfrac{1}{n^3} \\ 0 & 0 & 0 \\ \dfrac{1}{2^n} & 0 & 0 \end{pmatrix}.$$

Hence $\lambda_1 = \tfrac{1}{2}$, $\lambda_2 = 1$, and $\lambda_3 = n$. Thus for $n_0 = 2$, our system satisfies the hypotheses of Theorem 7.26. Consequently, there are three solutions

$$y_1(n) \sim \left(\frac{1}{2}\right)^n \begin{pmatrix} 1 \\ 0 \\ 0 \end{pmatrix},$$

$$y_2(n) \sim \begin{pmatrix} 0 \\ 1 \\ 0 \end{pmatrix},$$

$$y_3(n) \sim \left(\prod_{j=1}^{n-1} j\right) \begin{pmatrix} 0 \\ 0 \\ 1 \end{pmatrix} = (n-1)! \begin{pmatrix} 0 \\ 0 \\ 1 \end{pmatrix}.$$

Remark Before ending this section we make one further comment on the conditions in Theorem 7.26. This comment concerns the necessity for some condition on $B(n)$ such as (7.4.15). Certainly Condition (7.4.15) holds when $B(n) = 0(n^{-\alpha})$, $n \to \infty$, for some $\alpha > 1$ (i.e., $n^{\alpha}|B(n)| \leq L$ for all $n \geq n_0$). On the other hand, the condition $B(n) = 0(n^{-1})$ is not sufficient for Formula (7.4.16) to hold, and a simple example illustrates this point. Let us take $k = 1$, $D(n) = 1$, and $B(n) = \frac{1}{n}$. Then Eq. (7.4.1) takes the form $y(n + 1) = \left(\frac{n+1}{n}\right) y(n) = \left(1 + \frac{1}{n}\right) y(n)$ which has the general solution $y(n) = cn$, for some constant c. Hence no solution satisfies Formula (7.4.16).

Exercise 7.4.

In Problems 1 through 5 find asymptotic estimates (using Theorem 7.22) for a fundamental set of solutions of the given system.

1. $y(n + 1) = (D(n) + B(n))y(n)$, where

$$D(n) = \begin{pmatrix} \dfrac{3}{n+2} & 0 \\ 0 & n+1 \end{pmatrix}, \qquad B(n) = \begin{pmatrix} \dfrac{1}{n^2} & \dfrac{3}{n^3} \\ 0 & \dfrac{5}{n^{3/2}} \end{pmatrix}.$$

2. $y(n + 1) = (D(n) + B(n))y(n)$, where

$$D(n) = \begin{pmatrix} \cos \pi n & 0 & 0 \\ 0 & \dfrac{n}{n+1} & 0 \\ 0 & 0 & 3 \end{pmatrix}, \qquad B(n) = \begin{pmatrix} \dfrac{\sin n}{n^3} & \dfrac{n}{e^n} & 0 \\ 0 & 0 & \dfrac{n}{3^n} \\ \dfrac{1}{2^n} & 0 & \dfrac{n}{n^3+5} \end{pmatrix}.$$

3. $y(n + 1) = A(n)y(n)$, where

$$A(n) = \begin{pmatrix} 1 + \dfrac{1}{n} & 0 & \dfrac{1}{n(n+1)} \\ 0 & \dfrac{1}{n} & 0 \\ 0 & 0 & 1 + (-1)^n \cos n\pi \end{pmatrix}.$$

4. $y(n + 1) = A(n)y(n)$, where

$$A(n) = \begin{pmatrix} n & e^{-n} & 0 \\ 0 & 3 - e^{-2n} & 0 \\ 2^{-n} & 0 & 1 + n \end{pmatrix}.$$

5. Give an example of a two-dimensional difference system where Theorem 7.23 does not hold.

6. Prove that $\Phi_1(n)$ and $\Phi_2(n)$ are fundamental matrices for Eq. (7.4.2).

7. Define a diagonal matrix $P = \text{diag}(a_1, a_2, \ldots, a_k)$ where a_i

$$= \begin{cases} 0, & \text{if } \lambda_i \notin S \\ 1, & \text{if } \lambda_i \in S \end{cases}.$$

   Prove the following statements

   (a) $P^2 = P$ (a *projection* matrix).
   (b) $\Phi(n)P = \Phi_1(n)$.
   (c) $\Phi(n)(I - P) = \Phi_2(n)$.
   (d) $\Phi_1(n)P = \Phi_1(n)$, $\qquad \Phi_2(n)(I - P) = \Phi_2(n)$.

8. In Formula (7.4.4), let $x(n) = \Phi(n)\Phi^{-1}(n_0)\xi$.

   Using the notation of Problem 6, show that

   (i) if $\xi \in PR^k$, then $Py(n_0) = \xi$,
   (ii) if $\xi \in (I - P)R^k$, then $(I - P)y(n_0) = \xi$.

9. Prove Lemma 7.24.

10. Prove Formula (7.4.7) using mathematical induction on i.

11. Prove that the solution $y(n)$ of Eq. (7.4.1) defined by Eq. (7.4.5) is bounded for $n \geq n_0 \geq 0$.

12. Prove that, under assumption of Theorem 7.22, for each bounded solution $y(n)$ of Eq. (7.4.1) there exists a bounded solution $x(n)$ of Eq. (7.4.2).

13.* (Open Problem). Improve Theorem 7.22 by relaxing Condition (7.4.4), requiring only conditional covergence of $\sum_{n=n_0}^{\infty} B(n)$.

14.* (Hard) Extend Theorem 7.22 to the case when $D(n)$ is a constant matrix in a one-block Jordan form and then when $D(n)$ is a constant matrix in the general Jordan form.

15.* (Hard) Extend Theorem 7.22 to the case when $D(n)$ has an eigenvalue equal to zero.

16.* (Open Problem). Suppose that there are r distinct eigenvalues $\lambda_1(n)$, $\lambda_2(n)$, $\ldots, \lambda_r(n)$ with distinct moduli. Prove that, with the conditions of Theorem 7.22 holding for $1 \leq i \leq r$, there are solutions $y_i(n), 1 \leq i \leq r$ of system equation (7.4.1) that satisfy Formula (7.4.11).

17.* [10] Consider system equation (7.4.14). Prove that Assumption (H) is satisfied if $D_i(n)$ satisfies the following: there exist constants $\mu > 0$ such that for each λ_j, $i \neq j$, either

$$\prod_{1=0}^{n} \left| \frac{\lambda_i(1)}{\lambda_j(1)} \right| \to +\infty \text{ as } n \to \infty,$$

and

$$\prod_{1=n_1}^{n_2} \left| \frac{\lambda_i(1)}{\lambda_j(1)} \right| \geq \mu > 0, \qquad \text{for all } 0 \leq n_1 \leq n_2,$$

or

$$\prod_{1=n_1}^{n_2} \left| \frac{\lambda_i(1)}{\lambda_j(1)} \right| \leq K, \qquad \text{for all } 0 \leq n_1 \leq n_2.$$

7.5   High Order Difference Equations

In this section we turn our attention to the kth order scalar equations of the form

$$y(n + k) + (a_1 + p_1(n))y(n + k - 1) + \cdots + (a_k + p_k(n))y(n) = 0, \qquad (7.5.1)$$

where $a_i \in R$ and $p_i(n)$, $1 \leq i \leq k$, are real sequences. As we have seen in Chapter 3, Eq. (7.5.1) may be put in the form of a k-dimensional system of first order difference equations which is asymptotically constant. Thus we are led to the study of a special case of Eq. (7.4.1), namely, the asymptotically constant system

$$y(n + 1) = [A + B(n)]y(n), \qquad (7.5.2)$$

where A is a $(k \times k)$ constant matrix that is not necessarily diagonal. This system is, obviously, more general than the system induced by Eq. (7.5.1). The first asymptoticity result concerning system equation (7.5.2) is a consequence of Theorem 7.26.

Theorem 7.28. [11] Suppose that the matrix A has k linearly independent eigenvectors $\xi_1, \xi_2, \ldots, \xi_k$ and k corresponding eigenvalues $\lambda_1, \lambda_2, \ldots, \lambda_k$. If Condition (7.4.3) holds for $B(n)$, then system equation (7.5.2) has solutions $y_i(n)$, $1 \leq i \leq k$ such that

$$y_i(n) = [\xi_i + o(1)]\lambda_i^n. \qquad (7.5.3)$$

Proof In order to be able to apply Theorem 7.26 we need to diagonalize the matrix A. This may be accomplsiehd by letting

$$y = Tz \qquad (7.5.4)$$

in Eq. 7.5.2, where

$$T = (\xi_1, \xi_2, \ldots, \xi_k), \qquad (7.5.5)$$

that is, the ith column of T is ξ_i.

Then we obtain

$$Tz(n + 1) = [A + B(n)]Tz(n),$$

or

$$z(n + 1) = [D + \tilde{B}(n)]z(n), \tag{7.5.6}$$

where $D = T^{-1}AT = \text{diag}(\lambda_1, \lambda_2, \ldots, \lambda_k)$ and $\tilde{B}(n) = T^{-1}B(n)T$. It is easy to see that $\tilde{B}(n)$ satisfies Condition (7.4.10). Now Formula (7.5.3) follows by applying Theorem 7.26.

Example 7.29. Find an asymptotic estimate of a fundamental set of solutions of

$$y(n + 1) = [A + B(n)]y(n), \tag{$*$}$$

where

$$A = \begin{pmatrix} 2 & 2 & 1 \\ 1 & 3 & 1 \\ 1 & 2 & 1 \end{pmatrix}$$

$$B(n) = \begin{pmatrix} 1/n^2 + 1 & 0 & (.5)^n \\ 0 & (.2)^n & 0 \\ e^{-n} & 0 & \log n/n^2 \end{pmatrix}.$$

Solution The eigenvalues of A are $\lambda_1 = 5$, $\lambda_2 = 1$, and $\lambda_3 = 1$, and the corresponding eigenvectors are $\xi_i = \begin{pmatrix} 1 \\ 1 \\ 1 \end{pmatrix}$, $\xi_2 = \begin{pmatrix} 1 \\ 0 \\ -1 \end{pmatrix}$, and $\xi_3 = \begin{pmatrix} 1 \\ 0 \\ -2 \end{pmatrix}$. Furthermore, $B(n)$ satisfies Condition (7.4.3). Thus by Theorem 7.24, Eq. $(*)$ has the solutions

$$y_1(n) = (1 + o(1))(5^n) \begin{pmatrix} 1 \\ 1 \\ 1 \end{pmatrix} \sim \begin{pmatrix} 1 \\ 1 \\ 1 \end{pmatrix} (5^n),$$

$$y_2(n) = (1 + o(1)) \begin{pmatrix} 1 \\ 0 \\ -1 \end{pmatrix} \sim \begin{pmatrix} 1 \\ 0 \\ -1 \end{pmatrix},$$

$$y_3(n) = (1 + o(1)) \begin{pmatrix} 1 \\ 0 \\ -2 \end{pmatrix} \sim \begin{pmatrix} 1 \\ 0 \\ -2 \end{pmatrix}.$$

Next, we apply Theorem 7.26 to establish the following asymptoticity result for Eq. (7.5.1).

Corollary 7.30. Suppose that the polynomial

$$p(A) = \lambda^k + a_1\lambda^{k-1} + \cdots + a_k \tag{7.5.7}$$

has distinct roots $\lambda_1, \lambda_2, \ldots, \lambda_k$ and that

$$\sum_{n=1}^{\infty} |p_i(n)| < \infty, \text{ for } 1 \leq i \leq k. \tag{7.5.8}$$

Then Eq. (7.5.1) has k solutions $y_1(n), y_2(n), \ldots, y_k(n)$ with

$$y_i(n) = [1 + o(1)]\lambda_i^n. \tag{7.5.9}$$

Proof First we put Eq. (7.5.1) into the form of a k-dimensional system

$$z(n + 1) = [A + B(n)]z(n), \tag{7.5.10}$$

where

$$A = \begin{pmatrix} 0 & 1 & \cdots & 0 \\ 0 & 0 & 1 & 0 \\ \vdots & & & \vdots \\ -a_k & -a_{k-1} & \cdots & -a_1 \end{pmatrix},$$

$$B(n) = \begin{pmatrix} 0 & 0 & \cdots & 0 \\ 0 & 0 & \cdots & 0 \\ -p_k(n) & -p_{k-1}(n) & \cdots & -p_1(n) \end{pmatrix},$$

$$z(n) = (y(n), y(n+1), \ldots, y(n+k-1))^T.$$

Notice that Polynomial (7.5.7) is the characteristic polynomial of the matrix A. Furthermore, for each eigenvalue λ_i there corresponds the eigenvector $\xi_i = (1, \lambda_i, \lambda_i^2, \ldots, \lambda_i^{k-1})^T$. In addition, the matrix $B(n)$ satisfies Condition (7.4.3). Hence one may apply Corollary 7.30 to conclude that there are k solutions $z_1(n), z_2(n), \ldots, z_k(n)$ of Eq. 7.5.10 such that for $1 \leq i \leq k$

$$z_i(n) = \begin{pmatrix} y_i(n) \\ y_i(n+1) \\ y_i(n+2) \\ \vdots \\ y_i(n+k-1) \end{pmatrix} = (1 + o(1))\lambda_i^n \begin{pmatrix} 1 \\ \lambda_i \\ \lambda_i^2 \\ \vdots \\ \lambda_i^{k-1} \end{pmatrix}.$$

Hence $y_i(n) = [1 + o(1)]\lambda_i^n$.

Example 7.31. Find asymptotic estimates of fundamental solutions to the difference equation

$$y(n+3) - \left(2 + e^{-n-2}\right) y(n+2) - \left(1 + \frac{1}{n^2+1}\right) y(n+1) + 2y(n) = 0.$$

Solution The characteristic equation is given by $\lambda^3 - 2\lambda^2 - \lambda + 2 = 0$ with roots $\lambda_1 = 2, \lambda_2 = 1, \lambda_3 = -1$. Notice that $p_1(n) = -e^{-n-2}$, $p_2(n) = -\frac{1}{n^2+1}$, and

$p_3(n) = 0$ all satisfy Condition (7.5.8). Hence Corollary 7.30 applies to produce solutions $y_1(n)$, $y_2(n)$, and $y_3(n)$ defined as follows:

$$y_1(n) = [1 + o(1)]2^n, \qquad y_2(n) = 1 + o(1), \qquad y_3(n) = [1 + o(1)](-1)^n.$$

Corollary 7.30 is due to Evgrafov [12]. It says that, for each characteristic root of Polynomial (7.5.7), at least one solution behaves as in Formula (7.5.9), provided that the rate of convergence of the coefficients is not too slow.

What happens if all the roots of the characteristic Eq. (7.5.7) are equal? This same question was addressed by Coffman [13], where he obtained the following result.

Theorem 7.32. Suppose that the polynomial Eq. (7.5.7) has a k-fold root of 1 and

$$\sum_{n=1}^{\infty} n^{k-1}|p_i(n)| < \infty, \qquad \text{for } 1 \le i \le k. \tag{7.5.11}$$

Then Eq. (7.5.1) has k solutions $y_1(n)$, $y_2(n)$, ..., $y_k(n)$ with

$$y_i(n) = n^{i-1}(1 + o(1)), (n \to \infty). \tag{7.5.12}$$

We remark here that the actual result of Coffman is stronger than the statement of Theorem 7.32. Indeed he proved that

$$\Delta^m y_i(n) = \begin{cases} \dbinom{n}{i-m} + o\left(\dfrac{i-m}{n}\right) & \text{for } 1 \le m \le i \\[3mm] o\left(\dfrac{i-m}{n}\right) & \text{for } i \le m \le k-1 \end{cases}.$$

The curious reader might wonder whether Coffman's Theorem (7.31) applies if the polynomial Eq. (7.5.7) has a k-fold root not equal to 1. Luckily, by a very simple trick, one is able to do exactly that. Assume that the characteristic Eq. (7.5.7) has a k-fold root $\mu \ne 1$. Then Polynomial (7.5.7) may be written as

$$(\lambda - \mu)^k = 0. \tag{7.5.13}$$

Letting $y(n) = \mu^n x(n)$ in Eq. (7.5.1) we obtain

$$\mu^{n+k} x(n+k) + \mu^{n+k-1}(a_1 + p_1(n))x(n+k-1) + \cdots + \mu^n(a_k + p_k(n))x(n) = 0$$

or

$$x(n+k) - \frac{1}{\mu}(a_1 + p_1(n))x(n+k-1) + \cdots + \frac{1}{\mu^k}(a_k + p_k(n))x(n) = 0. \tag{7.5.14}$$

The characteristic Eq. (7.5.14) is given by

$$\lambda_k + \frac{a_1}{\mu}\lambda^{k-1} + \frac{a_2}{\mu^2}\lambda^{k-2} + \cdots + \frac{a_k}{\mu^k} = 0,$$

which has a k-fold root $\lambda = 1$. Moreover, if $p_i(n)$, $1 \le i \le k$, satisfies Condition (7.5.1), then so does $(1/\mu^i)p_i(n)$. Hence Theorem 7.32 applies to Eq. (7.5.14) to yield solutions $x_1(n), x_2(n), \dots, x_k(n)$ with

$$x_i(n) = n^{i-1}(1 + o(1)), \qquad (n \to \infty).$$

Consequently, there are solutions $y_1(n), y_2(n), \dots, y_k(n)$ of Eq. (7.5.1) such that

$$y_i(n) = n^{i-1}(1 + o(1))\mu^n.$$

We now summarize the above observations in the following corollary.

Corollary 7.33. Suppose that the Polynomial (7.5.7) has a k-fold root μ and that Condition (7.5.11) holds. Then Eq. (7.5.1) has k solutions $y_1(n), y_2(n), \dots, y_k(n)$ such that

$$y_i(n) = n^{i-1}(1 + o(1))\mu^n. \qquad (7.5.15)$$

Example 7.34. Investigate the asymptotic behavior of solutions of the difference equation

$$y(n + 3) - (6 + e^{-n-2})y(n + 2) + \left(12 - \frac{1}{(n+1)^4}\right) y(n + 1) - 8y(n) = 0.$$

Solution The characteristic equation is given by $\lambda^3 - 6\lambda^2 + 12\lambda - 8 = 0$ with roots $\lambda_1 = \lambda_2 = \lambda_3 = 2$. Also $p_1(n) = -e^{-n-2}$, $p_2(n) = -1/((n+1)^4)$, and $p_3(n) = 0$ all satisfy Condition (7.5.11). Hence, by Corollary 7.33 there are three solutions $y_1(n) = (1 + o(1))2^n$, $y_2(n) = n(1 + o(1))2^n$, and $y_3(n) = n^2(1 + o(1))2^n$.

Example 7.35. Consider the difference equation

$$x(n + 2) + p_1(n)x(n + 1) + p_2(n)x(n) = 0 \qquad (7.5.16)$$

where

$$p_1(n) \ne 0, \qquad \text{for } n \ge n_0 \ge 0$$

and that

$$\lim_{n \to \infty} 4p_2(n)/p_1(n)p_1(n - 1) = p \qquad (7.5.17)$$

exists. Let $\alpha(n)$ be defined by

$$\alpha(n) = (4p_2(n)/p_1(n)p_1(n - 1)) - p. \qquad (7.5.18)$$

Assume that $p \ne 0$, $p \le 1$ and

$$\sum_{j=n_0}^{\infty} |\alpha(j)| < \infty. \qquad (7.5.19)$$

Show that Eq. (7.5.16) has two solutions

$$x_{\pm}(n) \sim \left(-\frac{1}{2}\right)^n \prod_{j=n_0}^{n-1} (p_1(j)(1 \pm \nu \mp \alpha(j))/2\nu), \qquad (7.5.20)$$

where $v = \sqrt{1 - p}$.

Solution Let

$$x(n) = \left(-\frac{1}{2}\right)^n \left(\prod_{j=n_0}^{n-1} p_1(j)\right) y(n). \tag{7.5.21}$$

Then Eq. (7.5.16) is transformed to

$$y(n + 2) - 2y(n + 1) + (p + \alpha(n))y(n) = 0. \tag{7.5.22}$$

Let $z(n) = (z_1(n), z_2(n))^T = (y(n), y(n + 1))^T$. Then Eq. (7.5.22) may be put into a system of the form

$$\begin{pmatrix} z_1(n + 1) \\ z_2(n + 1) \end{pmatrix} = \begin{pmatrix} 0 & 1 \\ v^2 - 1 - \alpha(n) & 2 \end{pmatrix} \begin{pmatrix} z_1(n) \\ z_2(n) \end{pmatrix}. \tag{7.5.23}$$

Again we let

$$\begin{pmatrix} z_1(n) \\ z_2(n) \end{pmatrix} = \begin{pmatrix} 1 & 1 \\ -(v - 1) & v + 1 \end{pmatrix} \begin{pmatrix} u_1(n) \\ u_2(n) \end{pmatrix}.$$

Then Eq. (7.5.23) becomes

$$\begin{pmatrix} u_1(n + 1) \\ u_2(n + 1) \end{pmatrix} = \begin{pmatrix} (1 - v + \alpha(n))/2v & \alpha(n)/2v \\ -\alpha(n)/2v & (1 + v - \alpha(n))/2v \end{pmatrix} \begin{pmatrix} u_1(n) \\ u_2(n) \end{pmatrix}. \tag{7.5.24}$$

If we let $u(n) = (u_1(n), u_2(n))^T$, then we may write Eq. (7.5.24) in the form

$$u(n + 1) = (D(n) + B(n))u(n) \tag{7.5.25}$$

where

$$D(n) = \begin{pmatrix} (1 - v + \alpha(n))/2v & 0 \\ 0 & (1 + v - \alpha(n))/2v \end{pmatrix},$$

$$B(n) = \begin{pmatrix} 0 & \alpha(n)/2v \\ -\alpha(n)/2v & 0 \end{pmatrix}.$$

By Theorem 7.26, there are two solutions of Eq. (7.5.25) given by

$$u_+(n) \sim \left[\prod_{j=n_0}^{n-1} (1 - v + \alpha(j))/2v\right] \begin{pmatrix} 1 \\ 0 \end{pmatrix},$$

$$u_-(n) \sim \left[\prod_{j=n_0}^{n-1} (1 + v - \alpha(j))/2v\right] \begin{pmatrix} 0 \\ 1 \end{pmatrix}.$$

These two solutions produce two solutions of Eq. (7.5.23)

$$z_+(n) = \begin{pmatrix} y_+(n) \\ y_+(n + 1) \end{pmatrix} = \begin{pmatrix} 1 & 1 \\ -(v - 1) & v + 1 \end{pmatrix} \left(\prod_{j=n_0}^{n-1} (1 - v + \alpha(j))/2v\right).$$

Hence

$$y_+(n) \sim \prod_{j=n_0}^{n-1} (1 - v + \alpha(j))/2v.$$

Using Eq. (7.5.21) we obtain

$$x_+(n) = \left(-\frac{1}{2}\right)^n \prod_{j=n_0}^{n-1} p_1(j)(1 - v + \alpha(j))/2v.$$

Similarly, one may show that

$$x_-(n) = \left(-\frac{1}{2}\right)^n \prod_{j=n_0}^{n-1} p_1(j)(1 + v - \alpha(j))/2v. \tag{7.5.26}$$

(See Exercise 7.5, Problem 11.)

Exercise 7.5.

In Problems 1 through 4 find an asymptotic estimate of a fundamental set of solutions of the given equation $y(n + 1) = [A + B(n)]y(n)$.

1. $A = \begin{pmatrix} 2 & 0 \\ 0 & 3 \end{pmatrix}$, $B(n) = \begin{pmatrix} e^{-n} & 0 \\ \dfrac{1}{(n+1)^2} & (0.1)^n \end{pmatrix}$.

2. $A = \begin{pmatrix} 1 & 6 \\ 5 & 2 \end{pmatrix}$, $B(n) = \begin{pmatrix} 0 & e^{-n-1} \\ 2^{-n} & \dfrac{n}{e^n} \end{pmatrix}$.

3. $A = \begin{pmatrix} -1 & 0 & 0 \\ 0 & 1 & 0 \\ 0 & 0 & 4 \end{pmatrix}$, $B(n) = \begin{pmatrix} 3^{-n} & 0 & 2^{-n} \\ \dfrac{\sin n}{(n+1)^2} & 0 & 0 \\ 0 & e^{-n} & \dfrac{1}{(n+1)^3} \end{pmatrix}$.

4. $A = \begin{pmatrix} 5 & 4 & 2 \\ 4 & 5 & 2 \\ 2 & 2 & 2 \end{pmatrix}$, $B(n) = \begin{pmatrix} 0 & (0.2)^n & 0 \\ (0.1)^n & 0 & e^{-n^2} \\ 0 & \dfrac{1}{n^2+1} & 0 \end{pmatrix}$.

In Problems 5 through 10 investigate the asymptotic behavior of solutions of the given equation.

5. $y(n + 2) - (5 + e^{-n})y(n + 1) + \left(6 - \dfrac{1}{(n+1)^2}\right) y(n) = 0$.

6. $y(n + 2) - (4 + ne^{-n})y(n) = 0.$

7. $y(n + 2) + (4 + ne^{-n})y(n) = 0.$

8. $y(n + 3) - 6y(n + 2) + (11 + (\sin n)e^{-n})y(n + 1) - 6y(n) = 0.$

9. $y(n + 3) - (3 + 2^{-n})y(n + 2) + 3y(n + 1) - y(n) = 0.$

10. $y(n + 3) = 15y(n + 2) + 75y(n + 1) - (125 + (0.1)^n)y(n) = 0.$

11. Complete the proof of Example 7.35 by verifying Formula 7.5.26.

12.* Consider the second order difference equation

$$x(n + 2) + p_1(n)x(n + 1) + p_2(n)x(n) = 0. \qquad (7.5.27)$$

Assume that $p_1(n) \neq 0$ for $n \geq n_0$ and that

(i) $\lim\limits_{n \to \infty} 4p_2(n)/p_1(n)p_1(n - 1) = p.$ $\qquad (7.5.28)$

(ii) Let $\alpha(n) = [4p_2(n)/p_1(n)p_1(n - 1)] - p,$ $\qquad \sum\limits_{n=n_0}^{\infty} |\alpha(n)| < \infty.$
$\qquad\qquad\qquad\qquad\qquad\qquad\qquad\qquad\qquad\qquad\qquad (7.5.29)$

If p is neither 0 nor 1, show that Eq. (7.5.27) has two solutions

$$x_{\pm}(n) \sim \left(-\frac{1}{2}\right)^n \left(\prod_{j=n_0}^{n-1} p_1(j)\right)(1 \pm v)^n, \qquad (n \to \infty),$$

where $v = \sqrt{1 - p}.$

$\left[\text{Hint: Let } x(n) = \left(-\frac{1}{2}\right)^n \left(\prod_{j=n_0}^{n-1} p_1(j)\right) z(n).\right]$

13. In Problem 12, suppose that $p = 1$ and that all the assumptions there hold except that the condition $\sum_{n=n_0}^{\infty} |\alpha(n)| < \infty$ is replaced by $\sum_{n=n_0}^{\infty} n|\alpha(n)| < \infty.$

Show that there are two solutions $x_1(n) \sim \left(-\frac{1}{2}\right)^n \prod_{j=n_0}^{n-1} p_1(j)$ and $x_2(n) \sim n \left(-\frac{1}{2}\right)^n \prod_{j=n_0}^{n-1} p_1(j), (n \to \infty).$

14. Consider the difference Eq. (7.5.27) such that $p_1(n) \neq 0$ for $n \geq n_0$. Assume that $\lim_{n \to \infty}(p_2(n))/p_1(n)p_1(n-1) = 0$ and $\alpha(n) = (p_2(n))/p_1(n)p_1(n-1).$

(a) Use the transformation $x(n) = \left(-\frac{1}{2}\right)^n \prod_{j=n_0}^{n-1} p_1(j)z(n)$ to transform Eq. (7.5.27) to $z(n + 2) - 2z(n + 1) + \alpha(n)z(n) = 0.$

(b) Show that Eq. (7.5.27) has two solutions $x_1(n) \sim (-1)^n \prod_{j=n_0}^{n-1} p_1(j)$ and $x_2(n) = o(v^n|x_1(n)|)$ for any v, with $0 < v < 1.$

15.* Consider the difference Eq. (7.5.16) with Conditions (7.5.16) and (7.5.17) satisfied. If p is real and $p > 1$, show that Formula (7.5.20) remains valid if we assume that

$$\sum_{j=n_0}^{\infty} j|\alpha(j)| < \infty. \tag{7.5.30}$$

16.* Show that Formula (7.5.20) remains valid if one replaces Hypothesis (7.5.19) by

$$\sum_{j=n_0}^{\infty} |\alpha(j)|^{\sigma} < \infty \tag{7.5.31}$$

for some real number σ, with $1 \leq \sigma \leq 2$.

17.* Show that the conclusions of Problem 16 remain valid if Condition (7.5.31) is replaced by

$$\sum_{j=n_0}^{\infty} |\alpha(j)|^{\sigma} j^{\tau-1}$$

for some real numbers σ and τ such that $1 \leq \sigma \leq 2$ and $\tau > \sigma$.

7.6 Nonlinear Difference Equations

In this section we consider the nonlinearly perturbed system

$$y(n+1) = A(n)y(n) + f(n, y(n)) \tag{7.6.1}$$

along with the associated unperturbed system

$$x(n+1) = A(n)x(n), \tag{7.6.2}$$

where $A(n)$ is an invertible $(k \times k)$ matrix function on Z^+ and $f(n, y)$ is a function from $Z^+ \times R^k \to R^k$ which is continuous in y. Let $\Phi(n)$ be the fundamental matrix of System (7.6.2).

The first step in our analysis is to extend the variation of constants formula (Theorem 7.21) to System (7.6.1). Since $A(n)$ is not assumed here to be a diagonal matrix, we ned to replace Definition 7.20 by a more general definition of dichotomy.

Definition 7.36. System (7.6.2) is said to possess an ordinary dichotomy if there exists a projection matrix P and a positive constant M such that

$$
\begin{aligned}
|\Phi(n)P\Phi^{-1}(m)| &\leq M, &&\text{for } n_0 \leq m \leq n, \\
|\Phi(n)(I-P)\Phi^{-1}(m)| &\leq M, &&\text{for } n_0 \leq n \leq m.
\end{aligned}
\tag{7.6.3}
$$

Notice that if $A(n) = \text{diag}(\lambda_1(n), \ldots, \lambda_k(n))$, then this definition reduces to Definition 7.22 if we let $\Phi_1(n) = \Phi(n)P$ and $\Phi_2(n) = \Phi(n)(I-P)$.

Theorem 7.37. [11, 14, 15] Suppose that System (7.6.2) possesses an ordinary dichotomy. If, in addition,

$$\sum_{j=n_0}^{\infty} |f(j, 0)| < \infty \tag{7.6.4}$$

and

$$|f(n, x) - f(n, y)| \le \gamma(n)|x - y|, \tag{7.6.5}$$

where $\gamma(n) \in l^1([n_0, \infty))$ i.e., $\sum_{j=n_0}^{\infty} \gamma(j) < \infty$, then for each bounded solution $x(n)$ of Eq. (7.6.2) there corresponds a bounded solution $y(n)$ of Eq. (7.6.1) and vice versa. Furthermore, $y(n)$ is given by the formula

$$y(n) = x(n) + \sum_{j=n_0}^{n-1} \Phi(n) P \Phi^{-1}(j+1) f(j, y(j))$$

$$- \sum_{j=n}^{\infty} \Phi(n)(I - P)\Phi^{-1}(j+1) f(j, y(j)). \tag{7.6.6}$$

Proof The proof mimics that of Theorem 7.22 with some obvious modifications. Let $x(n)$ be a bounded solution of Eq. (7.6.2). Define a sequence $\{y_i(n)\}(i = 1, 2, \ldots)$ successively by letting $y_1(n) = x(n)$ and

$$y_{i+1}(n) = x(n) + \sum_{j=n_0}^{n-1} \Phi(n) P \Phi^{-1}(j+1) f(j, y_i(j))$$

$$- \sum_{j=n}^{\infty} \Phi(n)(I - P)\Phi^{-1}(j+1) f(j, y_i(j)). \tag{7.6.7}$$

We use mathematical induction to show that $y_i(n)$ is bounded on $[n_0, \infty)$ for each i. First we notice from assumption that $|y_1(n)| \le c_1$. Now suppose that $|y_i(n)| \le c_i$. Then it follows from Eqs. (7.6.4), (7.6.5), and (7.6.7) that

$$|y_{i+1}(n)| \le c_1 + M \sum_{j=n_0}^{\infty} \left[\gamma(j)|y_i(j)| + |f(j, 0)| \right]$$

$$\le c_1 + M \left[\sum_{j=n_0}^{\infty} c_i \gamma(j) + \tilde{M} \right] = c_{i+1},$$

where

$$\sum_{j=n_0}^{\infty} |f(j, 0)| = \tilde{M}.$$

Hence $y_i(n)$ is bounded for each i.

As in the proof of Theorem 7.22, one may show that the sequence $\{y_i(n)\}$ converges uniformly on $[n_0, \infty)$ to a bounded solution $y(n)$ of Eq. (7.6.1). Conversely, let $y(n)$ be a bounded solution of Eq. (7.6.1). Then one may verify easily that

$$\tilde{y}(n) = \sum_{j=n_0}^{n-1} \Phi(n)P\Phi^{-1}(j+1)f(j,\tilde{y}(j)) - \sum_{j=n}^{\infty} \Phi(n)(I-P)\Phi^{-1}(j+1)f(j,\tilde{y}(j))$$

is another bounded solution of Eq. (7.6.1). Hence $x(n) = y(n) - \tilde{y}(n)$ is a bounded solution of Eq. (7.6.2).

The preceding result does not provide enough information about the asymptotic behavior of solutions of system equation (7.6.1). In order to obtain such results we need one more assumption on Eq. (7.6.2).

Theorem 7.38. [11] Let all the assumptions of Theorem 7.37 hold. If $\Phi(n)P \to 0$ as $n \to \infty$, then for each bounded solution $x(n)$ of Eq. (7.6.2) there corresponds a bounded soution $y(n)$ of Eq. (7.6.1) such that

$$y(n) = (x(n) + o(1)) \tag{7.6.8}$$

or

$$y(n) \sim x(n).$$

Proof The proof is similar to the proof of Theorem 7.23 and is left to the reader as Exercise 7.6, Problem 7.

Example 7.39. Consider the equation

$$\begin{pmatrix} y_1(n+1) \\ y_2(n+1) \end{pmatrix} = \begin{pmatrix} 3 & 0 \\ 0 & 1/2 \end{pmatrix} \begin{pmatrix} y_1(n) \\ y_2(n) \end{pmatrix} + \begin{pmatrix} \sin y_1(n)/n^2 \\ [1 - \cos y_2(n)]/n^2 \end{pmatrix}.$$
$$\tag{7.6.9}$$

Here

$$A(n) = \begin{pmatrix} 3 & 0 \\ 0 & 1/2 \end{pmatrix}, \qquad f(n,y) = \begin{pmatrix} \sin y_1/n^2 \\ (1 - \cos y_2)/n^2 \end{pmatrix}.$$

Using the Euclidean norm we obtain $\sum_{j=1}^{\infty} |f(j,0)| = 0$. Moreover, for

$$x = \begin{pmatrix} x_1 \\ x_2 \end{pmatrix}, y = \begin{pmatrix} y_1 \\ y_2 \end{pmatrix}$$

$$|f(n,x) - f(n,y)| = \frac{1}{n^2} \begin{vmatrix} \sin x_1 - \sin y_1 \\ \cos x_2 - \cos y_2 \end{vmatrix}$$

$$= \frac{1}{n^2}\sqrt{(\sin x_1 - \sin y_1)^2 + (\cos x_2 - \cos y_2)^2}.$$
$$\tag{7.6.10}$$

By the mean value theorem

$$\frac{|\sin x_1 - \sin y_1|}{|x_1 - y_1|} = |\cos c - \sin c|, \text{ for some } c \in (x_1, y_1)$$

$$\leq 2,$$

and

$$\frac{|\cos x_2 - \cos y_2|}{|x_2 - y_2|} \le 2.$$

Hence substituting into Eq. (7.6.10) we obtain

$$|f(n, x) - f(n, y)| \le \frac{2}{n^2}|x - y|.$$

The associated homogeneous equation

$$x(n + 1) = A(n)x(n)$$

has a fundamental matrix $\Phi(n) = \begin{pmatrix} 3^n & 0 \\ 0 & (1/2)^n \end{pmatrix}$ and two solutions; one bounded $x_1(n) = \binom{0}{1}(1/2)^n$ and one unbounded $x_2(n) = \binom{1}{0}3^n$. If we let the projection matrix

$$P = \begin{pmatrix} 0 & 0 \\ 0 & 1 \end{pmatrix},$$

then

$$\Phi(n)P = \begin{pmatrix} 0 & 0 \\ 0 & (1/2)^n \end{pmatrix} \to 0$$

as $n \to \infty$.

Hence all the conditions of Theorem (7.38) hold. Thus corresponding to the bounded solution $x_1(n) = \binom{0}{1}(1/2)^n$ there corresponds a solution $y(n)$ of Eq. (7.6.9) such that

$$y(n) \sim \binom{0}{1}(1/2)^n.$$

Next we specialize Theorem (7.38) to the following kth order nonlinear equation of Poincaré's type

$$y(n+k) + (a_1 + p_1(n))y(n+k-1) + \cdots + (a_k + p_k(n))y(n) = f(n, y(n)). \quad (7.6.11)$$

Corollary 7.40. Suppose that the characteristic equation $\lambda^k + a_1\lambda^{k-1} + \cdots + a_k = 0$ has distinct roots λ_i, $1 \le i \le k$, and $\sum_{n=1}^{\infty} |p_j(n)| < \infty, 1 \le j \le k$. Assume further that Conditions (7.6.4) and (7.6.5) hold. Then for each λ_j with $|\lambda_j| \le 1$ there corresponds a solution y_j of Eq. (7.6.11) such that $y_j(n) \sim \lambda_j^n$.

Proof By Corollary 7.30, the homogeneous part of Eq. (7.6.11)

$$x(n + k) + (a_1 + p_1(n))x(n + k - 1) + \cdots + (a_k + p_k(n))x(n) = 0$$

has solutions $x_1(n), x_2(n), \ldots, x_k(n)$ with $x_j(n) \sim \lambda_j^n$. If $|\lambda_j| \le 1$, then $x_j(n)$ is bounded. Corresponding to this bounded solution $x_j(n)$ there is a solution $y_j(n)$ of Eq. (7.6.11) with $y_j(n) \sim x_j(n)$. Thus $y_j(n) \sim \lambda_j^n$.

Example 7.41. Investigate the asymptotic behavior of solutions of the equation

$$y(n+2) - \frac{3}{2}y(n+1) + \frac{1}{2}y(n) = e^{-n}/(1+y^2(n)). \qquad (7.6.12)$$

Solution The characteristic equation is given by

$$\lambda^2 - \frac{3}{2}\lambda + \frac{1}{2} = 0$$

with distinct roots

$$\lambda_1 = 1, \lambda_2 = \frac{1}{2}.$$

Now

$$\sum_{n=0}^{\infty} f(n,0) = \sum_{n=0}^{\infty} e^{-n} < \infty.$$

Moreover,

$$|f(n,x) - f(n,y)| = e^{-n}\left|\frac{1}{1+x^2} - \frac{1}{1+y^2}\right|$$

$$= e^{-n}\frac{|x+y|}{(1+x^2+y^2+x^2y^2)} \cdot |x-y|$$

$$\leq |x-y|.$$

Hence all the assumptions of Corollary 7.40 are satisfied. Consequently Eq. (7.6.12) has two solutions $y_1(n) \sim 1$ and $y_2(n) \sim (1/2)^n$.

Exercise 7.6.

In Problems 1 through 3 investigate the asymptotic behavior of solutions of the equation $y(n+1) = A(n)y(n) + f(n,y(n))$.

1. $A(n) = \begin{pmatrix} \frac{1}{2} & 0 \\ 0 & \frac{1}{n+2} \end{pmatrix}$, $f(n,y) = \begin{pmatrix} e^{-n}\cos y_1 \\ 0 \end{pmatrix}$.

2. $A(n) = \begin{pmatrix} 2 & 3 \\ 3 & -1 \end{pmatrix}$, $f(n,y) = \begin{pmatrix} \frac{1}{n^3} \\ 0 \end{pmatrix}$.

3. $A(n) = \begin{pmatrix} 3 & 2 \\ 2 & 1 \end{pmatrix}$, $f(n,y) = \begin{pmatrix} ne^{-n} & y_1 \\ e^{-n} & y_2 \end{pmatrix}$.

4. Study the asymptotic behavior of solutions of

$$y(n+2) + \left(\frac{3}{2} + \frac{1}{n^2}\right)y(n+1) - (1+e^{-n})y(n) = \frac{\sin y(n)}{n^2}.$$

5. Study the asymptotic behavior of solutions of

$$y(n + 2) - 4y(n + 1) + 3y(n) = \frac{1}{n^2 + y^2}, \qquad n \geq 1.$$

6. Study the asymptotic behavior of solutions of $y(n+2)+(1+e^{-n})y(n) = e^{-n}$.

7. Prove Theorem 7.38.

References

[1] P. Ribenboim, *The book of Prime Number Records*, Springer-Verlag, New York, 1988.

[2] H. Poincaré, Sur Les Equation Linéaires aux Differentielles Ordinaires et aux Différence Finies, *Amer. J. Math.*, **7** (1885), 203–258.

[3] O. Perron, "Über Summengleichungen und Poincarésche Differenzengleichungen," *Math. Annalen*, **84** (1921) 1.

[4] H. Meschkowski, H, *Differenzengleichungen*, Vandehoeck and Ruprecht, Göttingen, 1959.

[5] A. Drozdowicz, *On the Asymptotic Behavior of Solutions of the Second Order Difference Equations*, Glasnik Matematicki, to appear.

[6] ———— "Asymptotic Behavior of the Solutions of the Second Order Difference Equations," *Proc. Amer. Math. Soc.*, **99** (1987), 135–140.

[7] A. Drozdowicz and J. Popenda, *Asymptotic Behavior of the Solutions of the Second Order Difference Equations*, Fasciculi Mathematici, to appear.

[8] N. Levinson, "The Asymptotic Behavior of a System of Linear Differential Equations," *Amer. J. Math.* **68** (1946), 1–6.

[9] M.S.P. Eastham, *The Asymptotic Solutions of Linear Differential Systems*, Clarendon, Oxford, 1989.

[10] Z. Benzaid and D.A. Lutz, "Asymptotic Representation of Solutions of Perturbed Systems of Linear Difference Equations," *Studies Appl. Math.* **77** (1987), 195–221.

[11] S. Elaydi, "Asymptotic Theory for Linear and Nonlinear Difference Equations," *Proceedings of the Fourth Colloquium on Differential Equations*, Plovdiv, Bulgaria, 1993.

[12] M. Evgrafov, "The Asymptotic Behavior of Solutions of Difference Equations," *Dokl. Akad. Nauk SSSR*, **121** (1958), 26–29 (Russian).

[13] C.V. Coffman, "Asymptotic Behavior of Solutions of Ordinary Difference Equations," *Trans. Amer. Math. Soc.* **110** (1964), 22–51.

[14] J. Schinas, "Stability and Conditional Stability of Time-Difference Equations in Banach Spaces," *J. Inst. Math. Appl.* **14** (1974), 335–346.

[15] M. Pinto, "Discrete Dichotomies," *Comp. Math. Appl.* **28** (1994), 259–270.

8

Oscillation Theory

In previous chapters we were mainly interested in the asymptotic behavior of solutions of difference equations both scalar and nonscalar. In this chapter we will go beyond the question of stability and asymptoticity. Of particular interest is to know whether a solution $x(n)$ oscillates around an equilibrium point x^*, regardless of its asymptotic behavior. Since we may assume without loss of generality that $x^* = 0$, the question that we will address here is whether solutions oscillate around zero or whether solutions are eventually positive or eventually negative.

Sections 8.1 and 8.2 follow closely the paper of Erbe and Zhang [1] and the book of Gyori and Ladas [2]. In Section 8.3 we follow the approach in the paper of Hooker and Patula [3]. For more advanced treatment of oscillation theory the reader is referred to [4], [2], [5], [6].

8.1   Three-Term Difference Equations

In this section we consider the three-term difference equation (of order $k + 1$)

$$x(n + 1) - x(n) + p(n)\, x(n - k) = 0, \quad n \in Z^+, \qquad (8.1.1)$$

where k is a positive integer and $p(n)$ is a sequence defined for $n \in Z^+$.

A nontrivial solution $x(n)$ is said to be *oscillatory* (around zero) if for every positive integer N there exists $n \geq N$ such that $x(n)\, x(n + 1) \leq 0$. Otherwise, the solution is said to be *nonoscillatory*. In other words, a solution $x(n)$ is oscillatory if it is neither eventually positive nor eventually negative. The solution $x(n)$ is said to be oscillatory around an equilibrium point x^* if $x(n) - x^*$ is oscillatory around zero. The special case when $k = 1$ and $p(n) = p$ is a constant real number has been treated previously in Section 2.5. In this case Eq. (8.1.1) may be written in the more convenient form

$$x(n + 2) - x(n + 1) + p\, x(n) = 0. \tag{8.1.2}$$

The characteristic roots of Eq. (8.1.2) are given by

$$\lambda_{1,2} = \frac{1}{2} \pm \frac{1}{2}\sqrt{1 - 4p}.$$

Recall from Section 2.5 that all solutions of Eq. (8.1.2) oscillate if and only if λ_1 and λ_2 are not positive real numbers. Hence every solution of Eq. (8.1.2) oscillates if and only if $p > \frac{1}{4}$.

Let us now turn our attention back to Eq. (8.1.1). This equation is the discrete analogue of the delay differential equation

$$x'(t) + p(t)\, x(t - k) = 0. \tag{8.1.3}$$

The oscillatory behavior of Eq. (8.1.3) and its discrete analogue Eq. (8.1.1) is remarkably similar, with one exception, when $k = 0$. In this case, the equation

$$x'(t) + p(t)\, x(t) = 0$$

has the solution

$$x(t) = x(t_0)\, \mathrm{Exp}\left(-\int_{t_0}^{t} p(s)\, ds\right)$$

which is never oscillatory. However, the discrete analogue

$$x(n + 1) = (1 - p(n))\, x(n)$$

has the solution $x(n) = [\prod_{j=n_0}^{n-1} (1 - p(j))]x(n_0)$ which oscillates if $1 - p(j) < 0$ for all $j \geq n_0$.

To prepare for the study of the oscillatory behavior of Eq. (8.1.1) we first investigate the solutions of the following associated difference inequalities:

$$x(n + 1) - x(n) + p(n)\, x(n - k) \leq 0, \tag{8.1.4}$$

$$x(n + 1) - x(n) + p(n)\, x(n - k) \geq 0. \tag{8.1.5}$$

In the sequel we make use of the notions of the *limit superior* and the *limit inferior* of a sequence $\{a(n)\}$, denoted by $\lim_{n\to\infty} \sup a(n)$ and $\lim_{n\to\infty} \inf a(n)$, respectively. Let $\beta(n)$ be the least upper bound of the set $\{a(n), a(n + 1), a(n + 2), \ldots\}$. Then either $\beta(n) = \pm\infty$ for every n or the sequence $\{\beta(n)\}$ is monotonically decreasing sequence of real numbers, and thus $\lim_{n\to\infty} \beta(n)$ exits. Similarly, let $\alpha(n)$ be the greatest lower bound of the set $\{a(n), a(n + 1), a(n + 2), \ldots\}$.

Definition 8.1. Let $\{a(n)\}$ be a sequence of real numbers. Then

(i) $\lim_{n\to\infty} \sup\ a(n) = \lim_{n\to\infty} \beta(n)$.

(ii) $\lim_{n\to\infty} \inf\ a(n) = \lim_{n\to\infty} \alpha(n)$.

Note that $\lim_{n\to\infty} a(n)$ exists if and only if $\lim_{n\to\infty} \sup\ a(n) = \lim_{n\to\infty} \inf\ a(n) = \lim_{n\to\infty} a(n)$.

Example 8.2. Find the limit superior and the limit inferior for the following sequences:

$$S_1 :\quad 0, 1, 0, 1,\ \ldots\ .$$
$$S_2 :\quad 1, -2, 3, -4,\ \ldots,\ (-1)^{n+1}n,\ \ldots\ .$$
$$S_3 :\quad 3/2, -1/2, 4/3, -1/3, 5/4, -1/4, 6/5, -1/5,\ \ldots\ .$$

Solution

$$\lim_{n\to\infty} \sup\ S_1\ =\ 1,\qquad \lim_{n\to\infty} \inf\ S_2 = 0,$$
$$\lim_{n\to\infty} \sup\ S_2\ =\ \infty,\qquad \lim_{n\to\infty} \inf\ S_2 = -\infty,$$
$$\lim_{n\to\infty} \sup\ S_3\ =\ 1,\qquad \lim_{n\to\infty} \inf\ S_3 = 0.$$

Theorem 8.3. [1] Suppose that

$$\lim_{n\to\infty} \inf\ p(n) = p > \frac{k^k}{(k+1)^{k+1}}. \tag{8.1.6}$$

Then the following statements hold:

(i) Inequality (8.1.4) has no eventually positive solution;

(ii) Inequality (8.1.5) has no eventually negative solution.

Proof (i) To prove statement (i), assume the contrary, that is there exists a solution $x(n)$ of Inequality (8.1.4) which is eventually positive. Hence there exists a positive integer N_1, such that $x(n) > 0$ for all $n \geq N_1$. Dividing Inequality (8.1.4) by $x(n)$, we get for $n \geq N_1$

$$\frac{x(n+1)}{x(n)} \leq 1 - p(n)\frac{x(n-k)}{x(n)}. \tag{8.1.7}$$

If we let $z(n) = \frac{x(n)}{x(n+1)}$, then

$$\frac{x(n-k)}{x(n)}\ =\ \frac{x(n-k)}{x(n-k+1)}\ \frac{x(n-k+1)}{x(n-k+2)},\ \ldots,\ \frac{x(n-1)}{x(n)}$$
$$=\ z(n-k)\ z(n-k+1),\ \ldots,\ z(n-1).$$

Substituting into Inequality (8.1.7) yields

$$\frac{1}{z(n)} \leq 1 - p(n)z(n-k)z(n-k+1), \ \ldots, \ z(n-1), \quad n \geq N_1 + k. \quad (8.1.8)$$

Now Condition (8.1.6) implies that there exists a positive integer N_2 such that $p(n) > 0$ for all $n \geq N_2$. Put $N = \max\{N_2, \ N_1 + k\}$. Then for $n \geq N, x(n + 1) - x(n) \leq -p(n) x(n-k) \leq 0$. Consequently, $x(n)$ is nonincreasing and thus $z(n) \geq 1$. Let $\lim_{n\to\infty} \inf z(n) = q$. Then from Inequality (8.1.8) we have

$$\lim_{n\to\infty} \sup \frac{1}{z(n)} = \frac{1}{\displaystyle\lim_{n\to\infty} \inf z(n)}$$

$$= 1/q$$

$$\leq 1 - \lim_{n\to\infty} \inf [p(n)\, z(n-k)\, z(n-k-1), \ \ldots, \ z(n-1)]$$

or

$$1/q \leq 1 - p\, q^k$$

which yields

$$p \leq \frac{q-1}{a^{k+1}}. \quad (8.1.9)$$

Let $h(q) = (q-1)/q^{k+1}$. The $h(q)$ attains its maximum at $q = (k+1)/k$. Hence $\max_{q\geq 1} h(q) = (k^k)/(k+1)^{k+1}$. Hence from Inequality (8.1.9) we have $p \leq (k^k)/(k+1)^{k+1}$, a contradiction. This completes the proof of part (i) of the theorem. The proof of part (ii) is left to the reader as Exercise 8.1, Problem 6.

Corollary 8.4. If Condition (8.1.6) holds, then every solution of Eq. (8.1.1) oscillates.

Proof Suppose the contrary and let $x(n)$ be an eventually positive solution of Eq. (8.1.1). Then Inequality (8.1.4) has an eventually positive solution which contradicts Theorem 8.3. On the other hand if Eq. (8.1.1) has an eventually negative solution, then so does Inequality (8.1.5) which again violates Theorem 8.3.

The above corollary is sharp as may be evidenced by the following example where we let

$$p(n) = \frac{k^k}{(k+1)^{k+1}}.$$

Example 8.5. Consider the difference equation $x(n+1) - x(n) + (k^k/(k+1)^{k+1})\, x(n-k) = 0$. Then $x(n) = \left(\frac{k}{k+1}\right)^{n-1}$, $n > 1$, is a nonoscillatory solution of the equation.

Next we give a partial converse of Corollary (8.4).

Theorem 8.6. [1] Suppose that $p(n) \geq 0$ and

$$\sup p(n) < \frac{k^k}{k+1^{k+1}}. \quad (8.1.10)$$

Then Eq. (8.1.1) has a nonoscillatory solution.

Proof As in the proof of Theorem 8.3, we let $z(n) = x(n)/x(n+1)$ in Eq. (8.1.1) to obtain

$$1/z(n) = 1 - p(n) z(n-k) z(n-k+1), \ldots, z(n-1). \tag{8.1.11}$$

To complete the proof, it suffices to show that Eq. (8.1.11) has a positive solution. To construct such a solution we define

$$z(1-k) = z(2-k) = \ldots = z(0) = a = \frac{k+1}{k} > 1 \tag{8.1.12}$$

and

$$z(1) = [1 - p(1) z(1-k) z(2-k) \ldots z(0)]^{-1}. \tag{8.1.13}$$

Then $z(1) > 1$ also. Claim that $z(1) < a$. To show this we have

$$\begin{aligned}
\frac{z(1)}{a} &= \frac{1}{a[1 - p(1) z(1-k) \ldots z(0)]} \\
&\leq \frac{k}{(k+1)\left[1 - \frac{k^k}{(k+1)^{k+1}} \cdot \left(\frac{k+1}{k}\right)^k\right]} \\
&< 1.
\end{aligned}$$

Hence by induction, we may show that $1 < z(n) < a$, with $n = 1, 2, 3, \ldots$. Moreover, $z(n)$ is a solution of Eq. (8.1.11). Now let $x(1) = 1$, $x(2) = x(1)/z(1)$, $x(3) = x(2)/z(2)$, and so on. Then $x(n)$ is a nonoscillatory solution of Eq. (8.1.1).

For the special case when $p(n) = p$ is a constant real number we have the following stronger result.

Theorem 8.7. Consider the equation

$$x(n+1) - x(n) + p \, x(n-k) = 0, \tag{8.1.14}$$

where k is a positive integer and p is a nonnegative real number. Then every solution of Eq. (8.1.14) oscillates if and only if $p > k^k/(k+1)^{k+1}$.

Proof Combining the results of Corollary 8.4, Example 8.5, and Theorem 8.6 yields the proof.

Remark 8.8. Gyori and Ladas [2] showed that every solution of the kth order equation

$$x(n+k) + p_1 x(n+k-1) + \cdots + p_k x(n) = 0 \tag{8.1.15}$$

oscillates if and only if its characteristic equation has no positive roots. Based on this theorem (Exercise 8.1, Problem 8), they were able to show that every solution of the three-term Eq. (8.1.14), where $k \in Z - \{-1, 0\}$, oscillates if and only if $p > k^k/(k+1)^{k+1}$.

Exercise 8.1.

1. Find limit superior and limit inferior of the following sequences:

    (a) S_1: 2/3, 1/3, 3/4, 1/4, 4/5, 1/5, 5/6, 1/6, ...

    (b) S_2: $(-1)^{n+1}$.

    (c) S_3: $\alpha n/(1 + \beta n)$.

    (d) S_4: $1 + (-1)^{n+1}$.

2. Prove the following statements:

    (a) $\lim_{n\to\infty} \sup (1/a(n)) = 1/\lim_{n\to\infty} \inf\ a(n)$,

    (b) If $a(n) > 0$, then $\lim_{n\to\infty} \sup (-a(n)) = -\lim_{n\to\infty} \inf\ a(n)$,

    (c) $\lim_{n\to\infty} \inf\ a(n) \le \lim_{n\to\infty} \sup\ a(n)$.

3. Show that every solution of the equation

$$x(n + 1) - x(n) + p\, x(n) = 0$$

   oscillates if and only if $p \ge 1$, where $p \in R$.

4. Show that every solution of the equation

$$x(n + 1) - x(n) + p\, x(n + 1) = 0$$

   oscillates if and only if $p \le -1$.

5. Consider the difference equation

$$\Delta^2 x(n) + p(n)\, x(n + 1) = 0$$

   where $p(n) > a > 0$ for $n \in Z^+$. Show that every nontrivial solution of the equation is oscillatory. (Hint: Use Theorem 8.3.)

6. Prove part (ii) of Theorem 8.3.

7. The characteristic equation of Eq. (8.1.14) is given by

$$\lambda^{k+1} - \lambda^k + p = 0, \text{ where } p \ge 0.$$

   Show that the characteristic equation has no positive roots if and only if $p > (k^k)/(k + 1)^{k+1}$. (Hint: Consider the function $f(\lambda) = \lambda^{k+1} - \lambda^k + p$ and show that it attains its minimum when $\lambda = (k - 1)/k$.)

8. Use the result in Problem 7 to prove Theorem 8.7.

9.* [1] Assume that

$$\lim_{n \to \infty} \inf p(n) = q > 0,$$

and

$$\lim_{n \to \infty} \sup p(n) > 1 - q.$$

Prove that all conclusions of Theorem 8.3 hold.

In Problems 10 through 12 consider the equation with several delays

$$x(n + 1) - x(n) + \sum_{j=1}^{m} p_j(n) x(n - k_j) = 0 \qquad (8.1.16)$$

where k_j are positive integers.

10. Suppose that $p_i(n) \geq 0$ and

$$\sum_{i=1}^{m} \lim_{n \to \infty} \inf p_i(n) \left[\frac{(k_i + 1)^{k_i+1}}{(k_i)^{k_i}} \right] > 1.$$

Show that every solution of Eq. (8.1.16) oscillates.

11. Suppose that $p_i(n) \geq 0$ and

$$\lim_{n \to \infty} \inf \left(\sum_{j=1}^{m} p_i(n) \right) > \frac{(\bar{k})^{\bar{k}}}{(\bar{k} + 1)^{\bar{k}+1}},$$

where $\bar{k} = \min \{k_1, k_2, \ldots, k_m\} \geq 1$. Show that every solution of Eq. (8.1.16) oscillates.

12.* Suppose that

$$\lim_{n \to \infty} \inf \sum_{i=1}^{m} p_i(n) = c > 0$$

and

$$\lim_{n \to \infty} \sup \sum_{i=1}^{m} p_i(n) = 1 - c.$$

Prove that every solution of Eq. (8.1.16) oscillates.

8.2   Nonlinear Difference Equations

In this section we will investigate the oscillatory behavior of the nonlinear difference equation

$$x(n + 1) - x(n) + p(n) f(x(n - k)) = 0, \qquad (8.2.1)$$

where $k \in Z^+$ and $N \in Z^+$. The first theorem is due to Erbe and Zhang [1].

Theorem 8.9. Suppose that f is continuous on R and satisfies the following assumptions:

(i) $xf(x) > 0$, $x \neq 0$,

(ii) $\lim_{x\to 0} \inf \frac{f(x)}{x} = L$, $0 < L < \infty$,

(iii) $p L > \frac{k^k}{(k+1)^{k+1}}$ if $k \geq 1$ and $p L > 1$ if $k = 0$, where $p = \lim_{n\to\infty} \inf p(n) > 0$.

Then every solution of Eq. (8.2.1) oscillates.

Proof Assume the contrary and let $x(n)$ be a nonoscillatory solution of Eq. (8.2.1). Suppose that $x(n) > 0$ for $n \geq N$. This implies by Assumption (i) that $f(x(n)) > 0$. Hence $x(n + 1) - x(n) = -p(n) f(x(n - k)) < 0$ and thus $x(n)$ is decreasing. Hence $\lim_{n\to\infty} x(n) = c \geq 0$.

Taking the limit of both sides of Eq. (8.2.1) yields $f(c) = 0$ which by Assumption (i) gives $c = 0$. Hence $\lim_{n\to\infty} x(n) = 0$. Dividing Eq. (8.2.1) by $x(n)$ and letting $z(n) = x(n)/x(n + 1) \geq 1$ yields

$$1/z(n) = 1 - p(n) z(n - 1), \quad \dots, \quad z(n - k) \frac{f(x(n - k))}{x(n - k)}. \qquad (8.2.2)$$

Let $\lim_{n\to\infty} \inf z(n) = r$. By taking limit superior in Eq. (8.2.2) we obtain

$$1/r \leq 1 - pLr^k$$

or

$$pL \leq \frac{r - 1}{r^{k+1}}. \qquad (8.2.3)$$

It is easy to see that the function $h(r) = (r - 1)/r^{k+1}$ attains its maximum at $r = (k + 1)/k$ and its maximum value is thus $k^k/(k + 1)^{k+1}$. Hence Inequality (8.2.3) becomes

$$pL \leq \frac{k^k}{(k + 1)^{k+1}}$$

which contradicts Assumption (iii).

Remark If we let $\lim_{n\to\infty} \inf f(x)/x = 1$, then the linearized equation associated with Eq. (8.2.1), where $p(n)$ is equal to a constant real number p, is given by

$$y(n + 1) - y(n) + p\, y(n - k) = 0 \qquad (8.2.4)$$

which has been studied in Section 8.1. We may now rephrase Theorem 8.9 as follows:

Suppose that Assumptions (i) and (ii) hold with $L = 1$ and $p(n)$ is constant. If every solution of Eq. (8.2.4) oscillates then so does every solution of Eq. (8.2.1).

Gyori and Ladas [2] considered the more general equation with several delays

$$x(n + 1) - x(n) + \sum_{i=1}^{m} p_i\, f_i(x(n - k_i)) = 0, \qquad (8.2.5)$$

where $p_i > 0$, k_i is a positive integer, and f_i is a continuous function on R, with $1 \leq i \leq m$. They have obtained the following result.

Theorem 8.10. Suppose that the following hold:

(i) $x f_i(x) > 0$, for $x \neq 0$, $1 \leq i \leq m$,

(ii) $\lim_{x \to 0} \inf \frac{f_i(x)}{x} \geq 1$, $1 \leq i \leq m$,

(iii) $\sum_{i=1}^{m} p_i \frac{(k_i+1)^{k_i+1}}{k_i^{k_i}} > 1$.

Then every solution of Eq. (8.2.5) oscillates.

Proof (See Ref. 2, Theorem 7.4.1.)
Actually under additional conditions it was proved in [Ref. 2, Corollary 7.4.1] that every solution of the nonlinear equation oscillates if and only if every solution of the corresponding linearized equation oscillates.

We now apply the obtained results to study the oscillatory behavior of the Pielou logistic delay equation. The stability of this equation has been determined previously in Example 4.

Example 8.11. [2] Consider the Pielou logistic delay equation

$$y(n+1) + \frac{\alpha\, y(n)}{1 + \beta\, y(n-k)}, \quad \alpha > 1, \beta > 0, \quad k \text{ a positive integer} \quad (8.2.6)$$

Show that every positive solution of Eq. (8.2.6) oscillates about its positive equilibrium point $y^* = (\alpha - 1)/\beta$ if

$$\frac{\alpha - 1}{\alpha} > \frac{k^k}{(k+1)^{k+1}}. \quad (8.2.7)$$

Solution We follow Method 2 in Example 4.25 by letting $y(n) = ((\alpha - 1)/\beta)e^{x(n)}$ in Eq. (8.2.6). We obtain the equation

$$x(n+1) - x(n) + \frac{\alpha - 1}{\alpha} f(x(n-k)) = 0 \quad (8.2.8)$$

where

$$f(x) = \frac{\alpha}{\alpha - 1} \ln \left(\frac{(\alpha - 1)e^x + 1}{\alpha} \right).$$

It may be shown that the function f satisfies Conditions (i) and (ii) in Theorem 8.9 with $L = 1$. Hence by Theorem 8.9 every solution of Eq. (8.2.8) oscillates about 0. This implies that every solution of Eq. (8.2.6) oscillates about the equilibrium point $y^* = (\alpha - 1)/\beta$.

Exercise 8.2.

1. Consider the difference equation

$$\Delta x(n) + e^{x(n-1)} - 1 = 0.$$

   Determine the oscillatory behavior of all solutions.

2. Consider the difference equation

$$x(n+1) = x(n) \, \mathrm{Exp}\left[r\left(1 - \frac{x(n)}{\alpha} \right) \right], \quad r > 0, \quad \alpha > 0, \quad x(0) > 0.$$

   (a) Show that $x^* = \alpha$ is the only positive equilibrium point.

   (b) Show that every solution oscillates about α if $r > 1$.

   (c) Show that if $r = 1$, every solution converges monotonically to α.

3. Consider the difference equation

$$x(n+1) = x(n) \, \mathrm{Exp}\left[r\left(1 - \frac{x(n-1)}{\alpha} \right) \right], \quad r > 0, \quad \alpha > 0, \quad x(0) > 0.$$

   Show that every solution oscillates about $x^* = \alpha$ if $r > 1/4$. [Hint: Let $x(n) = \alpha \, e^{y(n)}$ and then use Theorem 8.9.]

4. Consider the difference equation

$$x(n+1) = x(n) \, \mathrm{Exp}\left[r\left(1 - \frac{x(n-1)}{\alpha} - \frac{x(n-2)}{\beta} \right) \right],$$

$$r > 0, \quad \alpha > 0, \quad \beta > 0, \quad x(0) > 0.$$

   Show that every solution oscillates about $x^* = (\alpha\beta)/\alpha + \beta$ if $r > (4(\alpha + \beta))/9\alpha + 16\beta$. [Hint: Let $x(n) = \alpha \, e^{y(n)}$ and then apply Theorem 8.10.]

5. Consider the difference equation

$$\Delta \, x(n) + p(1 + x(n)) \, x(n) = 0, \quad p > 0, \quad 1 + x(n) > 0.$$

   Show that every solution oscillates if $p > 1$.

6. Consider the difference equation

$$\Delta \, x(n) + p(1 + x(n)) \, x(n-1) = 0, \quad p > 0, \quad 1 + x(n) > 0.$$

   Show that every solution oscillates if $p > 1/4$. [Hint: Let $z(n) = x(n)/x(n+1)$, then mimic the proof of Theorem 8.9.]

7. [2] Consider the difference equation

$$\Delta\, x(n) + p(n)[1 + x(n)]\, x(n - k) = 0, \qquad p(n) > 0$$

for $n \geq 1$, $x(n) + 1 > 0$ for $n \geq -k$. Prove that every solution oscillates if $\lim_{n \to \infty} \inf p(n) = c > k^k/((k + 1)^{k+1})$. (Hint: Follow the hint in Problem 6.)

8. Consider the difference equation

$$x(n + 1) = \frac{\alpha\, x(n)}{1 + \beta\, x(n - k) + \gamma\, x(n - 1)}$$

with $\alpha > 1$, $\beta > 0$, $\gamma > 0$, k, $1 \in Z^+$. Find conditions under which all solutions oscillate.

8.3   Self-Adjoint Second Order Equations

In this section we consider second order difference equations of the form

$$\Delta[p(n - 1)\Delta x(n - 1)] + q(n)\, x(n) = 0 \tag{8.3.1}$$

where $p(n) > 0$, $n \in Z^+$. Equation (8.3.1) is called *self-adjoint*, a name borrowed from its continuous analogue

$$[p(t)x'(t)]' + q(t)\, x(t) = 0.$$

Equation (8.3.1) may be written in the more familiar form

$$p(n)x(n + 1) + p(n - 1)x(n - 1) = b(n)x(n) \tag{8.3.2}$$

where

$$b(n) = p(n - 1) + p(n) - q(n). \tag{8.3.3}$$

As a matter of fact any equation of the form

$$p_0(n)x(n + 1) + p_1(n)\, x(n) + p_2(n)\, x(n - 1) = 0 \tag{8.3.4}$$

with $p_0(n) > 0$ and $p_2(n) > 0$ can be written in the self-adjoint form (8.3.1) or (8.3.2). To find $p(n)$ and $q(n)$ from $p_0(n)$, $p_1(n)$ and $p_2(n)$, multiply both sides of Eq. (8.3.4) by a positive sequence $h(n)$. This yields

$$p_0(n)\, h(n)\, x(n + 1) + p_1(n)\, h(n)\, x(n) + p_2(n)\, h(n)\, x(n - 1) = 0. \tag{8.3.5}$$

Comparing Eq. (8.3.5) with Eq. (8.3.2) we obtain

$$\begin{aligned} p(n) &= p_0(n)\, h(n), \\ p(n - 1) &= p_2(n)\, h(n). \end{aligned}$$

Thus

$$p_2(n + 1) h(n + 1) = p_0(n) h(n)$$

or

$$h(n + 1) = \frac{p_0(n)}{p_2(n + 1)} h(n). \tag{8.3.6}$$

Hence

$$h(n) = \prod_{j=n_0}^{n-1} \frac{p_0(j)}{p_2(j + 1)}$$

is a solution of Eq. (8.3.6). This gives us

$$p(n) = p_0(n) \prod_{j=n_0}^{n-1} \frac{p_0(j)}{p_2(j + 1)}.$$

Also from Eq. (8.3.3) we obtain

$$q(n) = p_1(n) h(n) + p(n) + p(n - 1).$$

In [5] Hartman introduced the notion of *generalized* zeros in order to obtain a discrete analogue of Sturm's Separation Theorem in differential equations. Next we give Hartman's definition.

Definition 8.12. A solution $x(n), n \geq n_0 \geq 0$, of Eq. (8.3.1) has a generalized zero at $r > n_0$ if either $x(r) = 0$ or $x(r - 1)x(r) < 0$.

In other words, a generalized zero of a solution is either an actual zero or where the solution changes its sign.

Theorem 8.13. (Sturm Separation Theorem) Let $x_1(n)$ and $x_2(n)$ be two linearly independent solutions of Eq. (8.3.1). Then the following statements hold:

(i) $x_1(n)$ and $x_2(n)$ cannot have a common zero, that is if $x_1(r) = 0$, then $x_2(r) \neq 0$.

(ii) If $x_1(n)$ has a zero at n_1 and a generalized zero at $n_2 > n_1$, then $x_2(n)$ must have a generalized zero in $(n_1, n_2]$.

(iii) If $x_1(n)$ has generalized zeros at n_1 and $n_2 > n_1$, then $x_2(n)$ must have a generalized zero in $[n_1, n_2]$.

Proof

(i) Assume that $x_1(r) = x_2(r) = 0$. Then the Casoratian

$$C(r) = \begin{vmatrix} x_1(r) & x_2(r) \\ x_1(r + 1) & x_2(r + 1) \end{vmatrix} = 0.$$

It follows from Theorem 2.13 that $x_1(n)$ and $x_2(n)$ are linearly dependent, a contradiction.

(ii) Assume that $x_1(n_1) = 0, x_1(n_2 - 1) x(n_2) < 0$ (or $x_1(n_2) = 0$). We may assume that n_2 is the first generalized zero greater than n_1. Suppose that $x_1(n) > 0$ for $n_1 < n < n_2$ and $x_1(n_2) \leq 0$.

Now if $x_2(n)$ has no generalized zeros in $(n_1, n_2]$, then $x_2(n)$ is either positive in $[n_1, n_2]$ or negative in $[n_1, n_2]$. Without loss of generality let $x_2(n) > 0$ on $[n_1, n_2]$. Now pick a positive real number M and $r \in (n_1, n_2)$ such that $x_2(r) = Mx_1(r)$ and $x_2(n) \geq Mx_1(n)$ in $[n_1, n_2]$. By the principle of superposition, the sequence $x(n) = x_2(n) - Mx_1(n)$ is also a solution of Eq. (8.3.1) with $x(r) = 0$ and $x(r - 1) x(r + 1) \geq 0$, with $r > n_1$. Letting $n = r$ in Eq. (8.3.1) we obtain

$$\Delta[p(r - 1)\Delta x(r - 1)] + q(r) x(r) = 0.$$

Since $x(r) = 0$, we have

$$p(r - 1)\Delta^2 x(r - 1) + \Delta x(r)\Delta p(r - 1) = 0$$

or

$$p(r) x(r + 1) = -p(r - 1) x(r - 1). \qquad (8.3.7)$$

Since $x(r + 1) \neq 0, x(r - 1) \neq 0$ and $p(n) > 0$, Eq. (8.3.7) implies that $x(r - 1) x(r + 1) < 0$ which is a contradiction. This completes the proof of part (ii). The proof of part (iii) is left to the reader as Exercise 8.3, Problem 6.

Remark 8.14 Based on the notion of generalized zeros, we can give an alternate definition of oscillation. A solution of a difference equation is oscillatory on $[n_2, \infty)$ if it has infinitely many generalized zeros on $[n_0, \infty)$. An immediate consequence of the Sturm Separation Theorem (Theorem 8.13) is that if Eq. (8.3.1) has an oscillatory solution, then all its solutions are oscillatory. We caution the reader that the above conclusion does not hold in general for nonself-adjoint second order difference equations.

For example, the difference equation $x(n+1) - x(n-1) = 0$ has a nonoscillatory solution $x_1(n) = 1$ and an oscillatory solution $x_2(n) = (-1)^n$. Observe that this equation is not self-adjoint.

We are now ready to give some simple criteria for oscillation.

Lemma 8.15. If there exists a subsequence $b(n_k) \leq 0$, with $n_k \to \infty$ as $k \to \infty$, then every solution of Eq. (8.3.2) oscillates.

Proof Assume the contrary, that there exists a nonoscillatory solution $x(n)$ of Eq. (8.3.2). Without loss of generality, suppose that $x(n) > 0$ for $n \geq N$. Then

$$p(n_k) x(n_k + 1) + p(n_k - 1) x(n_k - 1) - b(n_k) x(n_k) > 0, \qquad \text{for } n_k > N,$$

which is a contradiction.

One of the most useful techniques in oscillation theory is the use of the so-called *Ricatti transformations*. We will introduce only one transformation that is needed in the development of our results. Two other transformations will appear in the exercises. In Eq. (8.3.2), let

$$z(n) = b(n + 1) \, x(n + 1) / (p(n) \, x(n)). \tag{8.3.8}$$

Then $z(n)$ satisfies the equation

$$c(n) \, z(n) + \frac{1}{z(n - 1)} = 1 \tag{8.3.9}$$

where

$$c(n) = \frac{p^2(n)}{b(n) \, b(n + 1)}. \tag{8.3.10}$$

Next we give a crucial result that relates Eq. (8.3.2) with Eq. (8.3.9).

Lemma 8.16. Suppose that $b(n) > 0$ for $n \in Z^+$. Then every solution $x(n)$ of Eq. (8.3.2) is nonoscillatory if and only if every solution $z(n)$ of Eq. (8.3.9) is positive for $n \geq N$, for some $N > 0$.

Proof Suppose that $x(n)$ is a nonoscillatory solution of Eq. (8.3.2). Then $x(n) \, x(n + 1) > 0$ for $n \geq N$. Equation (8.3.8) then implies that $z(n) > 0$. Conversely, assume that $z(n)$ is a positive solution of Eq. (8.3.9). Using this solution we construct inductively a nonoscillatory solution $x(n)$ of Eq. (8.3.2) as follows: Let $x(N) = 1$, $x(n + 1) = (p(n))/b(n + 1) \, z(n) \, x(n)$, with $n > N$. Then one may verify that $x(n)$, with $n \geq N$ is indeed a solution of Eq. (8.3.2) which is nonoscillatory. By Sturm Separation Theorem, every solution of Eq. (8.3.2) is thus nonoscillatory.

We need a comparison result concerning Eq. (8.3.9) that will be needed to establish the main result of this section.

Lemma 8.17. If $c(n) \geq a(n) > 0$ for all $n > 0$ and $z(n) > 0$ is a solution of the equation

$$c(n) \, z(n) + \frac{1}{z(n - 1)} = 1, \tag{8.3.11}$$

then the equation

$$a(n) \, y(n) + \frac{1}{y(n - 1)} = 1 \tag{8.3.12}$$

has a solution $y(n) \geq z(n) > 1$ for all $n \in Z^+$.

Proof Since $c(n) > 0$ and $z(n) > 0$, it follows from Eq. (8.3.1) that $1/(z(n - 1)) < 1$. This implies that $z(n - 1) > 1$ for all $n \geq 1$. We now define inductively a solution $y(n)$ of Eq. (8.3.12). Choose $y(0) \geq z(0)$ and let $y(n)$ satisfy Eq. (8.3.11). Now from Eqs. (8.3.12) and (8.3.12) we have

$$a(n) \, y(n) + \frac{1}{y(n - 1)} = c(n) \, z(n) + \frac{1}{z(n - 1)}.$$

So

$$a(1) \, y(1) + \frac{1}{y(0)} = c(1) \, z(1) + \frac{1}{z(0)}.$$

Since $y(0) \geq z(0)$, $1/y(0) \leq 1/z(0)$ and hence $a(1) \, y(1) \geq c(1) \, z(1)$ or

$$y(1) \geq \frac{c(1)}{a(1)} \, z(1) \geq z(1) > 1.$$

Inductively, one may show that

$$y(n) \geq z(n) > 1.$$

Theorem 8.18. If $b(n) \, b(n+1) \leq (4-\varepsilon) \, p^2(n)$ for some $\varepsilon > 0$ and for all $n \geq N$, then every solution of Eq. (8.3.2) is oscillatory.

Proof If $b(n) \, b(n-1) \leq (4-\varepsilon) \, p^2(n)$ for some $\varepsilon \geq 4$, then $b(n) \, b(n-1) \leq 0$. The conclusion of the theorem then follows from Lemma 8.15. Hence we may assume that $0 < \varepsilon < 4$. Now assume that Eq. (8.3.2) has a nonoscillatory solution. Then by Lemma 8.16, Eq. (8.3.9) has a positive solution $z(n)$ for $n \geq N$. Using the assumption of the theorem in Formula (8.3.10) yields

$$c(n) = \frac{p^2(n)}{b(n) \, b(n+1)} \geq \frac{p^2(n)}{(4-\varepsilon) \, p^2(n)} = \frac{1}{4-\varepsilon}.$$

Then it follows from Lemma 8.17 that the equation

$$\frac{1}{4-\varepsilon} \, y(n) + \frac{1}{y(n-1)} = 1 \qquad\qquad (8.3.13)$$

has a solution $y(n)$, $n \geq N$, such that $y(n) \geq z(n) > 1$ for all $n \geq N$. Define a positive sequence $x(n)$ inductively as follows: $x(N) = 1$, $x(n+1) = 1/\sqrt{4-\varepsilon} \, y(n) \, x(n)$ for $n \geq N$. Then

$$y(n) = \sqrt{4-\varepsilon} \left(\frac{x(n+1)}{x(n)} \right). \qquad\qquad (8.3.14)$$

Substituting for $y(n)$ for Eq. (8.3.14) into Eq. (8.3.13) yields $x(n+1) - \sqrt{4-\varepsilon} \, x(n) = x(n-1) = 0$, $n \geq N$, whose characteristic roots are

$$\lambda_{1,2} = \frac{\sqrt{4-\varepsilon}}{2} \pm i \, \frac{\sqrt{\varepsilon}}{2}.$$

Thus its solutions are oscillatory which gives a contradiction. The proof of the theorem is now complete.

It is now time to give some examples.

Example 8.19. Consider the difference equation

$$y(n+1) + y(n-1) = \left(2 + \frac{1}{2} (-1)^n \right) y(n).$$

Here $p(n) = 1$ and $b(n) = \left(2 + \frac{1}{2}\,(-1)^n\right)$. Now

$$
\begin{aligned}
b(n)\, b(n+1) &= \left(2 + \frac{1}{2}\,(-1)^n\right)\left(2 + \frac{1}{2}\,(-1)^{n+1}\right) \\
&= 3\frac{3}{4}.
\end{aligned}
$$

Thus $b(n)\, b(n+1) \le \left(4 - \frac{1}{5}\right) p^2(n)$. By Theorem 8.18, we conclude that every solution is oscillatory.

The following example will show the sharpness of Theorem 8.18 in the sense that if ε is allowed to be a sequence tending to zero, then the theorem fails.

Example 8.20. [3] Consider the equation

$$
x(n+1) + x(n) = b(n)\, x(n), \quad n = 1, 2, 3, \ldots
$$

where

$$
b(n) = \frac{\sqrt{n+1} + \sqrt{n-1}}{\sqrt{n}}.
$$

Now

$$
b(n)\, b(n+1) =
$$

$$
\frac{\sqrt{(n+1)(n+2)} + \sqrt{(n-1)(n+2)} + \sqrt{n(n+1)} + \sqrt{n(n-1)}}{\sqrt{n(n+1)}}
$$

$$
< 4.
$$

But $\lim_{n\to\infty} b(n)\, b(n+1) = 4$. Hence if one takes $\varepsilon_n = 4 - b(n)\, b(n+1)$, then $\varepsilon_n \to 0$ as $n \to \infty$. However, Theorem 8.18 fails since $x(n) = \sqrt{n}, n \ge 1$ is a nonoscillatory solution of the equation.

A partial converse of Theorem 8.18 now follows.

Theorem 8.21. If $b(n)\, b(n+1) \ge 4\, p^2(n)$ for $n \ge N$, then every solution of Eq. (8.3.2) is nonoscillatory.

Proof From Formula (8.3.10) and the assumption we obtain $c(n) \le \frac{1}{4}$. We now construct inductively a solution $z(n)$ of Eq. (8.3.9) as follows: Put $z(N) = 2$, and

$$
z(n) = \frac{1}{c(n)}\left(1 - \frac{1}{z(n-1)}\right), \quad n > N
$$

observe that

$$
z(N+1) = \frac{1}{c(N+1)}\left(1 - \frac{1}{z(N)}\right) \ge 4\left(1 - \frac{1}{2}\right) = 2.
$$

Similarly one may show that $z(n) \ge 2$ for $n \ge N$. Hence by Lemma 8.16, we conclude that every solution of Eq. (8.3.2) is nonoscillatory.

Example 8.22. Consider the difference equation

$$\Delta(n \,\Delta x(n-1) + \frac{1}{n}\, x(n) = 0.$$

Here $p(n) = n + 1$ and $q(n) = \frac{1}{n}$. Using Formula (8.3.3) we obtain

$$b(n) = 2n + 1 - \frac{1}{n}.$$

Now

$$
\begin{aligned}
b(n)\, b(n+1) \;&=\; \left(2n + 1 - \frac{1}{n}\right)\left(2n + 3 - \frac{1}{(n+1)}\right) \\
&=\; 4n^2 + 8n + 3 - \frac{2n+1}{(n+1)} - \frac{2n+3}{n} + \frac{1}{n(n+1)} \\
&\geq\; 4n^2 \text{ for all } n \geq 1.
\end{aligned}
$$

Hence by Theorem 8.21, every solution is nonoscillatory.

Exercise 8.3.

In Problems 1 through 5 determine the oscillatory behavior of all solutions.

1. $\Delta[(n-1)\, x(n-1)] + \frac{1}{n} x(n) = 0.$

2. $x(n+1) + x(n-1) = \left(2 - \frac{1}{n}\right) x(n).$

3. $x(n+1) + x(n-1) = \left(2 + \frac{1}{n}\right) x(n).$

4. $\Delta^2[x(n-1)] + \frac{1}{n \ln (n)}\, x(n) = 0, \quad n > 1.$

5. $\Delta[(n-1)\, x(n-1)] + x(n) = 0.$

6. Prove part (iii) of Theorem 8.13.

7.[7] Show that if $b(n) \leq \min \{p(n), p(n-1)\}$ for $n \geq N$, for some positive integer N, then every solution of Eq. (8.3.2) is oscillatory

8. Show that if $b(n) \leq p(n)$ and $p(n)$ is eventually nonincreasing, then every solution of Eq. (8.3.2) is oscillatory. (Hint: Use Problem 7.)

9. Show that if $b(n) \leq p(n-1)$ and $p(n)$ is eventually nondecreasing, then every solution of Eq. (8.3.2) is oscillatory. (Hint: Use Problem 7.)

10. (A second Riccati transformation). Let $z(n) = x(n+1)/x(n)$ in Eq. (8.3.2).

   (i) Show that $z(n)$ satisfies the equation

$$p(n)\, z(n) + \frac{p(n-1)}{z(n-1)} = b(n). \qquad (8.3.15)$$

(ii) Assuming $p(n) > 0$ show that every solution of Eq. (8.3.2) is nonoscillatory if and only if Eq. (8.3.15) has a positive solution $z(n), n \geq N$, for some $N > 0$.

11.* Use the second Riccati transformation in Problem 10 to show that if $b(n) \leq p(n-1)$ and $\lim_{n \to \infty} \sup (p(n))/p(n-1) > \frac{1}{2}$, then every solution of Eq. (8.3.2) oscillates.

12.³ Show that if $b(n) \geq \max \{p(n-1), 4p(n)\}$, for all $n \geq N$, for some $N > 0$, then every solution of Eq. (8.3.2) is nonoscillatory. (Hint: Use Theorem 8.21.)

13. Show that if $p(n_k) \geq b(n_k) b(n_k + 1)$ for a sequence $n_k \to \infty$, then every solution of Eq. (8.3.2) is oscillatory.

14. As in Formula (8.3.10), let

$$c(n) = \frac{p^2(n)}{b(n) \, b(n+1)}, \quad n \geq 0.$$

Show that either one of the following implies that every solution of Eq. (8.3.2) oscillates.

(i) $\lim_{n \to \infty} \sup c(n) > 1$.

(ii) $\lim_{n \to \infty} \sup \frac{1}{n} \sum_{j=1}^{n} c(j) > 1$.

(Hint: Use Problem 13.)

References

[1] L.H. Erbe and B.G. Zhang, *Oscillation of Discrete Analogues of Delay Equations, Differential and Integral Equations, vol. 2*, 1989, pp. 300–309.

[2] I. Gyori and G. Ladas, *Oscillation Theory of Delay Differential Equations with Applications*, Claredon, Oxford 1991.

[3] J.W. Hooker and W.T. Patula, "Riccati Type Transformation for Second Order Linear Difference Equations," *J. Math. Anal. Appl.* **82** (1981), 451–462.

[4] R.P. Agarwal, *Difference Equations and Inequalities, Theory, Methods and Applications*, Marcel Dekker, New York, 1992.

[5] P. Harman, "Difference Equations: Disconjugacy, Principal Solutions, Green's Functions, Complete Monotonicity," *Trans. Amer. Math. Soc.* **246** (1978), 1–30.

[6] W.G. Kelley and A.C. Peterson, *Difference Equations, An Introduction with Applications*, Academic, New York 1991.

[7] W.T. Patula, "Growth and Oscillation Properties of Second Order Linear Difference Equations," *Siam J. Math. Anal.* **19** (1979), 55–61.

Answers to Selected Problems

Exercises 1.1 and 1.2

1.   (a) $cn!$

      (b) $c3^{\frac{n(n-1)}{2}}$

      (c) $ce^{n(n-1)}$

      (c) $\dfrac{c}{n}$

2.   (a) $(c-4)\,2^{-n}+4$

      (b) $cn^{-1}+2n^{-1}\,(n-1)(n+2)$

3.   (a) $n!\,(2^{n}+c-1)$

      (b) $c+\dfrac{e^{n}-1}{e-1}$

6. $x(n+1)=x(n)+\frac{1}{2}\,n(n+1)+1$

9.   38 payments + final payment $52.29

11.   (a) $A(n+1)=(1+r)\,A(n)+T$

       (b) $25,000\,[(1.008)^{n}-1]$

13. $136,283.

15.   (a) $r=1-(0.5)^{1/5700}$

       (b) (b) 2,933 years

Exercise 1.3

3.   (a) $\dfrac{\alpha - 1}{\beta}$

5.   (b) $\mu = 3.3$

7.   (i) $D(n) = -p(n) + 15$
           $S(n + 1) = 2p(n) + 3$

      (iii) $p* = 4$, unstable

11.   (a) $p(n + 1) = -\dfrac{1}{2}p^2(n) + 1$

      (b) $p* = -1 + \sqrt{3}$

      (c) asymptotically stable

17. $y^2(t) = 4 - 3e^{-1/2t}$

Exercise 1.4

1. $\begin{cases} 0: & \text{asymptotically stable} \\ \pm 1: & \text{unstable} \end{cases}$

3. 0: asymptotically stable

5. 0: unstable

7. 0: unstable

17.   (a) from the left

      (b) from the right.

Exercise 1.5

5. $\{0, 1\}$: asymptotically stable

7. $|b^2 - 3ab| < 1$

9. $\left\{ \dfrac{1}{3}, \dfrac{2}{3} \right\}$

11.   (b) unstable

13. $f(x) = -x$

Exercise 2.1

13.
$$\begin{cases} x^3 = x^{(1)} + 3x^{(2)} + x^{(3)} \\ x^4 = x^{(1)} + 7x^{(2)} + 6x^{(3)} + x^{(4)} \\ x^5 = x^{(1)} + 15x^{(2)} + 25x^{(3)} + 10x^{(4)} + x^{(5)} \end{cases}$$

15. $\dfrac{1}{2} n(n-1) + n(n-1)(n-2) + \dfrac{1}{4} n(n-1)(n-2)(n-3) + c$

Exercise 2.2

1. (a) 0, linearly dependent

   (b) $2(5^{3n+3})$, linearly independent

   (c) $(-1)^{n+1} (27) 4^n$, linearly independent

   (d) 0, linearly dependent

3. (a) $c_1 + c_2 n + c_3 n^3$

   (b) $c_1 \cos\left(\dfrac{n\pi}{2}\right) + c_2 \sin\left(\dfrac{n\pi}{2}\right)$

   (c) linearly dependent

   (d) need one more solution.

15. $\dfrac{2^n}{n!} \displaystyle\sum_{r=0}^{n-1} \dfrac{r!}{2^{r+1}}$

16. $(n+1) \displaystyle\sum_{r=0}^{n-1} \dfrac{(-1)^r}{(r+2)!}$

Exercise 2.3

1. (a) $x(n+2) - 7x(n+1) + 10x(n) = 0$

   (c) $x(n+2) - 5\sqrt{2}\, x(n+1) + 25x(n) = 0$

   (e) $x(n+4) - 4x(n+3) + 6x(n+2) - 4x(n+1) + x(n) = 0$

2. $M(n+2) = 2M(n+1) + M(m)$

3. $c_1 4^n + c_2 (-4)^n$

5. $c_1 3^n + c_2 n\, 3^n + c_3 2^n \cos\dfrac{n\pi}{2} + c_4 2^n \sin\dfrac{n\pi}{2}$

7. $c_1 2^{n/2} \cos\dfrac{n\pi}{2} + c_2 2^{n/2} \sin\dfrac{n\pi}{2} + c_4 2^{n/2} n \cos\dfrac{n\pi}{2} + c_5 2^{n/2} n \sin\dfrac{n\pi}{2}$

11. (d) $\begin{array}{ll} T_n: & 1, x, 2x^2 - 1 \\ U_n: & 1, 2x, 4x^2 - 1 \end{array}$

23. $a^n \dfrac{\sin(n+1)\theta}{\sin\theta}$, $\theta = \cos^{-1}\left(\dfrac{b}{2a}\right)$.

Exercise 2.4

1. $\dfrac{1}{2} n + 5/4$

2. $e^n/(e^n + 8e + 12)$

3. $\dfrac{1}{12} n\, 4^n + \dfrac{7}{54} n - \dfrac{1}{18} n^2 + \dfrac{1}{9} n^3$

5. $\dfrac{1}{2} \cos\left(\dfrac{n\pi}{2}\right) - \dfrac{1}{2} n \cos\left(\dfrac{n\pi}{2}\right)$

7. $2 - 7N + 8n^2$

9. $\dfrac{6}{25}(3^n) - \dfrac{6}{25}\cos\left(\dfrac{n\pi}{2}\right) + \dfrac{9}{50}\sin\left(\dfrac{n\pi}{2}\right) + \dfrac{1}{30} n\, 3^n$

11. $c_1(-7)^n + c_2(-1)^n + \dfrac{1}{27} n 2^n - \dfrac{8}{243} 2^n$

15. $y(2) = 1$

$y(3 + 4n) = y(4 + 4n) = 0$

$y(5 + 4n) = y(6 + 4n) = -1$

$n = 0, 1, 2, 3, \ldots$

16. (d) $y(n) = \displaystyle\sum_{r=0}^{n-1} \left[\dfrac{y_1(r+1)\, y_2(n) - y_2(r+1)\, y_1(n)}{C(r+1)}\right] g(r)$

17. $y(n) = \dfrac{-n^2}{10} + \dfrac{3n}{50} - \dfrac{1}{125} + a + b(6^n)$

18. $y(n) = \left(\dfrac{2}{3}\right) 3^n - \left(\dfrac{1}{2}\right) 2^n - \left(\dfrac{1}{2}n\right) 2^n$

Exercise 2.5

1. repelling, oscillatory

3. attracting, oscillatory

13. (i) oscillatory,

    (ii) oscillate to ∞,

    (iii) solutions converge to $y* = \alpha c$

Exercise 2.6

1. $c_1(3)^n + c_2(-1)^n$

2. $c_1 2^n + c_2(n-1)!$

3. $1/(c-n)$

4. $\dfrac{5/6 + 2c(-6)^n}{1 + c(-6)^{n+1}}$

5. $\dfrac{3(1 + c(2/3)^{n+1})}{1 + c(2/3)^n}$

7. $x_0\, e^{2^n}$

9. $\sin^2(c\, 2^n)$

11. $e^{c(2^{n-1})}\, y(0)$

13. $1 - \cot(c\, 2^n)$

15. $\sin(c\, 2^n)$

Exercise 2.7

1. (a) $s_1(n+2) - \sigma\gamma\alpha s_1(n+1) - \sigma^2\,\gamma\,\beta(1-\alpha)\,s_1(n) = 0$
   (b) $\gamma \geq 50$

3. (a) $F(n+2) = F(n+1) + F(n)$
   (b) $2, 3, 5$

5. (i) 0.88
   (ii) 0.55
   (iii) 0.51

7. $p(n+3) + 3p(n+2) - 10p(n+1) + 6p(n) = 0$, $p(0) = 1$, $p(N) = p(N+1) = 0$

11. (a) $Y(n+2) - \alpha(1+\beta)\,Y(n+1) + \alpha\beta Y(n) = (1.05)^n$
    (b) $Y* = \dfrac{(1.05)^n}{1 - \alpha}$
    (c) $2^{-n/2}\left[c_1 \cos\dfrac{n\pi}{4}\right] + 0.1025(1.05)^n$

12. (a) $Y(n+2) - c_1\,Y(n+1) - c_2\,Y(n) = nh + K + I(0)$
    (b) $Y(n) = c_1\left(\dfrac{1+\sqrt{5}}{4}\right)^n + c_2\left(\dfrac{1-\sqrt{5}}{4}\right)^n + \alpha + \beta n$

13. (a) $Y(n + 2) - 2\beta Y(n + 1) + \beta Y(n) = V_0$

15. (b) $\dfrac{1}{2}$

17. (c) $u(n) = \dfrac{1}{2} s \alpha \dfrac{(n - 2)}{2} \left[(\alpha - 1) \cos \dfrac{n\pi}{2} + 2\sqrt{\alpha} \sin \dfrac{n\pi}{2} \right]$

$V(n) = \dfrac{1}{2} s \alpha \dfrac{(n - 2)}{2} (\alpha + 1) \cos \dfrac{n\pi}{2}$

Exercise 3.1

1. $\begin{bmatrix} 2^{n+1} - 3^n & 3^n - 2^n \\ 2^{n+1} - 2(3^n) & 2(3^n) - 2^n \end{bmatrix}$

3. $\begin{pmatrix} 2^{n+1} - 3^n & -2 + 2^{n+1} & \dfrac{1}{2} - \dfrac{1}{2} 3^n \\ (-2)^n + 3^n & 2 - 2^n & -\dfrac{1}{2} + \dfrac{1}{2} 3^n \\ -2^{n+2} + 4(3^n) & 4 - 2^{n+2} & -1 + 2(3^n) \end{pmatrix}$

5. $\begin{pmatrix} \dfrac{1}{3}(2^{n+1} + (-1)^n) \\ 2^{n+1} \end{pmatrix}$

7. $\begin{pmatrix} 3 - 2^{n+1} \\ 2(1 - 2^{n-1}) \\ 2(-1 + 2^n) \end{pmatrix}$

13. (i) $\begin{pmatrix} 2^{n+1} - 3^n & 3^n - 2^n \\ 2^{n+1} - 2(3^n) & 2(3^n) - 2^n \end{pmatrix}$

   (ii) same as Problem 3.

15. (a) $\begin{pmatrix} 0 & 1 & 0 \\ 0 & 0 & 1 \\ \dfrac{1}{2} & \dfrac{1}{2} & 0 \end{pmatrix}$

   (b) (2/5, 1/5)

Exercise 3.2

11. $\begin{pmatrix} \dfrac{11}{16} + \dfrac{3}{4} n - \dfrac{11}{16} 5^n \\ \dfrac{-5}{16} - \dfrac{1}{4} n - \dfrac{11}{16} 5^n \end{pmatrix}$

17. $a_1(-2)^n + a_2(-6)^n$

19. $a_1 + a_2 \, 4^n + \frac{1}{12} n \, 4^n$

21. $a_1 \left(\dfrac{1 - \sqrt{5}}{2} \right)^n + a_2 \left(\dfrac{1 + \sqrt{5}}{2} \right)^n$

Exercise 3.3

1. $\begin{pmatrix} 2^{n+1} & -4^n \\ 2 & 4^n \end{pmatrix}$

3. $\begin{pmatrix} \dfrac{3}{7}[(-1)^{n+1} + 6^n] \\ \dfrac{1}{7}[3(-1)^n + 4 \, 6^n] \\ 0 \end{pmatrix}$

5. $c_1 \, 2^n \begin{pmatrix} 0 \\ 1 \\ 0 \end{pmatrix} + c_2 \begin{pmatrix} 1 \\ -1 \\ 0 \end{pmatrix} + c_3 \, 3^n \begin{pmatrix} 1 \\ n \\ 2 \end{pmatrix}$

9. $\begin{pmatrix} 2^{n/2} \left[-c_2 \sin \dfrac{n\pi}{4} + c_3 \cos \dfrac{n\pi}{4} \right] \\ 2^{n/2} \left[-c_2 \cos \dfrac{n\pi}{4} - c_3 \sin \dfrac{n\pi}{4} \right] \\ c_1 + 2^{n/2} \left[c_2 \cos \dfrac{n\pi}{4} + c_3 \sin \dfrac{n\pi}{4} \right] \end{pmatrix}$

11. (a) $\begin{pmatrix} 3^n & n3^{n-1} \\ 0 & 3^n \end{pmatrix}$

(b) $\begin{pmatrix} 2^n & n2^{n-1} & n(n-1)2^{n-3} \\ 0 & 2^n & n2^{n-1} \\ 0 & 0 & 2^n \end{pmatrix}$

(c) $\begin{pmatrix} 2 & 1 & -2 \\ -1 & 0 & 1 \\ 0 & 0 & -\dfrac{2}{3} \end{pmatrix} \begin{pmatrix} 3^n & n3^{n-1} & \dfrac{n(n-1)}{2} 3^{n-2} \\ 0 & 3^n & n3^{n-1} \\ 0 & 0 & 3^n \end{pmatrix}$

$\begin{pmatrix} 0 & -1 & 3/2 \\ 1 & 2 & 0 \\ 0 & 0 & 3/2 \end{pmatrix}$

(d) $\begin{pmatrix} 2^n & 0 & 0 & 0 \\ 0 & 2^n & n2^{n-1} & n(n-1)2^{n-3} \\ 0 & 0 & 2^n & n2^n \\ 0 & 0 & 0 & 2^n \end{pmatrix}$

13. $\begin{pmatrix} c_1 2^n(1-n/2) + c_2 n2^n + c_3(3n^2 2^{n-3} + 9n2^{n-3}) \\ -c_1 n2^{n-2} + c_2 2^n(1-n) - 3c_3 n(n-1)2^{n-4} \\ c_3 2^n \end{pmatrix}$

Exercise 3.5

5. $\begin{pmatrix} 1 \\ 0 \\ 0 \end{pmatrix}$

7. $\begin{pmatrix} 1/2 \\ 1/2 \end{pmatrix}$

9. $\begin{pmatrix} 0 & 1/3 & 0 & 0 & 0 \\ 2/3 & 0 & 1/3 & 0 & 0 \\ 0 & 2/3 & 0 & 0 & 0 \\ 1/3 & 0 & 0 & 1 & 0 \\ 0 & 0 & 2/3 & 0 & 1 \end{pmatrix} \begin{pmatrix} 1/5 \\ 4/5 \end{pmatrix}$

11. 177.78; 272.22

13. 0

Exercise 4.1

1. (a) 3,3,3

   (b) $6,4,3\frac{3}{4}$

   (c) 6,7,5.21

Exercise 4.2

1. (a) unstable;

   (b) asymptotically stable;

   (c) asymptotically stable;

   (d) stable

5. (a) uniformly stable;

   (b) no conclusion;

(c) asymptotically stable;

(d) no conclusion

Exercise 4.4

1. (a) asymptotically stable,

    (b) unstable,

    (c) unstable,

    (d) asymptotically stable

3. unstable

5. stable, but not asymptotically stable.

Exercise 4.5

1. exponentially stable

3. (a) $\begin{pmatrix} 0 \\ 0 \\ 0 \end{pmatrix}$, $\begin{pmatrix} 1 \\ 1 \\ 0 \end{pmatrix}$

    (b) no conclusion, unstable.

5. unstable

7. $|a| < 1, |b| < 1.$

Exercise 4.6

9. (i) a: $a^2 \le 1,\quad b^2 \le 1;\quad$ b: $a^2 < 1,$
       b: $b^2 < 1$ or $a^2 \le 1,\quad b^2 \le 1, a^2 + b^2 \ne 2.$

    (ii) $\{(x, 0), (0, y)|x, y \in R\}$

Exercise 5.1

1. (a) $\dfrac{z(z - \cos \omega)}{z^2 - 2z \cos \omega + 1},\quad |z| > 1$

    (b) $\dfrac{z(z^2 - 1) \sin 2}{(z^2 - 2z \cos 2 + 1)^2},\quad |z| > 1$

    (c) $\dfrac{z}{(z - 1)^2},\quad |z| > 1$

3. $\dfrac{-z - a^2 + a}{z(z - a)},\quad |z| > |a|$

5. $\dfrac{(z+1)^2 z^{n-3}}{z^n - 1}$

9. $\dfrac{1}{(z-a)^3}$

17. (a) $\dfrac{z^2 \sin \omega}{(z-a)(z^2 - 2z \cos \omega + 1)}$

   (b) $\dfrac{z^2(z - \cos \omega)}{(z-1)(z^2 - 2z \cos \omega + 1)}$

Exercise 5.2

1. (a) $2/3[2^{-n} - 1]$
   (b) $-1/7(-2)^n + 1/7n(-2)^n + 6/7$

2. (a) $0, 1, -4, 10, -28, \ldots$
   (b) $x(n) = n\, e^{-na}$

3. (a) $(-2)^{n-3}\,(3n^2 - n)$
   (b) $2^{-n+1} + 2 \sin\left(\dfrac{(n-1)}{2}\,\pi\right)$

5. $\dfrac{1}{\sqrt{5}}\left[\left(\dfrac{1+\sqrt{5}}{2}\right)^n - \left(\dfrac{1-\sqrt{5}}{2}\right)^n\right]$

7. $\dfrac{1}{2}(n+1)$

11. $\dfrac{1-e}{2-e} + \left(\dfrac{1}{2-e}\right)(e-1)^n$

Exercise 5.3

1. $x(n) = \dfrac{1}{3}x(0)[1 + 2(4^n)]$
   unstable

3.   1. unstable
     2. uniformly stable

   5. unstable

Exercise 5.4

1. asymptotically stable

3. not asymptotically stable

Exercise 5.5

3. asymptotically stable

5. uniformly stable

Exercise 5.6

5. (a) $x(n) = -\dfrac{1}{7}(-3)^n + \dfrac{1}{7}(4^n)$

7. (a)
$$\begin{pmatrix} (1 + \sqrt{2})\, 2^{n-1} + \dfrac{(1 - \sqrt{2})}{2}(-1)^n \\ 0 \\ 0 \\ \left(\dfrac{3 - \sqrt{6}}{5}\right) 3^n + \left(\dfrac{2 + \sqrt{6}}{5}\right)(-2)^n \end{pmatrix}$$

(b)
$$\begin{pmatrix} \dfrac{1 + \sqrt{2}}{2}(2^n - n - 1) + \dfrac{1 - \sqrt{2}}{8}[(-1)^n + 2n - 1] \\ 0 \end{pmatrix}$$

11.
$$\begin{pmatrix} -2 + \dfrac{1}{12}(2^n) + \dfrac{3}{2}(3^n) \\ -1 + \dfrac{1}{12}(2^n) + \dfrac{1}{2}(3^n) \end{pmatrix}$$

Index

Undergraduate Texts in Mathematics

(continued from page ii)

Macki-Strauss: Introduction to Optimal Control Theory.

Malitz: Introduction to Mathematical Logic.

Marsden/Weinstein: Calculus I, II, III. Second edition.

Martin: The Foundations of Geometry and the Non-Euclidean Plane.

Martin: Transformation Geometry: An Introduction to Symmetry.

Millman/Parker: Geometry: A Metric Approach with Models. Second edition.

Moschovakis: Notes on Set Theory.

Owen: A First Course in the Mathematical Foundations of Thermodynamics.

Palka: An Introduction to Complex Function Theory.

Pedrick: A First Course in Analysis.

Peressini/Sullivan/Uhl: The Mathematics of Nonlinear Programming.

Prenowitz/Jantosciak: Join Geometries.

Priestley: Calculus: An Historical Approach.

Protter/Morrey: A First Course in Real Analysis. Second edition.

Protter/Morrey: Intermediate Calculus. Second edition.

Ross: Elementary Analysis: The Theory of Calculus.

Samuel: Projective Geometry. *Readings in Mathematics.*

Scharlau/Opolka: From Fermat to Minkowski.

Sigler: Algebra.

Silverman/Tate: Rational Points on Elliptic Curves.

Simmonds: A Brief on Tensor Analysis. Second edition.

Singer/Thorpe: Lecture Notes on Elementary Topology and Geometry.

Smith: Linear Algebra. Second edition.

Smith: Primer of Modern Analysis. Second edition.

Stanton/White: Constructive Combinatorics.

Stillwell: Elements of Algebra: Geometry, Numbers, Equations.

Stillwell: Mathematics and Its History.

Strayer: Linear Programming and Its Applications.

Thorpe: Elementary Topics in Differential Geometry.

Troutman: Variational Calculus and Optimal Control. Second edition.

Valenza: Linear Algebra: An Introduction to Abstract Mathematics.

Whyburn/Duda: Dynamic Topology.

Wilson: Much Ado About Calculus.